The Solar System 2

SCIENCES

Universe, Field Director – Fabienne Casoli

Solar System, Subject Head – Thérèse Encrenaz

The Solar System 2

External Satellites, Small Bodies, Cosmochemistry, Dynamics, Exobiology

Coordinated by
Thérèse Encrenaz
James Lequeux

iSTE

WILEY

First published 2021 in Great Britain and the United States by ISTE Ltd and John Wiley & Sons, Inc.

ISTE Ltd
27-37 St George's Road
London SW19 4EU
UK

www.iste.co.uk

John Wiley & Sons, Inc.
111 River Street
Hoboken, NJ 07030
USA

www.wiley.com

Library of Congress Control Number: 2021940272

British Library Cataloguing-in-Publication Data
A CIP record for this book is available from the British Library
ISBN 978-1-78945-034-7

ERC code:
PE9 Universe Sciences
PE9_1 Solar and interplanetary physics
PE9_4 Formation of stars and planets

MIX
Paper from responsible sources
FSC® C013604

Contents

Preface

Thérèse ENCRENAZ[1] and James LEQUEUX[2]
[1] *LESIA, Paris Observatory, PSL University, Paris, France*
[2] *LERMA, Paris Observatory, PSL University, Paris, France*

The aim of this book is to present a global and synthetic vision of planetology to the reader, in other words, the study of the objects of the Solar System. This is an ambitious objective, because planetology has undergone considerable development over the last decades and, today, it presents interfaces with multiple disciplines. In our approach, we have chosen to prioritize the study of physico-chemical processes, in order to shed light on the mechanisms that are at the origin of the formation of the objects of the Solar System, or that are responsible for their evolution.

This work is a continuation of the book *Le Système solaire* (T. Encrenaz, J.-P. Bibring, M. Blanc, M.-A. Barucci, F. Roques, P. Zarka), published in 2003 by EDP-Sciences and CNRS Editions, in the "Savoirs Actuels" collection. This work was itself the third edition of *Le Système solaire* (T. Encrenaz, J.-P. Bibring, M. Blanc), first published in 1987 in co-edition with InterEditions and CNRS Editions. More than 30 years after this first edition, a complete revision of the work was essential: during this period, planetology has undergone several revolutions. The first ones, both realized thanks to ground-based telescopes, were the discovery of the first trans-Neptunian objects other than Pluto in 1992, and then, in 1995, the discovery of the first extrasolar planets around solar-type stars. The 1990, 2000 and 2010 decades saw the deep exploration of the planets Jupiter and Saturn and their system, with the Galileo and Cassini space missions. The beginning of the 21st century saw the resumption of the Mars exploration program with the launch of numerous probes, both in orbit and on the surface of the planet. Venus and Mercury were also visited by space probes, as well as several asteroids including Ceres and

The Solar System 2,
coordinated by Thérèse ENCRENAZ and James LEQUEUX.

Vesta; lastly, the comet Churyumov-Gerasimenko was the subject of extended exploration thanks to the spectacular European Rosetta mission. Meanwhile, the first detections of exoplanets paved the way to a new field of research in full development, that of exoplanetology. The extraordinary variety of objects discovered around other stars, both in terms of their orbits and their physical parameters, has raised new questions about the formation scenarios of planetary systems; the discovery of the phenomenon of migration in these systems has encouraged new research on the origin and evolution of our own Solar System. The discovery of a large proportion of rocky exoplanets – the "super-Earths" – has also stimulated work in exobiology with the ultimate goal of finding life on one of these exoplanets. Over the last 30 years, planetary scientists have forged increasingly close links with other astronomers (who are at the origin of the discovery of exoplanets), but also with geophysicists, chemists and biologists interested in the problem of the emergence of life, on Earth or elsewhere.

The plan of the work follows the main course of the previous works. The first volume opens with three general chapters. Chapter 1 presents a general description of the Solar System. Chapter 2 describes the environment of the Solar System, the formation of stars and planets within the disks as well as the extrasolar planets. Chapter 3 deals with the interaction of the Solar System with the interplanetary medium. Chapters 4 and 5 present a description of telluric and giant planets. The second volume presents the satellites and rings of giant planets (Chapter 1), comets, asteroids and dwarf planets (Chapter 2), meteorites and cosmochemistry (Chapter 3), the formation and dynamics of the Solar System (Chapter 4), the origin of life and extraterrestrial life (Chapter 5), and lastly, the methods of study of the Solar System (Chapter 6). Almost all of these chapters are composed of original texts. In the few cases where we have relied on texts from the last edition of the Solar System (2003), we have explicitly mentioned it. We thank J.-P. Bibring, M. Blanc, M.-A. Barucci and P. Zarka for their contributions in the previous book.

This book is intended primarily for students of planetology, astrophysics and geophysics at the undergraduate and graduate level, but we hope that it can be read with benefit by all scientists, researchers and engineers wishing to deepen their knowledge of planetology.

June 2021

1

Satellites and Rings of the Giant Planets

Athena COUSTENIS, Marcello FULCHIGNONI and Françoise ROQUES
LESIA, Paris Observatory, PSL University, Paris, France

1.1. Introduction

Ground-based measurements as well as measurements from the Voyager, Galileo, and Cassini-Huygens probes have shown that the icy satellites around the giant planets (see Figure 1.1) are not the dead and lifeless objects that scientists imagined until the beginning of the 2000s. Recent observations have in fact revealed that they are unique and active worlds with, for some, surprisingly diverse characteristics on the surface and in the thin atmospheres (Schenk 2010; Lang 2011; Catling and Kasting 2017).

Until the 1990s, our knowledge of the Solar System was based on the data collected by the Voyager missions, two NASA probes launched in 1977 that flew by the Solar system's giant planets in the 1980s as well as by 48 of their satellites out of the 200 or so that have been observed to date (Atreya *et al.* 1989).[1]

The Galileo mission, developed by NASA, had the goal of studying the planet Jupiter and its satellites. Launched in 1989 and in operation until 2003, Galileo allowed us to better characterize Jupiter's atmosphere, its magnetosphere, and its

1 See also: https://voyager.jpl.nasa.gov/.

large satellites.[2] Most notably, it provided new information about volcanism on Io, as well as substantial evidence about the presence of a liquid water ocean under the frozen surface of Europa and the presence of a magnetic field on Ganymede.

The ESA-NASA-ASI Cassini-Huygens mission, launched in 1997, included an orbiter (Cassini) for Saturn and a lander (Huygens) for Titan. It arrived in the Saturn system in 2004 (Coustenis 2015). For 13 years (2004–2017), it studied the Saturnian environment and unveiled details about the planet, its satellites, and its rings. The satellite Titan was the target of the European Huygens probe, which landed in January 2005. The mission instruments provided an impressive wealth of data that revolutionized our knowledge of the Saturn[3] system.

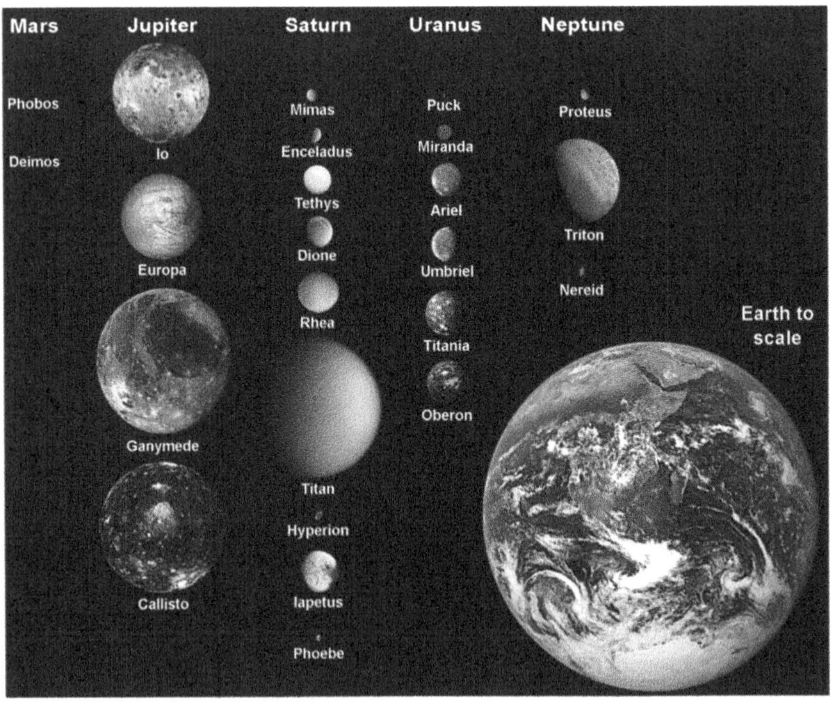

Figure 1.1. *Main satellites of the Solar system at scale (source: Wikimedia Commons, Bricktop, Deuar, KFP, TotoBaggins, City303, JCPagc2015). For a color version of this figure, see www.iste.co.uk/encrenaz/solar2.zip*

2 See also: https://www.jpl.nasa.gov/missions/galileo/.

3 See also: https://www.jpl.nasa.gov/missions/cassini-huygens/.

Most satellites are made up of a mixture of water and other kinds of ice, including, often, ammonia ice and rocky material (Atreya *et al.* 1989). A surprising variety of unusual and spectacular geological characteristics have been observed on these satellites, including patterns that bear witness to tectonic activity, volcanism, resurfacing, and more. Manifestations of geological and atmospheric processes that occur on our planet have also been observed, helping us to improve our understanding of the formation, evolution, and conditions for life in our Solar System and beyond. The latter presuppose the presence of liquid water for around a billion years, as well as the presence of chemical elements that are essential for life (referred to as "CHNOPS," for carbon C, hydrogen H, nitrogen N, oxygen O, phosphorus P, and sulfur S) and of sources of energy in a stable environment (Lissauer and de Pater 2019). According to a categorization proposed several years ago, there are two kinds of possible habitats: those in which underground oceans of water interact directly with a core rich in silicate, as on Europa, the smallest of Jupiter's Galilean satellites, or on Enceladus, a small satellite orbiting Saturn, while another class of habitats includes satellites that have bodies of liquid water sandwiched between two layers of ice, as we expect to be the case for Ganymede, the largest satellite in the Solar system, in orbit around Jupiter or Titan, Saturn's largest satellite (Coustenis and Encrenaz 2013; Coustenis 2014; Hand *et al.* 2020).

We know that the large satellites of the gas giants, which are located beyond the "ice line" (that is, the distance from the Sun corresponding to the limit for which condensation of water molecules is possible on the surface of a celestial body), contain large amounts of water. Indeed, with an average density of around 1.8 g/cm^3, the largest satellites (like Titan) contain almost 45% water by mass. Among these satellites, some have a core rich in silicates and can contain underground oceans of liquid water in contact with the core (as is probably the case for Europa and Enceladus). This is important, because in such environments on Earth, we find life (for instance, in the hydrothermal vents located at the bottom of the oceans) and therefore generate the heat from the chemical sources located under their ice crust. Other satellites can have oceans or pockets of liquid water trapped between two layers of ice (as suggested by measurements on Ganymede and Titan). In the study of the emergence of the elements of life on such satellites, the time scale is essential. If this liquid environment lasted for long enough, it could see the emergence of life; but, on the other hand, this could be inhibited if the ocean was so isolated from the environment that it was impossible for the concentration of ingredients necessary for life or the appropriate chemical inventory for the pertinent biochemical reactions to occur. What is important is to realize that these satellites contain large volumes of water, from whence they get their nickname of "ocean worlds" (see Figure 1.2). Sometimes the amount of water contained in these satellites exceeds that found on

Earth, sometimes by factors that can even reach an order of magnitude, as is the case for Titan and Ganymede.

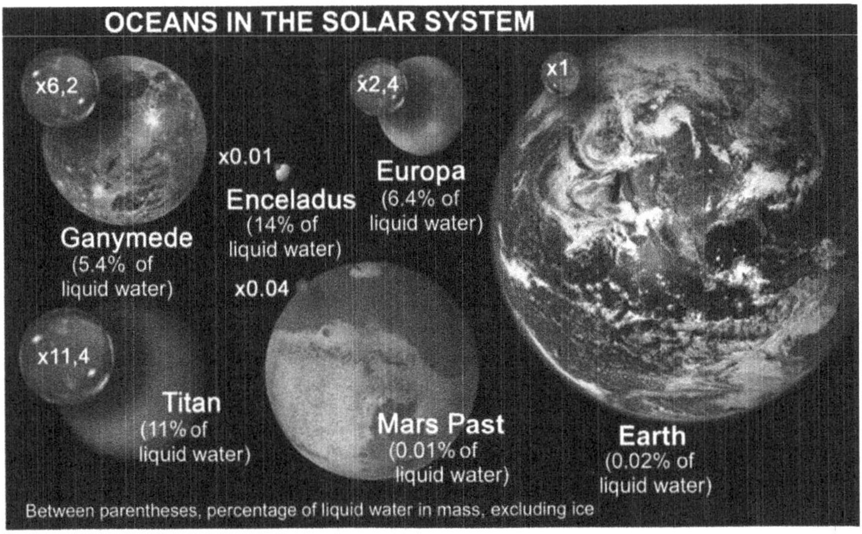

Figure 1.2. *Ocean worlds in the Solar system. The mass of the water in liquid or solid form is indicated, the mass of water on Earth taken as a unit. The percentage of liquid water in relation to the total mass of the planet or satellite is also given (source: PHL@UPR Arecibo, NASA). For a color version of this figure, see www.iste.co.uk/encrenaz/solar2.zip*

However, the exploration of these sub-surface oceans isn't easy, because the outer layers of ice on satellites can be very thick. We suppose that their thickness is on the order of several hundreds of kilometers on Ganymede, Callisto, and Titan. Europa may have a thinner layer of ice (possibly several dozens of km) because its proportion of water is only 10% of its mass, and it also has "chaotic" regions where the surface layer of ice may be even thinner. To understand the internal structure of the water layer, it is necessary to know the phase diagram of water under certain conditions of pressure and temperature.

On the edge of our Solar System, we find Uranus and Neptune surrounded by satellites that are unique in many ways, such as Triton and its geysers. Triton (in orbit around Neptune) and other objects beyond Neptune's orbit are thought to harbor liquid water oceans under the surface (still present today, even at the very low temperatures of their environments). The satellites of Uranus and Neptune have only been explored to date by the Voyager 2[4] mission.

Below, the most important discoveries about the outer Solar System's satellites from spatial and ground-based observations are described.

1.2. Jupiter's satellites

As of 2020, we are aware of 79 natural satellites around Jupiter (see Table 1.3, Volume 1, and Figure 1.1 in this volume) with diverse characteristics (Burns *et al.* 2004; Schenk 2010). The four largest satellites in the system, called the "Galileans" (Io, Europa, Ganymede, and Callisto) were discovered by Galileo in 1610.

1.2.1. *The Galilean satellites*

Among the Galilean satellites, Ganymede, Europa, and Io are caught in a 1:2:4 orbital resonance, which means that these objects periodically exercise strong gravitational influence on each other (see section 1.2.3, Volume 1, and section 4.5 in this volume). The three ice-covered Galilean satellites, Callisto, Ganymede, and Europa, present a wide diversity of characteristics on their surfaces, probably because of their different evolutions, even if they are situated in the same environment (see Figure 1.1). The unique characteristics of their geology bear witness to the parameters of their formation such as composition, density, and temperature, as well as evolving factors such as geophysical processes and the stage of differentiation (Schenk 2010). The Galilean satellites show signs of increased geological activity as their distance from Jupiter decreases (see Figure 1.3). As early as the 1980s, the observations from the Voyager probe showed large disparities between these four satellites, suggesting a strong influence of gravitational tidal forces on each of them. Thus, Io, the closest to Jupiter, is subjected to enormous tidal forces that lead to very intense volcanism. Europa may still be very active tectonically and volcanically today, because the small number of craters detected indicates a relatively young surface, while the larger Galilean satellites, Ganymede and Callisto, which are also the farthest, have old and densely cratered surfaces with a large range of sizes and internal structures (see Figure 1.3). Furthermore, the interactions with the Jovian magnetosphere and stellar winds, as well as the effects of the tides, have left traces on the surfaces that we see today in the form of æolian processes, cryovolcanism, and impact craters.

---

4 See also: https://solarsystem.nasa.gov/missions/voyager-2/in-depth/.

Understanding the gravitational interactions between Jupiter and the Galilean satellites is essential for a number of aspects of the science of the system of Jupiter, including habitability, because of their influence on the evolution of the interior and the surface of a satellite ((Burns *et al.* 2004; Lissauer and de Pater 2019); see also Figure 1.3). In particular, the effect of Laplace resonance can be decisive for the subterranean oceans of Europa and Ganymede, as well as for the future of volcanism on Io, where tidal dissipation is a significant source of heat, as it is for the other satellites.

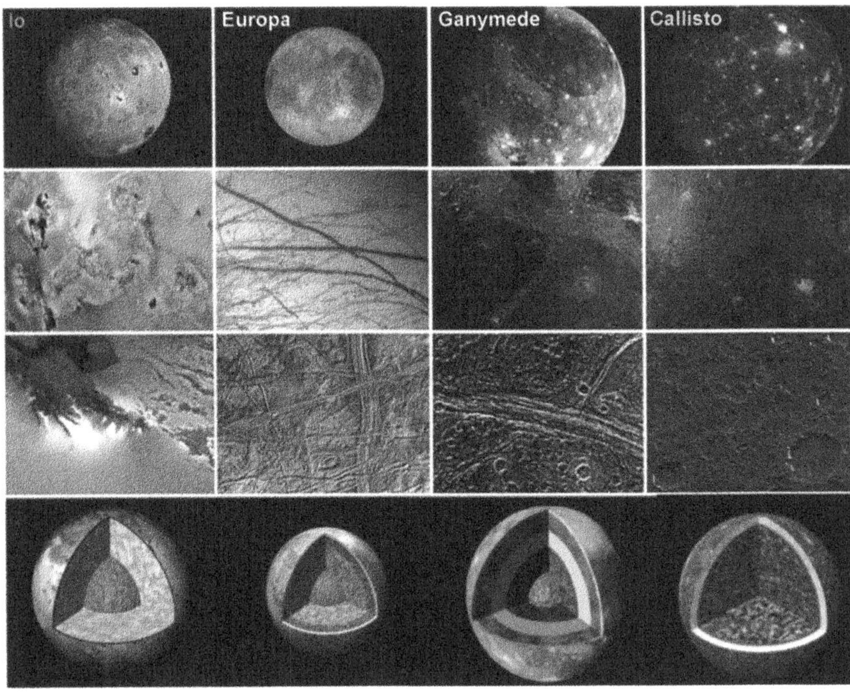

Figure 1.3. *The Galilean satellites of Jupiter, aspects of their surface and their internal structure; from left to right (as a function of their distance from Jupiter): Io, Europa, Ganymede, and Callisto, with their sizes at scale (source: JPL NASA). For a color version of this figure, see www.iste.co.uk/encrenaz/solar2.zip*

There is substantial proof that layers of ice and liquid exist on these satellites, but there is no certainty about their composition or their exact extent (Schenk 2010). On Ganymede, gravimetric data indicates a completely differentiated structure. A liquid layer as thick as 100 km in some places is trapped between the icy crust above and a layer below composed of ice at high pressure, which does not exist in natural environments on Earth. Europa has a similar internal structure. We think it has a metallic core and a layer of silicates, about the same size as the one on Ganymede, but a thinner hydrosphere that is no thicker than 100 km at its maximum. The biggest difference here is that, in the case of Europa, the liquid layer is almost certainly in direct contact with the silicate mantle. Callisto is very similar to Ganymede, although it is not completely differentiated and does not have an intrinsic magnetic field. This lack of differentiation, which is surprising for a body of its size, may be due to the fact that Callisto is not in resonance with the other Galilean satellites (see section 1.2.3, Volume 1 and section 4.4 in this work), which is one of the most striking pieces of evidence of the complexity of the interaction process in the Jovian system. Its surface is very old and highly cratered and, finally, it has a deep ocean similar to the one on Ganymede.

Observations indicate that the water under the surface is very probably mixed with other components such as salt hydrates or carbon dioxide, which we find on the surface of most of the Galilean satellites. Apart from the biochemical components that are necessary for life, other elements such as sodium, magnesium, calcium, iron, and potassium play an important role. However, while these components may be present on a surface, the conditions for their survival on Europa are not the same as on Ganymede, because their distance from the planet and hence the quantity of radiation of ions and electrons they receive are very different. Close to Jupiter, as in the case of Europa, this radiation is extremely harmful and probably deadly for any organic product present, preventing the survival of oxidants and organics on the surface for very long. To preserve Europa's potential habitability, these molecules would have to be able to enter the liquid water ocean under the surface. The composition of the surface of the Galilean satellites is largely reflected in their thin atmospheres (and exospheres), observed primarily in the UV from space and from the ground, atmospheres that result from the pulverization and sublimation of the substances on the surface. Conversely, the atmospheric components influence the surface composition. Europa has a dioxygen atmosphere with traces of sodium and potassium. Ganymede also has a thin oxygen atmosphere and a hydrogen exosphere. The intrinsic magnetic field around Ganymede is precious for habitability, because (as on Earth), it prevents high-energy particles from hitting the satellite.

1.2.1.1. *Io*

Io is the third largest of the Galilean satellites; a bit larger than our Moon, it is the closest to Jupiter (orbit at 6 Jovian radii), which exerts enormous tidal forces that lead to impressive geological and tectonic activity. One way in which this activity manifests is the presence of at least 250 volcanoes whose eruptions eject the gases, primarily sulfur dioxide, that create its thin atmosphere, which extends more than

300 km into space. Ionized particles are torn from Io because of its interaction with Jupiter's magnetosphere, forming a fairly substantial torus that can be detected in the UV (Catling and Kasting 2017).

The surface of Io shows a great variety of terrains, with many volcanic vents as well as non-volcanic mountains, calderas that are several kilometers deep, lakes of molten sulfur, and other patterns that give it its highly contrasted coloration, ranging from white to red and from green to orange (see Figure 1.4). The absence of impact craters on the surface indicates that it is young and is constantly being renewed by this intense activity (Schenk 2010).

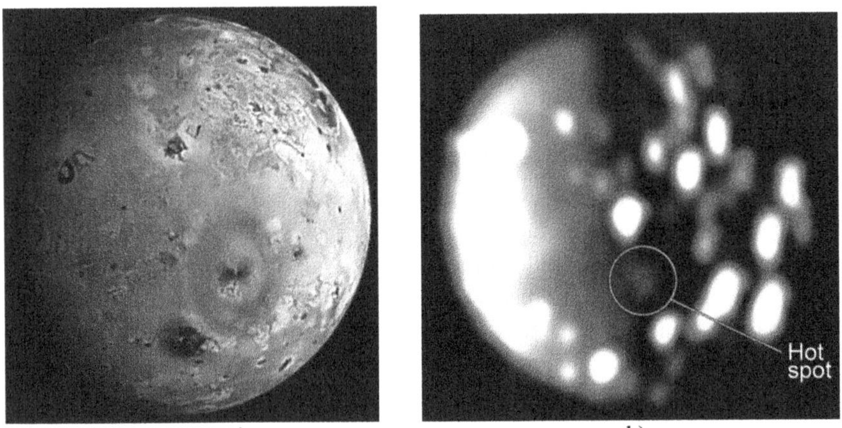

a) b)

Figure 1.4. *Io. For a color version of this figure, see www.iste.co.uk/encrenaz/solar2.zip*

COMMENTS ON FIGURE 1.4.– *(a) Composite color image of Io taken by the Galileo probe, showing the satellite's topography: volcanoes such as Loki Patera (the large dark horseshoe near the terminator), mountains, and plateaus can be seen. The large orange ring in the lower right is formed by matter deposited during eruptions of the largest volcano, Pele (source: NASA/JPL/University of Arizona). (b) Taken in 2018 by the Juno probe, this infrared image shows the location of a new hot spot on Io's southern hemisphere (source: NASA/JPL-Caltech/SwRI/ASI/INAF/JIRAM).*

The interior of Io is composed of a liquid iron core surrounded by a mantle made up of molten silicates enriched with sulfuric matter (see Figure 1.3). An extremely hot ocean of magma (1,500 K, approximately) that is 50 km thick is located at a depth of 30 or 50 km under Io's surface. Above this, the satellite's crust is very thin, and its morphology varies as a function of geologic activity.

1.2.1.2. *Europa*

Europa is the smallest of the four Galilean satellites in orbit around Jupiter ((Schenk 2010); see also Table 1.3, Volume 1). With a diameter of 3,140 km, Europa is slightly smaller than our Moon. Its distance from Jupiter is about 9.5 Jovian radii. Its density is 3.0 g/cm^3, which is typical for a mixture of rocks and ice. Its high albedo (0.67) suggests that its surface is primarily made up of water ice. Europa's surface temperature ranges from about ~125 K at the equator to ~50 K at the poles.

The spatial exploration of Europa began with the Voyager 2 encounter in 1979. Voyager's camera revealed an interesting surface topography made up of low-relief crests and chaotic terrains with few impact craters (Pappalardo *et al.* 2009). The pattern of crests already seemed to indicate that the icy crust was detached from the interior and moves slightly in relation to it. Europa thus became a privileged target for the Galileo mission, which approached Jupiter in 1995 and explored the Jovian system until 2003. Information was collected about the surface characteristics, the internal structure, and Europa's interaction with Jupiter's magnetosphere ((Burns 2004); see also Figure 1.3).

Among the most striking features of the observed surface of Europa are the dark streaks, called *linae*, which encircle the entire globe (see section 5.3.3 and Figure 5.9). They are countless, overlapping, and measure up to 1,000 km long and 1 km deep, but their edges never rise more than a few hundred meters high. In many cases, the crests are doubled, often with dark exterior borders and a central band. The images show that on each side of the lines, the edges are shifted with respect to each other (see Figure 1.3). These *linae*, probably of cryovolcanic origin, are interpreted as being the result of eruptions of hot water in a scenario that is similar to the one for terrestrial ocean ridges. Moreover, the fracture structures on the surface of Europa are different from those we would expect in the case of a completely solid object. Both recent and ancient fractures present different orientations, the more varied the older they are. This scheme suggests that the surface of Europa turns slightly faster than the sub-surface substances, which implies a liquid or viscous subsurface.

The surface of Europa also shows fractured and chaotic terrains called *lenticulae*, whose diameter varies from 10 to 100 km. There are many domes that seem to have been pushed out from the interior. These domes can be formed by the ascent of hot water between the cold ice of the exterior crust, in a scenario reminiscent of Earth's magma chambers. There are relatively few impact craters, which indicates that the surface of Europa is young. Its age can be estimated on the basis of the number of comets and asteroids in Jupiter's vicinity. This assessment results in an age of about

100 to 300 million years, which is a very short period of time compared to the satellite's history. Quite probably, Europa is still geologically active today.

It is now generally accepted that there is a water ocean under the icy crust of Europa ((Pappalardo *et al.* 2009; Hand *et al.* 2020); see also Figure 1.5). This ocean supplies the best explanation of the characteristics observed on the surface: *linae*, lenticular domes, fracture structures. In 2000, another discovery supported the ocean hypothesis: Galileo's magnetometer detected a weak magnetic moment, induced by the variable portion of Jupiter's magnetic field. The existence of this magnetic field requires the presence of a substance that is highly electrically conductive in the satellite's interior. The most plausible candidate is a large layer of liquid salt water under the surface. Spectral analysis of the dark surface characteristics should make it possible to detect the presence of salts in this ocean. Magnesium sulfate and sulfuric acid are possible candidates; sulfur components could explain the reddish color of the stripes.

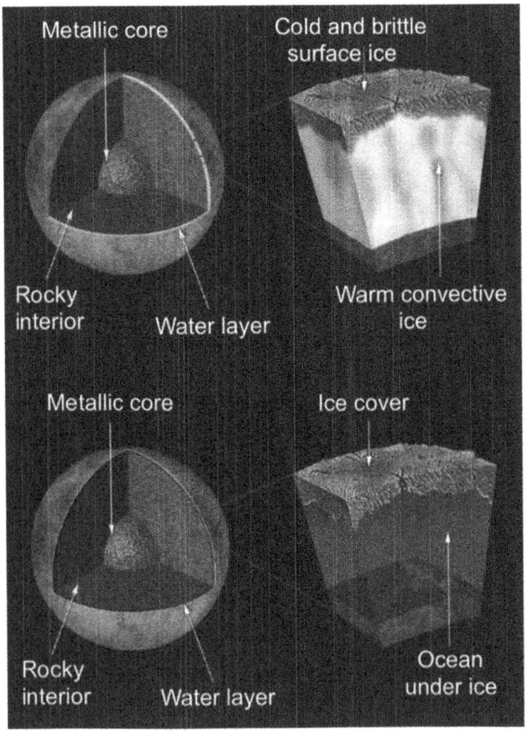

Figure 1.5. *Models of the interior of Europa with the presence of a liquid or ice ocean (source: NASA). For a color version of this figure, see www.iste.co.uk/encrenaz/solar2.zip*

The supposed existence of an underground ocean clearly depends on the capability of maintaining liquid water in the interior. What source of internal energy maintains the water at the right temperature on Europa? Radioactive heating is not sufficient because the volume/radius relationship is much smaller than on Earth. On the other hand, tidal heating is a plausible source of energy. Europa is known for being strongly differentiated, with a central core of iron and iron sulfide, surrounded by a mantle of silicate and then water (see Figure 1.5). Without the tidal effect, the water would be in its solid state. We have seen that Europa is in orbital resonance with Io and Ganymede. As a result, although it is in synchronous rotation with Jupiter, the bulge raised by Jupiter changes slightly in size and orientation under the effect of the other satellites when Europa moves along its orbit, causing a difference of 3 km between the radii in the radial and perpendicular directions, which leads to energy dissipation. According to models, this reaches its maximum effect in the lower part of the ice layer. In 2008, scientists proposed that another source of potentially more efficient energy could come from large planetary tides, generated by Rossby waves, as a result of the satellite's non-zero axis obliquity (Schenk 2010).

Furthermore, in 1995, observations from the Hubble Space Telescope revealed that Europa had a very tenuous atmosphere composed primarily of molecular oxygen, O_2, with a surface pressure of about 0.1 μPa, or 10^{-12} bars (Catling and Kasting 2017). Molecular oxygen is the primary atmospheric component because it does not freeze. Two years later, the Galileo probe also identified a tenuous ionosphere resulting from the interaction of Europa's surface with the high-energy particles of the Jovian magnetosphere. This magnetosphere contains a dense, high-energy plasma of oxygen and sulfur ions that escaped from Io. Thus, the surface of Europa is subjected to an intense bombardment of energetic particles from Jupiter's radiation belts, which leads to the rupture of water molecules and the generation of hydrogen peroxide, H_2O_2, a powerful oxidant that can be a chemical source of energy. Indeed H_2O_2 has a tendency to decompose exothermically into water and oxygen. It has been detected on the surface of Europa in the infrared and ultraviolet spectra. Europa's trailing hemisphere is particularly exposed to this effect, as shown by the black substance on the surface, interpreted as being a result of sulfur polymers altered by radiolysis. Colliding ions eject water molecules from the surface of Europa (this is the *pulverization effect*). Thus, the surface is slowly eroding: the rate of erosion is estimated at around 5–20 cm per million years. This provides additional evidence that the surface of Europa is relatively young. The Galileo probe's infrared spectrometer also detected mixtures of hydrated sulfuric acid and hydrogen peroxide, which could explain some of the surface's spectral characteristics. Molecular hydrogen cannot exist on Europa's surface because the satellite's gravitational field is too weak. It escapes, along with atomic and molecular oxygen, to form a neutral torus along the length of Europa's orbit. This torus was identified by the Galileo and Cassini probes. The question of the presence of convection in Europa's icy crust, which is likely to permit an exchange of matter between the surface and the ocean, remains open in the thin crust model because of the large number of tectonic characteristics such as fissures and visible faults on the surface (see Figure 1.5).

In 2016, water plumes were detected episodically on Europa by the HST, which may indicate cryovolcanic activity, as on Saturn's satellite Enceladus. In 2018, the results were confirmed by reanalyzing Galileo data from 1997, which provided new proof of the presence of an ocean of water below the icy crust.

How thick could the ice crust on Europa be? According to models, it could vary between several kilometers and a few dozen kilometers. The study of the largest craters on Europa, surrounded by concentric rings and filled with flat, fresh ice, suggests an icy crust that is 10 to 30 km thick, including a layer of warm, ductile ice above a liquid ocean that is about 100 km deep. We have seen that, according to some models of the internal structure, the liquid ocean should be in direct contact with the silicate mantle. The reanalysis of images from Galileo of chaotic terrains suggests that these areas form above pockets of liquid water located in an icy shell that is 3 km deep. While drilling 3 km down into Europa's surface is currently beyond our capabilities, this result, if confirmed, would have major consequences for astrobiology (Coustenis and Encrenaz 2013; Hand *et al.* 2020).

The determination of the topography of the surface and the internal structure at the interface between the silicate core and the liquid, as well as the detection of potential anomalies of mass at this location, will allow us to confirm the hypothesis that volcanism and/or hydrothermal activity could exist or have existed on Europa, as with the central ocean ridges on Earth. Such activity would release a variety of organic components into the ocean that are essential to maintaining simple forms of life at the bottom of the ocean, like those discovered around deep hydrothermal springs on Earth three decades ago. Concerning energy sources, the presence of an oxidizing agent on the surface, H_2O_2, and a reducing agent at the bottom of the ocean in the presence of hydrothermal activity introduces the possibility of a redox couple active on an oceanic scale. But this only works if the surface oxidants can supply the ocean through an icy crust that could be partially permeable. Subsequent exploration of Europa *in situ* by an orbiter and, even better (when the technology becomes available), by a lander, is necessary to confirm or to disprove these ideas.

1.2.1.3. *Ganymede*

Ganymede is in several ways a unique satellite in the Solar System because of its size (5,628 km in diameter), its intrinsic magnetic field, and its geological characteristics (Atreya 1989; Burns *et al.* 2004; Schenk 2010). Orbiting at about 15 Jovian radii, the largest satellite in the Solar system (larger than the planet Mercury) has a very thin atmosphere, which was detected by the HST in 1995. It is essentially composed of O_2, which is ionized by impacts of electrons to form atomic oxygen O, and probably also ozone O_3, as well as a small proportion of atomic hydrogen. The surface pressure is less than 1.5 μPa.

Its albedo (0.43) is compatible with the presence of water ice everywhere on the surface, on which two main types of terrain are distinguished. Ganymede thus occupies a key position in terms of geological evolution because it presents an old and dense terrain, like many of the terrains on Callisto, but also wide regions of tectonic resurfacing, similar to most of the surface of Europa (see Figure 1.3). One third of its surface is composed of old, dark regions that contain many craters formed four billion years ago. The other two thirds are more recent, lighter regions that probably indicate a process of resurfacing, forming a network and crisscrossed by large grooves and ridges (see Figure 1.6). The reason for this difference is not well understood; it may be the result of tectonic activity due to tidal heating. Hydrated minerals have been detected on Ganymede's trailing hemisphere (itself in synchronous rotation around Jupiter). Furthermore, various substances other than water ice have been identified on Ganymede's surface based on data from Galileo and from ground-based observations: carbon dioxide, sulfur dioxide, molecular oxygen, ozone, and various organic components probably formed locally thanks to the same process of radiolysis as on Europa (see Figure 1.6).

a) b) c)

Figure 1.6. *(a) Global geographical map of Ganymede; (b) mosaic of images of Ganymede; (c) illustration of the auroras on Ganymede (source: USGS Astrogeology Science Center/Wheaton/NASA/JPL-Caltech, ESA, G. Bacon (STScI)). For a color version of this figure, see www.iste.co.uk/encrenaz/solar2.zip*

The satellite's density (1,936 g/cm^3) indicates that it is made up of equal parts of water ice and rock. The observations show that it is a completely differentiated body, with a hydrosphere of at least 500 km representing 50% of the satellite's mass, a silicate mantle, and an iron core. Ganymede is one of only three solid bodies in the entire Solar System that possess an intrinsic magnetic field generated by a dynamo effect (the others are Mercury and Earth). Ganymede's magnetic field is in turn anchored in the enormous Jovian magnetosphere (Burns *et al.* 2004). Finally, it also possesses an induced magnetic field generated by currents circulating in a conductive liquid, which suggests the presence of a salty liquid ocean under the surface (see Figure 1.3). As a result of Ganymede's movement through Jupiter's magnetosphere, the auroral ovals that encircle the satellite should oscillate with an amplitude of 6° (see Figure 1.6). But if there is a large ocean in the depths of Ganymede, acting as a conductive environment, induction should produce an electrical field that in turn generates the satellite's magnetic field observed by Galileo and that could also reduce the amplitude of the oscillations of the auroral ovals. These ovals and their positions were measured by Hubble's UV instruments in 2017 and were found to be 2° instead of the 6° that was expected, leading to a strong argument in support of the existence of a liquid salty ocean in Ganymede's interior (Hand *et al.* 2020). We do not currently know if exchange processes are possible between the silicates and the salt ocean, which is presumed to lie almost 200 km below the surface of Ganymede, sandwiched between layers of ice.

With the various processes that leave their marks on its atmosphere (notably the presence of O, O_2, and possibly O_3) and its surface (such as tectonics, crater formation, and cryovolcanism), Ganymede is an archetype that allows us to understand a number of the processes that affect icy satellites throughout the outer Solar System and in some exoplanetary systems, those of the "ocean worlds," with very different properties from terrestrial planets but that nevertheless qualify for habitability.

1.2.1.4. *Callisto*

Callisto has a diameter of about 4,800 km and orbits Jupiter at an average distance of about 27 Jovian radii. Its density (1.83 gcm^{-3}) indicates that Callisto is made up of about half rock and half ice. Unlike the other Galilean satellites, its surface is dark, uniform, and completely covered with impact craters ((Schenk 2010); see also Figure 1.3). Callisto is the Galilean satellite that is farthest from Jupiter, and it is thus subjected to weaker tidal forces than the others, leading to a surface that has not been renewed by internal processes. The density of craters indicates that they were produced about 4 billion years ago, when all of the Solar System's bodies were subjected to heavy bombardment by comets and meteors. In addition to its large number of intermediate-sized craters, the surface of Callisto shows structures surrounded by rings that measure between hundreds and thousands of kilometers in diameter, likely created by major impacts. The largest, named

Valhalla, contains about 10 concentric rings with a maximum diameter of about 3,000 km. The preservation of this evidence of intense bombardment on the surface of Callisto is compatible with the lack of internal differentiation, which confirms the spatial measurements of its gravity field. Its interior is thus made up of well-mixed rock and ice, rather than presenting the core-mantle structure found on Io, Europa, and Ganymede. This satellite, alone among the Galilean satellites, has never been trapped in the orbital resonances responsible for tidal heating, which were so important in the evolution of Ganymede, Europa, and Io. Internal activity therefore did not noticeably modify Callisto's surface as it did on the other Galilean satellites. Nevertheless, Callisto has a weak magnetic field induced by Jupiter's field, which raises the possibility that a conductive layer of liquid salt water exists somewhere under its surface.

Unlike Ganymede, Callisto does not show large amounts of ice on its surface. The near-infrared spectra only contain weak indications of water ice, and the surface is much too dark to be made up exclusively of ice. The detailed images from Galileo reveal that the deposits of dark matter erased the smallest craters in certain areas (see Figure 1.3), and spectroscopic observations show that the substance is a mix of hydrated minerals that resemble clay and solid carbon dioxide. Spectroscopic studies have also shown the existence of a thin atmosphere composed primarily of carbon dioxide constantly escaping from the ground, traces of sulfur compounds (which may come from Io via volcanic eruptions), hydrogen peroxide (which is probably produced from water ice by photochemical reactions), and organic compounds potentially brought there by comets.

1.2.1.5. *Current and future exploration of the Galilean satellites*

The value of more thorough exploration of the giant planets and their systems became obvious following the discoveries made by the Voyager and Galileo missions. We currently have a mission exploring the Jovian system, the Juno probe from NASA, which was launched on August 5, 2011 and which has been on site since 2016, but this mission is essentially dedicated to the study of the giant gas planet, its magnetosphere, its formation, and its evolution, while the study of the satellites is *a priori* limited to their interactions with the planet's magnetic field.

In the preceding discussion, we demonstrated the interest of exploring the unique worlds that are the satellites of the giant planets and their potential habitats. One of the main objectives of the coming missions will be to determine the quantity of water present in the Jovian system, its distribution, and its potential contents in terms of organic material on and under the surface of the icy satellites, as well as the method of transport for substances between satellites through volcanism, pulverization, and impacts.

The Juice mission (*JUpiter ICy moons Explorer*) was chosen by ESA in 2013 as the first large mission in the 2015–2025 Cosmic Vision program. It was designed as a follow-up to the measurements taken by the Galileo mission, aiming to study in depth the system of Jupiter and its satellites, by concentrating on the largest, Ganymede. Juice's main theme is "the emergence of habitable worlds around gas giants", taking into account the necessary conditions implying the simultaneous presence of organic compounds, oligo-elements, liquid water, sources of energy, and relative stability of the environment over time.

The Juice mission is expected to launch in 2022 and arrive at Jupiter in 2031. The nominal mission should last for about 3.5 years. The mission scenario is designed to permit Juice to carry out an exploration of the Jovian system with the help of gravimetric assistance from the Galilean satellites to determine its trajectory. This exploration will include continuous surveillance of the magnetosphere and atmosphere of Jupiter, two flybys of Europa, and a flyby phase of Callisto on a trajectory that will reach 30° of latitude above Jupiter, culminating with a phase in orbit around Ganymede at 500 km from the surface, and possibly then even lower. The expected end of mission scenario involves crashing the spacecraft into Ganymede. The craft includes 10 instruments[5] and radio and Doppler experiments. The Juice mission will thus provide important information by establishing the existence of oceans below the surface of the crust of Ganymede, Europa, and Callisto, and by defining their main characteristics. For Europa, the Juice mission has planned two flybys above what we think are regions that were recently or are currently active and covered with thin layers of ice (regions referred to as "chaotic"), with the goal of extracting information about the underlying liquid ocean, its extent, and its composition. Observations will also be carried out on Io and other smaller satellites.

For technical and budgetary reasons, a surface lander or penetrator able to carry out a characterization *in situ* of the surface ice on Europa is not possible for now, nor is a dedicated orbiter, because of the high level of radiation generated by Jupiter's magnetic field. As a result, the characterization of the depth and breadth of Europa's ocean and of the boundary of its interface with the silicate mantle will be undertaken by Juice and a NASA mission specifically designed to study Europa, called Europa-Clipper. The latter is expected to launch in the mid-2020s and to arrive in

5 See also: https://sci.esa.int/web/juice/-/61110-juice-instruments.

the Jovian system about six years later, at about the same time as Juice, in order to carry out some 40 near flybys of the satellite.

1.2.2. *The minor Jovian satellites*

Besides the Galilean satellites, Jupiter harbors a number of other satellites of small size. The primary ones are classified into three distinct families:

– the inner satellites (Metis, Adrastea, Amalthea, and Thebe), located inside the orbit of Io, with quasi-circular and equatorial orbits;

– four satellites located outside the orbit of Callisto, on orbits with high inclination and prograde movement, with a semi-major axis of 150 and 170 Jovian radii (Elara, Lysithea, Himalia, and Leda);

– four satellites that are even farther away, located on orbits with high inclination but retrograde movement, with a semi-major axis of more than 300 Jovian radii (Sinope, Pasiphae, Carme, and Ananke).

In the latter two cases, these are most likely asteroids captured by Jupiter, or fragments of larger parent bodies that broke up near Jupiter because of tidal forces.

In addition to these three families, a number of Jupiter's small satellites were discovered in the 2000s by ground-based observations.[6]

1.3. Saturn's satellites

Other than ground-based observations, our current understanding of the system of Saturn is based on observations made *in situ* by instruments on-board the four missions that conducted different phases of exploration: Pioneer 11 (1979), the first probe to fly over Titan and Saturn, Voyager 1 (1980) and Voyager 2 (1981), which provided a first look at the entire system of Saturn (see Figure 1.7), and finally Cassini-Huygens (2004–2017), which studied the system of Saturn as a whole, revealing the characteristics of the planet, its rings, and its satellites.[7]

6 For more, see the list at: https://solarsystem.nasa.gov/moons/jupiter-moons/overview/?page=0&per_page=40&order=name+asc&search=&placeholder=Enter+moon+name&condition_1=9%3Aparent_id&condition_2=moon%3Abody_type%3Ailike and the images at: https://www.jpl.nasa.gov/spaceimages/details.php?id=PIA02530.

7 See also: https://solarsystem.nasa.gov/moons/saturn-moons/overview/.

In the Saturnan system, the Cassini orbiter has produced a wide range of new results about the 82 satellites known today, showing that these small bodies are far from being dead, icy, dull worlds, but rather active bodies with characteristics of resurfacing and cryovolcanism. Cassini's observations brought new surprises about several of these satellites. In particular, the combined research with the orbiter and the Huygens probe in the case of Titan, as well as the very close flybys of Enceladus, made it possible to reveal the satellites' distinctive characteristics and their potential in terms of exobiology (Coustenis and Encrenaz 2013; Coustenis and Raulin 2015; Schenk *et al.* 2018; Hand *et al.* 2020). Titan, with its thick atmosphere, in which we find methane clouds, and its surface covered with dunes, rivers, and lakes of liquid hydrocarbons, is a rich laboratory for studying chemical and physical processes similar to those on our planet, but with a different solvent and different substances. Furthermore, the measurements made by Cassini-Huygens indicated the very likely presence of a liquid ocean in the interior (Coustenis 2015). This is also the case for Enceladus: enormous plumes of icy particles that bear witness to geological activity and the discovery of simple organic compounds therein make it a privileged target for exobiology. Below, we will describe the major results from the Cassini-Huygens mission on Titan and Enceladus.

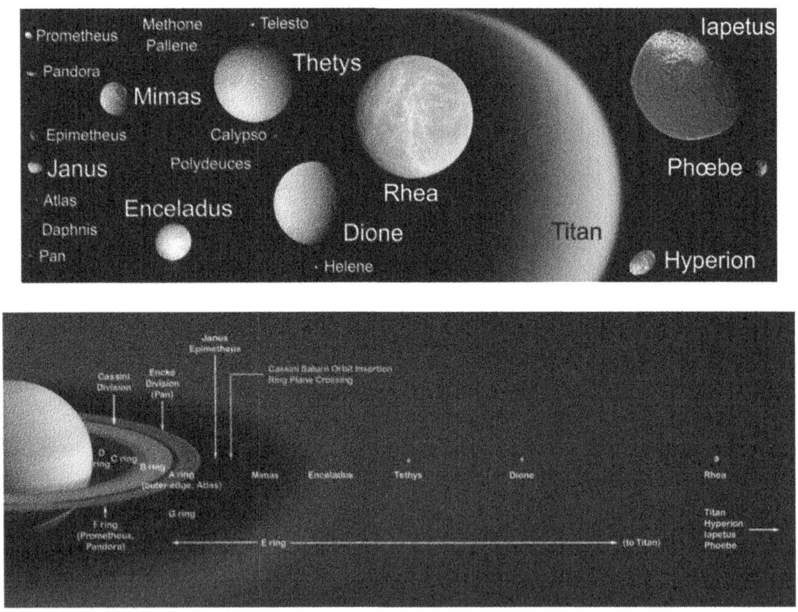

Figure 1.7. *A diagram of the lunar system of Saturn and its system of rings. Above, the various satellites at scale. Below, the rings and the main inner satellites (source: Wikimedia Commons, based on an image from NASA). For a color version of this figure, see www.iste.co.uk/encrenaz/solar2.zip*

1.3.1. *Titan*

Discovered in 1655 by Christiaan Huygens, Titan is the largest of Saturn's satellites and the second largest in the Solar System by size (5,150 km in diameter) after Ganymede ((Coustenis and Taylor 2008; Coustenis 2014); see also Figure 1.7 and Table 1.3, Volume 1). Titan, in synchronous rotation around Saturn (see Chapter 1, Volume 1, and Chapter 4 in this volume), moves along an orbit located 20 radii from the planet in about 16 days, which is also its period of rotation along its own axis. A recent study based on data from the Cassini probe shows that Titan is moving away from Saturn 100 times faster than previously supposed (11 cm per year). Before this rapid migration, 4.5 billion years ago, the satellite must have been much closer to the planet. This implies that the system of Saturn spread out much faster than we thought.

Titan's rotation period around the Sun is obviously the same as Saturn's: about 29.5 years. With an obliquity (angle between the axis of rotation and a line perpendicular to its orbital plane) of 27°, Titan, like Saturn, has seasons that each last a little longer than seven Earth years (Lorenz and Mitton 2008). Because of its almost synchronous rotation around Saturn, Titan's solid surface turns very slowly but, because of strong zonal winds, its atmosphere exhibits super-rotation. Titan's average distance from the Sun is 9.5 au. This corresponds to a solar flux above its atmosphere barely greater than 1% of the terrestrial flux. Moreover, Titan is sufficiently distant from the giant planet to avoid interaction with the rings. However, it is close enough to Saturn to allow its atmosphere to interact with the electrons in Saturn's magnetosphere. This interaction plays a key role, along with solar photons, in the chemical evolution of Titan.

Ground and space-based exploration of Titan has been providing basic information about the satellite for more than 350 years, particularly since the flybys by the Voyager 1 (1980) and Voyager 2 (1981) probes, which revealed that the surface of Titan is invisible to on-board cameras because it is enveloped by a thick atmospheric fog. Titan was found to be the only body in the Solar system other than the Earth to possess a dense atmosphere essentially made up of nitrogen (~98%), which combines with methane (~1.4% in the atmosphere) to give rise to a multitude of organic compounds such as hydrocarbons and nitriles (see Figure 1.8). Already at that time, this composition implied that Titan was a world that could resemble a primitive Earth, where prebiotic molecules might be found. This argument was one of the pillars on which the Cassini-Huygens mission was designed, which carried out the longest *in situ* exploration of a planet and its system of satellites to date (2004–2017).[8] Titan was at the heart of this exploration: 127 flybys were conducted,

8 See also: https://www.jpl.nasa.gov/missions/cassini-huygens/.

but it was especially the target for the European Huygens probe, which successfully descended into the atmosphere on January 14, 2005. All of the observations carried out by the instruments on the Cassini orbiter and the Huygens probe (Dougherty *et al.* 2010; Lorenz 2020) have allowed us to obtain a very detailed understanding of this far-off world.[9]

1.3.1.1. *The atmosphere of Titan: a biological factory*

Even before the Cassini-Huygens mission began, spectroscopic measurements of Titan's atmosphere from the ground but especially from the Voyager 1[10] mission had indicated the presence of a number of organic gases produced by photochemistry and photodissociation of dinitrogen, methane, and also oxygenated molecules ((Coustenis and Taylor 2008; Lorenz and Mitton 2008); see also Figure 1.8).

The Cassini-Huygens mission provided new information about the satellite's temperature and chemical composition (Dougherty *et al.* 2010; Coustenis 2014; Coustenis 2021).[11] The temperature and density profile measured by the HASI instrument on Huygens (see Figure 1.8) and by other instruments on Cassini showed a structured atmosphere that is the site of gravity waves above 400 km. We distinguish, in succession, the thermosphere (1,200–600 km), the mesosphere between 600 and 260 km, the stratosphere, where the temperature reaches its maximum of 187 K, while its minimum of 70 K is reached at the top of the troposphere at an altitude of 40 km, where a first temperature inversion occurs. The temperature on the surface of Titan is 93.6 K for a pressure of 1.5 bar (see Figure 1.8). The pressure and temperature values on the surface are therefore close to the triple point value of methane, which can therefore exist in gas, liquid, or solid state. In comparison, on the Earth's surface, environmental conditions are near the triple point of H_2O, so water cycles between the atmosphere and the surface, for example, through rain and dew evaporation (the hydrologic cycle). Titan is thus the only other place in the Solar System known to have a fluid cycle similar to that of water on Earth (Encrenaz and Coustenis 2017). However, on Titan, this happens with methane, and most of the methane resides in the atmosphere, while on Earth, most of the water is on the surface. The thermal profile changes depending on the region and the season (Horst 2017).

Cassini's ion and neutral mass spectrometer (INMS) described for the first time the ionic composition of the thermosphere where molecules are formed (see Figure 1.8). This data is complemented by measurements lower in the

9 Also see: https://www.esa.int/Science_Exploration/Space_Science/Cassini-Huygens/Cassini_instruments/.

10 See also: https://solarsystem.nasa.gov/missions/voyager-1/in-depth/.

11 See also: https://sci.esa.int/s/AM9ZGXA.

stratosphere acquired by the Cassini's Composite Infrared Spectrometer (CIRS) and detected by the Huygens's gas chromatograph mass spectrometer (GCMS) in the troposphere and during the descent toward Titan's equator on January 14, 2005. In the upper atmosphere, molecules of methane and nitrogen are dissociated by ultraviolet light from the Sun and by energetic particles accelerated by Saturn's magnetic field. The association of these parent molecules (N_2 and CH_4) produces a series of organic compounds (hydrocarbons like ethane, acetylene, and propane, and nitriles like hydrogen cyanide, acetonitrile, cyanoacetylene, etc.) which are present in small amounts (a few ppb (Coustenis 2021); see Figure 1.8). Some of these chemical components, such as hydrogen cyanide, HCN, are prebiotic molecules of interest. The atmosphere also contains small amounts of CO, CO_2, and H_2O of exogenous origin.

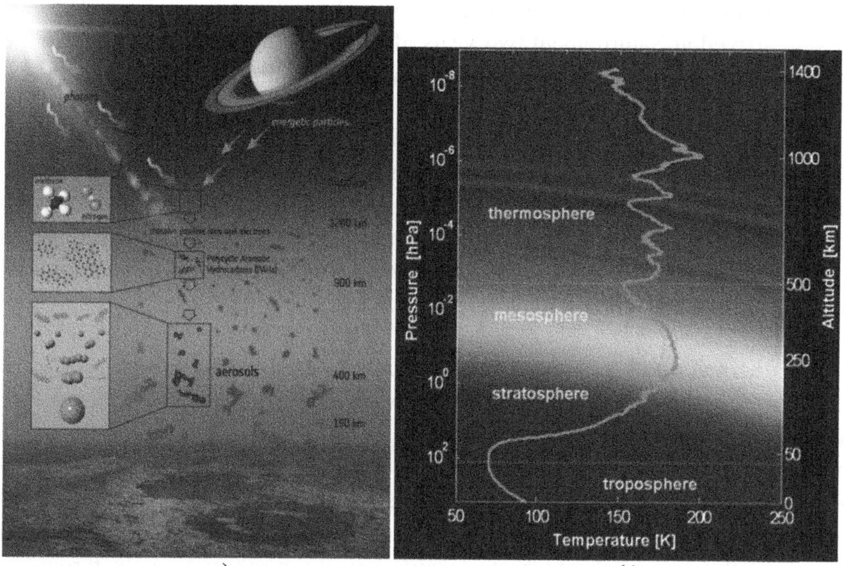

a) b)

Figure 1.8. *(a) The atmosphere of Titan; (b) the temperature profile measured by the HASI instrument during the descent of the Huygens probe through the satellite's atmospheric layers (source: ESA/ATG medialab; ESA/HASI). For a color version of this figure, see www.iste.co.uk/encrenaz/solar2.zip*

Complex refractory organic compounds combine and aggregate in the atmosphere, diffusing downward by creating aerosols and other condensates that form the clouds that hide the surface (see Figure 1.8). These clouds, made up primarily of methane, ethane, and other simple organic matter, form in the global fog and are dispersed and variable over long periods of time (Horst 2017). The clouds produce rain that deposits organic particles on the surface, where they accumulate and eventually create the lakes that are currently observable in the north (see section 1.3.1.4). This may be one of the most convincing pieces of evidence in support of a "methane" cycle on Titan that mimics the water cycle on Earth (Encrenaz and Coustenis 2017; Coustenis *et al.* 2018).

1.3.1.2. *Evolution of Titan's atmosphere during the Cassini mission*

The chemical species in the atmosphere of Titan evolve over time as they are subjected to the effects of solar radiation and circulation (Coustenis and Encrenaz 2013; Horst 2017; Coustenis *et al.* 2021; Teanby *et al.* 2019). After 13 years of thorough exploration of the Saturnian system by the Cassini mission, we have had the opportunity to witness atmospheric changes through two seasons on Titan. The mission arrived at Saturn about two years after the northern winter solstice in 2004 and ended in September 2017, just after the northern summer solstice.

During the mission, Cassini was thus able to glimpse for the first time the transition from autumn to winter at Titan's south pole and to witness the arrival of summer in the north. Titan's spatial and ground-based observations through this period show important variations of temperature and chemical composition in the neutral atmosphere before and after the northern spring equinox (NSE) in mid-2009 provoked by the change in sunlight exposure.

The thermal field and the chemical contents of minor gaseous species and condensates in the stratosphere and the mesosphere of Titan were determined during the Cassini mission by applying radiative transfer analysis to the Cassini/CIRS spectra, which cover the far infrared range of 10 to 1,500 cm^{-1} and were recorded during Titan's flybys from 2004 to 2017. The results show relatively calm equatorial evolution ((Teanby *et al.* 2019, Coustenis 2021); see also Figure 1.9). Conversely, this is not the case at the highest elevations in the north and south: through the seasons, strong temperature changes are observed near the poles of Titan, accompanied by a decrease or an increase in radiative emitters.

Furthermore, seasonal effects play an important role in the circulation in the atmosphere. Hadley cells dominate the meridional circulation of terrestrial atmospheres, like the ones on Venus, Earth, Mars, and Titan, with a great variety in strength, amplitude, and seasonality (Encrenaz and Coustenis 2017). Simulations with zonal wind models based on wind speed data captured by Huygens suggested that Titan's atmosphere circulates in a single enormous Hadley cell. At the time of Huygens's descent, during Titan's summer, hot gas was rising in the southern hemisphere and downwelling in the northern hemisphere, swept along by a high-altitude gas flow from south to north and a low-altitude gas flow from north to south. Such a large Hadley cell is only possible on a world in slow rotation such as Titan. The wind circulation cell from pole to pole seems to be centered in the stratosphere. Over the course of one Titan year (~30 Earth years), models predict that it should change every 12 years, with a transition period of three years (Coustenis 2021; Horst 2017; Teanby *et al.* 2019).

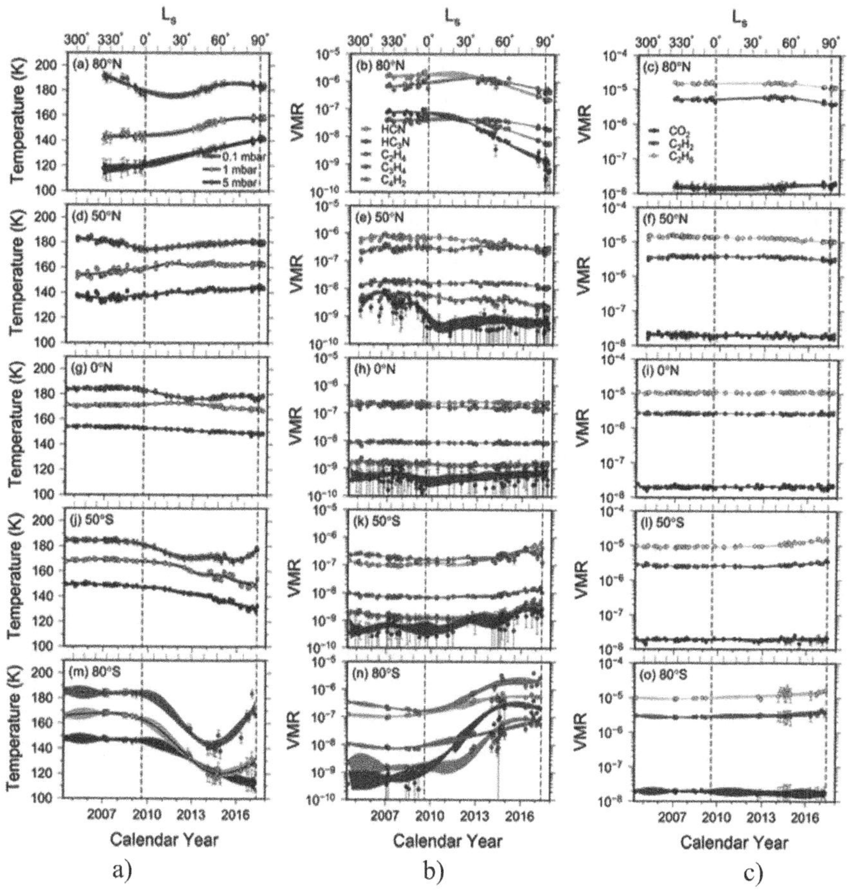

Figure 1.9. *Seasonal variations in the temperature and composition of Titan during the Cassini mission. (a) Change over time and latitude of the temperature at different levels of pressure; (b) and (c) change in the composition of the atmosphere for several gases (source: Teanby* et al. *2019)). For a color version of this figure, see www.iste.co.uk/encrenaz/solar2.zip*

During the first years of the Cassini mission, the northern latitudes exhibited an increased abundance until the NSE, which then decreased (from 2010 to 2012), returning to the values found during the V1 period, which took place one of Titan's years earlier in 1980, but remained still more elevated than at the south pole (see Figure 1.9). The increase observed in the north was associated with colder temperatures than in the south, between 15 and 29 K in the stratosphere at high latitudes. Things changed at the NSE, and until around 2014, with a strong drop in temperature observed near the south pole (adding up to a drop of about 40 K in four years after 2010), while a slight warming signaled the arrival of summer in the north after 2014 (about +6 K from February to September 2014, see Figure 1.9).

Furthermore, after the NSE, a large increase in several rare gases such as HC_3N and C_6H_6 began at Titan's south pole, while this was only previously observed at high northern latitudes, with an additional increasing trend for several other gases (C_4H_2, C_3H_4 and HCN). Confirmed over time (Coustenis 2021; see Figure 1.9), this trend indicates a seasonal atmospheric inversion on a large scale in Titan's unipolar circulation cell immediately after the equinox, with an increase in gases in the summer hemisphere and a corresponding decrease in the winter hemisphere.

As predicted by global climate models (GCM), a dynamic descending branch was observed for the first time at Titan's south pole, near the NSE, provoking reheating of the mesosphere (starting at around 400 km) via adiabatic heating and an increase in haze for latitudes above 80°S: indeed, it forced the air from the high atmosphere, where molecules form, to move toward the lower levels, leading to an enrichment in molecules at altitudes above 400 km, observed near the south pole in mid-2011 and later at even lower altitudes as the autumn progressed in the south. However, while the temperatures in the upper atmosphere should increase via adiabatic heating because of the expected strengthening of the vertical speed of the descending branch, it was instead radiative cooling that was observed in early 2012 in the stratosphere. This cooling (see Figure 1.9) is due to the decrease in solar flux during meridional autumn and to radiative cooling via highly enriched molecules at high altitude, which exceeds the adiabatic heating due to the descending branch. Net total cooling at high meridional latitudes causes the temperature to drop to values as low as 115 K in the low stratosphere, which leads to condensation of the gases at the poles at altitudes higher than those normally observed for other latitudes (Horst 2017).

Later on, after 2014, a significant and sudden increase in the temperature of the polar meridional stratosphere (from 10 to 50 K in the pressure range from 0.5 mbar to 0.05 mbar) was recorded with a change in the shape of the temperature profile. An inverse effect was expected in the north, but it was only observed with certainty after 2015 (Coustenis 2021). While the concentration of chemical products has continued to decrease near the north pole since that time, their abundance finally also decreased at the south pole, but only after a delay of two to three years. Indeed, in the stratosphere over Titan's south pole, some molecules remained abundant until around 2016, before experiencing a sudden and significant drop in 2017. In 2017, during the winter, Titan's south pole was losing its strong chemical growth, while the north pole had also slowly continued its decrease in gaseous opacities that had not picked up (see Figure 1.9). This indicates an asymmetrical response to the seasons in Titan's stratosphere (Coustenis 2019; Teanby *et al.* 2019).

In addition to the gases, observations of the southern stratospheric region of Titan in the autumn, while stratospheric temperatures are the coldest since the beginning of the Cassini mission, made possible the detection of condensates which had not previously been found in the north, including some new compounds such as benzene ice (C_6H_6). Several other unidentified signatures were observed in the spectra probing the deep stratosphere in a similar way; they could be spectral signatures of ice, probably mixtures of nitrile ice.

Data from Cassini impose constraints on photochemical and circulation models, which must take into account the asymmetry found in the behavior of the poles and the general environmental response of Titan to seasonal and dynamical forcing.

1.3.1.3. *Change in the atmosphere during a year on Titan*

With the conclusion of the Cassini mission, we have an almost complete picture of the evolution of the neutral atmosphere over a year on Titan by combining measurements from the IRIS infrared spectrometer on V1 acquired during the flyby of November 1980, with the CIRS and other continuous recordings from Cassini and Huygens from 2004 to 2017. Moreover, we have sparse intermediate data from ground-based and spatial observations such as with ISO or HST (Coustenis *et al.* 2013).

The composition values extracted from all of this data help us to search for inter-annual stratospheric thermal and chemical variations and provide the opportunity to evaluate the relative role of different physical processes in the long-term evolution of this complex environment. In mid-2010, when the season was exactly the same as during the V1 flyby, spectral analysis from Cassini/CIRS only showed limited inter-annual variations (Coustenis *et al.* 2013). A return to the temperatures and stratospheric abundance of 1980 was generally reached from 50°N to 50°S, which indicates that solar radiation is the dominant source of energy at 10 au, as on Earth, as expected by the general circulation and photochemical models. The exception concerned the more complex hydrocarbons (such as C_4H_2 and C_3H_4) and some nitriles in the north, where they were enriched at that time (Coustenis

2021). These results tell us more about seasonal, photochemical, and circulation effects on Titan, but remain rare and imprecise for the two seasons not covered by Cassini (summer to winter in the north).

1.3.1.4. *The surface of Titan*

Despite some indications of an inhomogeneous surface on Titan with light and dark regions already available before 2004, it was only with Cassini-Huygens that we were able to characterize the satellite's complex soil ((Coustenis 2014; Lorenz 2020); see also Figure 1.10)). One of the most effective applications of the synergy between the Cassini orbiter and the Huygens probe is the cartography of Titan's surface. While Cassini provided detailed views of Titan's surface with its camera, its spectrometer, and its radar, Huygens, descending through the atmosphere, sent back extremely detailed images with resolutions between 10 meters and 10 km. The Huygens Atmospheric Structure Instrument (HASI) provided the conditions of pressure and temperature for Titan's surface (1.5 bar and 93.7 K, see Figure 1.8).

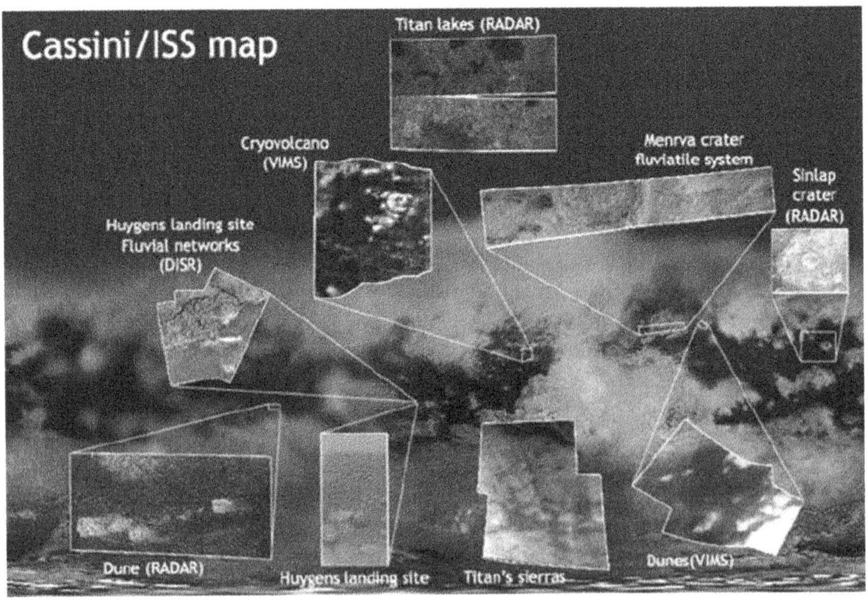

Figure 1.10. *The surface of Titan and some of the major discoveries made by the various instruments aboard the Cassini-Huygens mission: transitory atmospheric phenomena (clouds) and several geomorphological characteristics (craters, dunes, cryovolcanoes, channels, mountains, and lakes) are presented in this composite image (source: Université de Nantes and LESIA/SIGAL). For a color version of this figure, see www.iste.co.uk/encrenaz/solar2.zip*

Images of the surface of Titan taken by Cassini-Huygens show reliefs that resemble terrestrial landscapes, such as hills, dunes, and lakes ((Dougherty *et al.* 2010); see also Figure 1.10). But the components are completely different: a mixture of ice and organic substances. The ambient conditions and the direct measurement of the methane that evaporated under the probe after landing imply that the active erosive agent is liquid methane. The diversity of terrains on Titan observed by the Cassini-Huygens instruments is associated with multiple geological processes.

The erosion and deposit of solid and liquid matter by wind and rain shaped the surface of Titan. The liquids evaporate, forming clouds from which re-liquefied gases fall in the form of rain, which sculpts the surface of the satellite with rivers, seas, and lakes (depending on their size). In 2007, Cassini's radar discovered hydrocarbon lakes near Titan's north pole. The area covered by these structures ranges from less than 10 km^2 to more than 100,000 km^2. They are essentially confined to the region above latitude 55°N, but other small lakes have been discovered further south, such as Ontario Lacus. To date, several hundred areas of liquid water have been identified and mapped, and about 30 of them have been named. Lakes often form in depressions that are 600 to 800 meters deep. The three largest seas on Titan, Ligeia, Punga, and Kraken, extend over hundreds of kilometers and have shredded banks (see Figure 1.10) such as those observed on Earth in regions where reliefs have been filled by liquid (such as lakes in Alpine valleys). Cassini's radar measurements supplied bathymetric profiles that indicate depths of 100 to 200 meters for liquid methane lakes in the north. On the other hand, Lake Ontario, one of the only ones observed in the southern hemisphere, is primarily composed of ethane (Dougherty *et al.* 2010; Coustenis 2014). These lakes are found around 100 m above seas, and are similar to terrestrial karstic lakes, which implies the existence of subterranean rivers. The hydrological difference between the north pole regions indicates processes of formation that are probably different: it has been suggested that some lakes formed after explosions of heated nitrogen that created high barriers. Depressions at the south pole could be vestiges of ancient seas. This asymmetry between the two poles may be the result of long-term variations in climate (Coustenis *et al.* 2018; Lorenz 2020).

Radar observations suggest that the final fate of this precipitation is the production of dunes covered with organic matter, which were observed around Titan's equator by Cassini's radar. These dunes are the same size and form as terrestrial linear dunes, like those in the Namibian and Sahara deserts. This type of dune is formed by variable winds, which on Titan can be generated by atmospheric tides due to Saturn whose distance to Titan is variable, since its orbit is relatively eccentric. A vast sea of sand covers most of Titan's surface: elsewhere, we can see a deeply eroded terrain, at an altitude below that of the neighboring regions, called

Xanadu, which does not contain dunes. Xanadu seems to be one of the oldest regions on Titan: we find degraded circular structures there, which may be eroded vestiges of impact craters ((Coustenis 2015); see also Figure 1.10).

Drainage channels and dendritic networks observed by the Cassini radar, which are probably trenches like those at the Huygens landing site, have been observed at low and medium latitudes, although canyons of several hundreds of meters exist elsewhere. At high latitudes, we see meanders that suggest a deposit of small-grain sediments. The fluvial modification of the surface is obvious at the Huygens landing site, confirmed in Cassini's radar and near infrared images at much larger scales (Coustenis 2014; Lorenz 2020).

The almost complete absence of impact craters indicates that the surface is continually modified by active geologic processes such as æolian deposits (winds) and fluvial erosion (streams and rivers), which may affect a larger scale tectonic activity like the one, in some places, that formed mountain chains more than 1 kilometer high. Several zones may be active on the surface of Titan: domes, potential cryovolcanic flows and bright spots, mountain chains. The temperature on Titan is so low that water ice plays the role of rocky substrate, and water plays the role of lava. Although proof of active volcanism is not yet completely forthcoming, there are apparent changes on the surface that strongly suggest it. The region called Sotra Patera, for example, shows traces of possible cryovolcanic activity with flows of liquid water. Furthermore, the detection of argon 40 suggests that Titan's interior is geologically active. Theoretical calculations suggest a heat flow of about 8% of Earth's, which is enough to initiate cryovolcanism.

Cassini-Huygens thus found that the geological processes – impact, tectonic, fluvial, æolian – are similar to those on Earth, more so than those on Venus or Mars (Lorenz and Mitton 2008; Coustenis 2014; Encrenaz and Coustenis 2017). Titan could well be the best analog of an active telluric planet, although with different surface materials (Coustenis and Raulin 2015; Lorenz 2020).

1.3.1.5. *The interior of Titan*

The global density of Titan (1.88 g cm^{-3}) indicates that rocks and ice exist in equal proportions and that the body is differentiated into several layers. Titan may have an icy crust that is 50 to 150 km thick, above a liquid salt ocean that is a few hundred kilometers deep, with 10% to 30% dissolved ammonia acting as antifreeze. Below we find an ice layer under high pressure, followed by several layers of different forms of crystalline water ice mixed with ammonia ice and hydrocarbon ice. At the center is a rocky core that is more than 3,000 km in diameter (Dougherty *et al.* 2010; Coustenis and Raulin 2015).

Measurements by Cassini of a variation, small but significant, in Titan's rotation period have been interpreted to be the result of decoupling of the crust and the deep interior, separated by a liquid layer. In support of this argument, some of the surface features shifted by up to 30 km between October 2005 and May 2007, which suggests that the satellite's crust is mobile. An additional indication of the existence of such an ocean was discovered by the Huygens probe, which detected very low frequency radio waves in Titan's atmosphere; this is a Schumann resonance between the surface and the ionosphere, which implies the presence of a highly conductive surface under the successive layers of ice. Indeed, we think that the satellite's surface is a poor reflector of this type of wave, and that therefore they are reflected instead by the liquid-ice transition of an internal ocean.

Furthermore, measurements of variations in Titan's rotation and the amplitude of tides on its surface by Cassini in 2012 showed that the icy crust deforms in step with the orbital period, which is in agreement with the presence of a global ocean within a few hundred kilometers of the crust, as suggested by Huygens's measurements.

1.3.1.6. *Origin of the atmosphere and interactions with the surface and the interior*

Cassini-Huygens also provided important information about the origin and evolution of Titan's atmosphere by measuring the concentrations of rare gases (such as argon, observed for the first time) and their isotopic abundances, as well as that of nitrogen and the ratios of the stable isotopes of carbon. These measurements provide important indications about the global role of atmospheric escape, chemical conversion, outgassing, and recycling in the evolution of Titan's atmosphere (Catling and Kasting 2017; Schenk *et al.* 2018).

As we saw above, Titan's atmosphere and its surface are tightly linked. The organic matter falling from the atmosphere collects on the surface (see Figure 1.8) where it contributes to the formation of the observed geological forms. Analysis of Cassini's near-infrared spectro-imaging and radar data looking at the terrains at low and medium latitudes shows significant variations with time of the albedo and the composition of different areas. In this latitudinal dependence of the composition of Titan's surface, water ice seems to be the primary constituent at latitudes above 30°N and 30°S, while Titan's equatorial region is dominated in part by a very dark unknown substance and by another substance that has the properties of tholins deposited through the atmosphere (see Figure 1.8). The similarities and differences of the albedo between the different geomorphological units may indicate cryovolcanism (volcanoes of ice that spew water and other volatile products in the liquid or vapor phases on the surfaces of the giant planets' icy satellites), which thereby add methane into Titan's atmosphere and explain its reconstitution.

Indeed, the persistent presence of methane in Titan's atmosphere is a mystery because the Sun's ultraviolet radiation is capable of dissociating CH_4 molecules, and, therefore, since the satellite's formation 4.5 billion years ago, primary methane should have been completely destroyed. As a result, the presence of a source re-injecting it into the atmosphere must exist. In addition to cryovolcanism, other theories have been proposed: the sublimation of layers of frozen hydrocarbons, or the presence of oceans or large lakes of liquid methane on Titan's surface. It should be noted that the mass of carbon contained in the methane and its by-products in Titan's atmosphere, seas, and dunes exceeds by more than an order of magnitude that of the carbon present on Earth in the ocean, the biosphere, and fossil fuels (Dougherty *et al.* 2018).

The origin of Titan's atmosphere as a whole has been a source of debate since its discovery. According to the main theory, cometary impacts would have brought ammonia ice to Titan, and the ammonia would then have decomposed into the nitrogen that we see today. However, recent data about the comet 67P/Churyumov-Gerasimenko from the Rosetta mission and the constraints brought by isotopic ratios and rare gases reveal that a wealth of complex cometary organic material including nitrogen has been brought to Titan's interior. The measurements of the $^{36}Ar/N$ ratio by Cassini-Huygens as well as recent theories indicate that Titan's hot interior is responsible for the outgassing of N and other organic matter from the outer core within 300 km of the ice-rock interface, forming around 50% of today's atmosphere (Coustenis 2021). Subsequent measurements and internal heating models should allow us to better understand the nature and extent of cryovolcanism on Titan and the origin of its atmosphere.

1.3.2. *Enceladus*

Enceladus is one of Saturn's smaller satellites, with a diameter of 500 km (see Table 1.3, Volume 1). Enceladus is in 2:1 orbital resonance with Dione, which generates tidal forces. The energy produced by this process heats the interior and may govern its geological activity. Enceladus has one of the highest albedos of any object in the Solar system (Schenk *et al.* 2018). Covered with water ice, it reflects almost 80% of the Sun's light that it receives, which explains its extremely low surface temperature (around 73 K).

Enceladus presents five different types of terrain. Some parts are covered with craters whose diameter can be as large as 35 km, while other zones are devoid, which implies resurfacing events in a very recent geological past. There are fissures, plains, undulating terrains, and geysers. The region of Enceladus's south pole, in particular, is sprinkled with blocks of ice that can be as much as ten meters high, and the surface is sculpted by tectonic forms that are specific to the region: deep crevasses and fissures where the temperature is dozens of degrees higher than on the rest of the surface, indicating that below the icy crust, there may be liquid water (see section 5.3.3).

When Cassini performed its first encounter with Enceladus on February 17, 2005, the magnetometer saw that something was affecting Saturn's magnetosphere, with its plasma slowing and deviating in the vicinity of Enceladus. The data collected during the March 9, 2005 flyby provided additional evidence of the existence of an atmosphere around Enceladus's south pole. The cosmic dust analyzer recorded thousands of impacts from minuscule particles of dust and ice coming from a cloud around the satellite or from the adjacent Saturn's E ring. Indeed, several mathematical models show that this ring is unstable, with a lifetime between 10,000 and 1,000,000 years. Consequently, the particles that make it up must be constantly replenished. Enceladus is in orbit inside this ring, in its most narrow and dense area (see Figure 1.7), which has led us to suspect since the 1980s that Enceladus is the primary source of the E ring's particles. This hypothesis was confirmed during Cassini's close flybys in 2005 (Schenk *et al.* 2018).

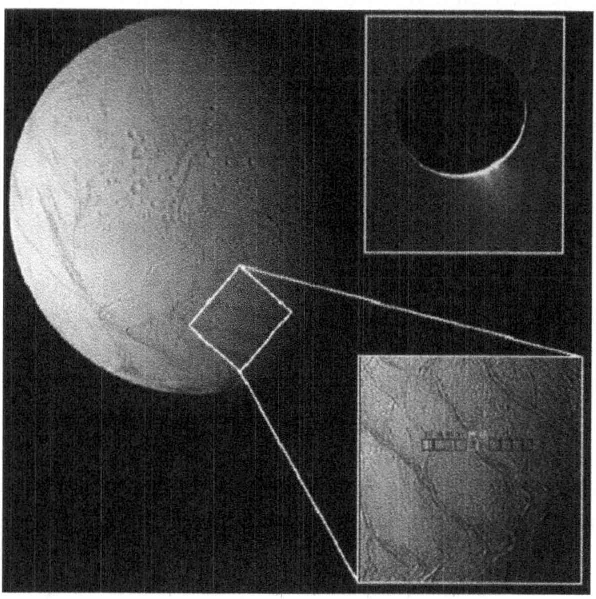

Figure 1.11. *The surface of Enceladus seen by Cassini, with the "tiger stripes" in the south, from which emanate plumes of water vapor (lower right zoom) (source: NASA). For a color version of this figure, see www.iste.co.uk/encrenaz/solar2.zip*

Cassini also provided definitive proof that Enceladus is geologically active, when several of Cassini's instruments detected plumes of gas and ice particles emanating from a series of hot fractures centered on the south pole, nicknamed the "tiger stripes" ((Schenk *et al.* 2018); see also Figures 1.11, 5.10, and 5.11). Moreover, the density measured by Cassini is 1.6 10^3 kg/m^3, higher than Saturn's other satellites, suggesting an internal composition that is richer in silicates and iron. The largest abundance of heavy elements may explain a higher degree of heating by radioactive elements than on the other satellites.

Several pieces of data from Cassini showed that reservoirs of liquid water under Enceladus's surface have produced geysers. In March 2008, Cassini swept Enceladus's south pole at an altitude of 52 km and dove into one of the plumes, collecting particles and gas and determining their composition. The gas analyzed by the INMS instrument closely resembled gases emitted by comets, showing the presence – in addition to water – of CO_2, methane, and other organic species (see Figure 1.12). The CIRS instrument on Cassini established that Enceladus's south pole was the hottest part of the satellite, surprisingly hotter than the equator. During its most recent flybys, Cassini detected complex molecules in the plumes, with atomic masses greater than 200 AMU. Like Titan, Enceladus therefore seems to be an organic chemical factory.

Enceladus is also the second icy satellite known to exhibit cryovolcanism. The other is Triton, but its cryovolcanism is not *a priori* connected to an internal source of heat. Enceladus can therefore be used to study the active processes that formed the surfaces of the other icy satellites. These processes include tidal heating, cryovolcanism, and ice tectonics (Hand *et al.* 2020).

Enceladus ejects plumes of salty water intermingled with grains of sand rich in silica, nitrogen (in the ammonia), nutrients, and organic molecules. The spectra also include traces of simple hydrocarbons such as methane (CH_4), propane (C_3H_8), and acetylene (C_2H_2) as well as formaldehyde (CH_2O), which are carbon compounds (see Figure 1.12). In particular, the presence of silica nanoparticles in Enceladus's plumes confirms that the liquid layer is in contact with a porous, rocky core: the water temperature to form these particles must be very high (360 K), suggesting that Enceladus possesses hydrothermal springs on its seabed.

In terrestrial oceans, hydrothermal sources host burgeoning life despite the extreme environment (Coustenis and Encrenaz 2013; Schenk *et al.* 2018; Hand *et al.* 2020). While other satellites in the Solar System have oceans of liquid water covered by kilometers of icy crust (Ganymede, for example), in the case of Enceladus, as on Jupiter's satellite Europa, pockets of liquid water may be only a few dozen meters under the surface (see Figure 1.12b). Enceladus is undoubtedly one of the places in the Solar System where exploration is most likely to find a habitable environment.

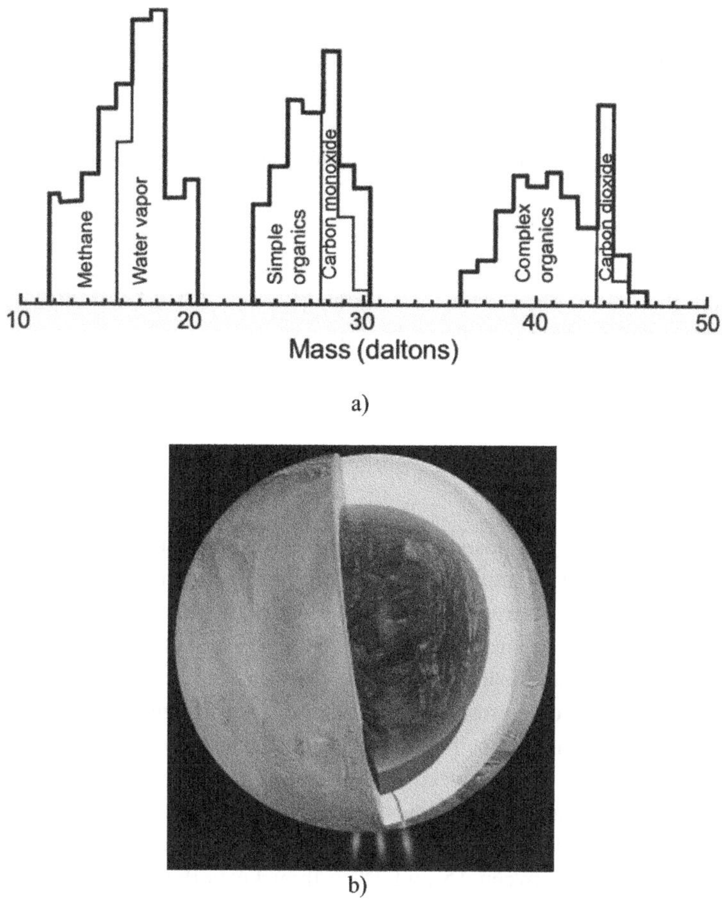

Figure 1.12. *(a) Mass spectrum of a plume on Enceladus and (b) a model of its interior (source: NASA/SWRI). For a color version of this figure, see www.iste.co.uk/encrenaz/solar2.zip*

1.3.3. *The other icy satellites*

This category includes Saturn's satellites other than Titan and Enceladus, whose main properties we will present (from outer to inner, Phoebe, Iapetus, Hyperion, Rhea, Dione, Tethys, and Mimas, Figure 1.7). All of these satellites, except for

Phoebe and Hyperion (which are probably captured objects), are locked by Saturn's tides in a 1:1 spin-orbit resonance, and they therefore always show the same side toward the planet. The recent discoveries by Cassini showed that these worlds are active, with the possible presence of liquid oceans as well under Dione and Mimas, and they interact strongly with the rings (Schenk 2010).[12] Furthermore, Cassini also discovered the presence of organic compounds (primarily hydrocarbons) on several of Saturn's satellites, including Iapetus, Phoebe, and Dione, thanks to their signatures in the spectro-imaging data collected by Cassini, showing that the dark substance in Saturn's system could be of organic origin.

1.3.3.1. *Regular and irregular satellites*

We say that several celestial objects constitute a "dynamic group" when the parameters that define their orbits are similar (Schenk 2010; Lang 2011). Saturn's most distant satellites belong to three dynamic groups:

– the Norse group;

– the Inuit group;

– the Celtic group.

The Norse group is composed of 29 outer satellites whose distances from Saturn range between 12 and 24 million km, with orbital inclinations between 135° and 175° and eccentricities between 0.13 and 0.7. All of these satellites have retrograde orbits, which indicates that they are objects captured by Saturn that came from the Centaur region or from beyond Neptune. Because of its orbital characteristics, Phoebe is part of the Norse group and may be the largest fragment resulting from the collision that created the group. The Inuit group is composed of five satellites whose distances from Saturn range between 11 and 18 million km, with orbital inclinations between 40° and 50° and eccentricities between 0.12 and 0.48. The Celtic group is composed of four outer satellites that share very similar orbital parameters: a distance from Saturn between 15 and 18 million km, orbital inclinations between 35° and 40°, and eccentricities between 0.45 and 0.53.

Closer to Saturn, we find *Rhea,* its second largest satellite, with a mean radius of 764 km. Rhea orbits the planet in 4.5 Earth days and is extensively cratered, much more so than Tethys and Dione, which are inner satellites. Rhea is the only satellite in the Solar System around which a disk of debris and a ring have been observed. Surface temperatures range from 99 K in sunny areas, dropping to 53 K in unlit areas. Rhea has a high albedo, showing that most of its surface is made up of water

12 See also: https://en.wikipedia.org/wiki/Moons_of_Saturn#/media/ File:Saturnmoonsdiagram.

ice, which, at Rhea's temperatures, acts like a rocky substrate. Rhea's density (1,233 kg/m^3) suggests that the satellite's interior is composed of three-quarters ice and one-quarter rock, with a homogeneous mixture of these two components, as indicated by its moment of inertia, which is larger than it would be if Rhea had a rocky core. A thin exosphere of oxygen and carbon dioxide has been detected around Rhea: during its diurnal rotation, Saturn's magnetic field periodically sweeps the satellite's surface, and the energy particles trapped in its lines interact with the surface by dissociating those organic molecules that are present and releasing oxygen and carbon dioxide. Its distance is 527,000 km from Saturn, such that the planetary tides do not lead to sufficient internal heating to produce flows of liquid water on the surface, which would have the effect of renewing it by wiping away pre-existing craters. However, the existence of regions with craters of different sizes indicates that a significant event renewed the surface in Rhea's distant past, given that there is no trace of young craters and that the average age of the plains is estimated to be about four billion years. The images obtained by the Cassini orbiter show deep canyons whose walls expose water ice. These cliffs also seem to bear witness to the existence of past tectonic activity.

Phoebe is one of the largest irregular satellites in the Norse group, as well as being the most distant of Saturn's satellites. Recently, thanks to infrared observations from the Spitzer satellite, a ring/disk around Saturn with very low density but extremely large (6 million km in diameter) was discovered, and its outer edge touches Phoebe's orbit: this is the E ring. Phoebe is therefore, with Enceladus and possibly Thetis and Dione, at the origin of this ring, which is supplied with dust ejected during meteor impacts. Contrary to most large satellites orbiting Saturn, Phoebe has a very dark albedo (0.06). Its surface is cratered and there is water ice on the wall of one of the largest craters. Its weak albedo and its retrograde orbit suggest that Phoebe came from the outer Solar system, a region where there is a lot of dark material, and that it was captured by Saturn. The hypothesis has been put forth that Phoebe could be a captured Centaur; in other words, an object from the Kuiper belt that migrated to the inner Solar System, providing planetologists with the opportunity to study a primitive object.

Iapetus, which was discovered in 1671 by Jean-Dominique Cassini, is the only regular satellite not to orbit in Saturn's equatorial plane (orbital inclination of 7.6°). It is in a 3:1 orbital resonance with Titan. Its main characteristic is the strong asymmetry of its two hemispheres: one is dark as coal (albedo 0.03–0.05) and the other is very bright (albedo 0.5–0.6). Several scenarios have been proposed to explain this asymmetry: dark particles from faraway Phoebe sweeping Iapetus; cryovolcanism, with eruptions of hydrocarbons that could form the dark surfaces after solar radiation has provoked chemical reactions in the ejected matter; the effect of thermal segregation due to the accumulation of dark matter on the hot side while volatiles migrate toward the cold side, which the ice makes even brighter. It is possible that in the distant past Iapetus was subjected to a small influx of dark matter from an external source (Phoebe, for example), which initiated this process. Another striking characteristic of Iapetus is an equatorial ridge, a mountain chain that is more than 10 km high, encircling the satellite's equator. On the side that never faces Saturn, the ridge seems to be broken, and even brighter isolated mountains have been observed there. There are two theories about the formation of this ridge: according to some, the ridge accumulated at a time when Iapetus had a much faster rotation than today; others think the mountain range is the result of material from the collapse of a ring.

Hyperion has a very eccentric orbit ($e = 0.10$) and is in 3:4 orbital resonance with Titan, which results in a rather chaotic 21-day orbit. Hyperion's density is a little more half that of water, which is probably due to a composition dominated by water ice and lighter substances, such as carbon dioxide or methane, with a porosity of more than 40%. This description is compatible with the hypothesis of a "catastrophic" origin for Hyperion, which would have been the result of the accretion of small icy and rocky bodies from the destruction of a primordial satellite by a major impact. The *rubble pile* thus formed has never attained sufficient mass to compact. Hyperion's surface is riddled with particularly deep craters that show no signs of ejecta, but whose interior shows signs of sliding terrain and wall collapse. The satellite's high porosity and low density suggests that these craters were formed by compression rather than excavation and that the ejected material tended to leave the satellite immediately. Many craters on Hyperion are bright, a sign of the presence of water ice, but the floors of the oldest of these craters have a lower albedo and a redder color as a result of an accumulation of dark matter left over from the sublimation of volatiles by heating due to extended daytime exposure to solar radiation.

Dione is a medium-sized icy satellite (1,124 km in diameter) in orbit around Saturn with a period of 2.7 days and a distance of 377,400 km. Dione has gravitationally locked two small satellites in its orbit around Saturn, called, by analogy with Jupiter's Trojan asteroids, the "Trojans" of Dione: *Helene* and *Polydeuces*. Dione is in orbital resonance with Mimas and Enceladus. Its density is 1,480 kg/m^3, implying that about a third of the satellite is composed of a dense core (probably silicates), with the rest being water ice. A very fine ice powder coming from Saturn's E ring, itself supplied by Enceladus's geysers, constantly bombards Dione's surface, where the average temperature is 87 K. The main geological characteristics of Dione's surface are heavily cratered terrains, with craters of up to 100 km in diameter, plains with a moderate or low number of craters, and fractured areas. The heavily cratered areas are, for the most part, on the trailing hemisphere. Logically, the leading hemisphere should be the one most frequently hit by impacts, but it seems that Dione was recently subjected to one or more impacts that turned it by 180°, which inverted the roles of the hemispheres. The fractured areas are dominated by very long subsidence fractures, similar to those observed on Rhea.

Tethys is composed almost entirely of water ice, making it very bright: its average albedo is 0.8. It looks a lot like Dione and Rhea except that it is less cratered, because its proximity to Saturn results in strong tidal heating effects that keep the layer of ice under the surface partially molten, allowing for effusions of liquid water that have erased or smoothed pre-existing surface structures. However, two giant structures persist on Tethys's surface: an immense impact crater and a very long valley. The Odysseus crater (400 km in diameter) dominates the satellite's western hemisphere, while the Ithaca Chasma is a valley (2,000 km long and 100 km wide, with a depth of between 2 and 3 km) that circles Tethys from north to south pole, at Odysseus's antipodes. Two hypotheses have been proposed to explain the origin of this valley by surface fracturing:

a) ice expansion, a consequence of the solidification of internal liquid water;

b) the impact that created Odysseus, generating seismic waves that interacted destructively at the crater's antipodes.

At the L4 and L5 Lagrange points of its orbit, Tethys harbors two small "Trojans" that are gravitationally locked in its own system – *Telesto* and *Calypso*.

Mimas is the smallest and closest of Saturn's icy satellites. Its low density suggests that it is composed essentially of water ice, the only substance that has been detected on its surface. In addition to its spin-orbit resonance with Saturn, Mimas is linked by its orbital resonances with Tethys (2:1), Dione (3:1), and Enceladus (3:2). Mimas interacts with particles from Saturn's B ring, with which it is in 1:2 resonance, creating divisions in the B ring and density waves in the A ring. Most of Mimas's surface is saturated with craters that are 40 km in diameter, but at the south pole the craters are generally smaller (20 km in diameter or less). This suggests that episodes of fusion of the icy crust or other processes that have renewed the surface occurred later than on the rest of the satellite. Mimas's distinctive characteristic is the giant Herschel crater (130 km in diameter), which covers a third of the surface, with walls that are around 5 km high and a central peak that is 6 km high. The shock waves generated by the impact probably caused the fractures, called *chasmata*, which are seen on the side of Mimas, located at Herschel's antipodes. The presence of such a heavily cratered surface, without any melting, implies that Mimas's interior is made up of solid ice. This is a paradox: being so close to Saturn and having a much more eccentric orbit than Enceladus's, Mimas should show stronger tidal effects and even more activity. Another surprise came from observations of Mimas in February 2010, which showed large differences of temperature on the surface that could not be explained by differences in the visible albedo. Mimas causes major disturbances in the orbits of the smaller satellites – *Methone* (D = 3 km), *Pallene* (D = 4 km), and *Anthe* (D = 2 km) – that orbit between Mimas and Saturn; they form the group called the *Alkyonides*.

1.3.3.2. *The small satellites near the rings and their interactions*

We have seen that Enceladus and Mimas have very significant interactions with the rings. The gases and microscopic particles emitted by geysers on Enceladus's meridional hemisphere constantly supply the E ring. Orbital resonances between Mimas and the rings of Saturn are responsible for the Huygens division, which delimits the B ring from the Cassini division, as well as several density waves in the A ring.

At 151,000 km from Saturn, *Janus* and *Epimetheus* are found between the weak F and G rings located between the A ring and the E ring (see Figure 1.7) and share the same orbit, which they travel in 17 hours. These satellites are both shaped like a large potato, with dimensions of 196 × 192 × 150 km for Janus and 135 × 108 × 105 km for Epimetheus; their low density (700 kg/m^3) seems to indicate that they are made up of a group of fragments held together by gravity. Their surface is cratered, with several craters larger than 30 km, and show grooves similar to those that characterize the surface of Phobos, one of Mars's two satellites. All of these elements seem to indicate that the fragmentation of a single satellite, shortly after the birth of the system of Saturn, led to the formation of this pair of satellites.

Six other small satellites have more limited contact with the rings; they are referred to as *ring shepherds* because their gravity sculpts one or more edges of the ring to which they are closest or in which they are immersed, sending grains and dust backward, which tends to escape. *Pan*, which is the innermost of Saturn's known satellites, has a mean radius of 14.1 km. It orbits at 134,000 km from Saturn in 13.8 hours, within the A ring where it formed and where it keeps the 325 km-wide Encke gap empty. *Daphnis* has a mean radius of almost 4 km and completes its orbit in 14 hours, at 136,500 km from Saturn, which creates the Keeler gap. *Atlas* orbits around the exterior edge of Saturn's A ring, which it maintains empty. Like Pan, Atlas's shape is characterized by a significant equatorial crest. It has a mean radius of 15.1 km and orbits at 137,670 km from the planet, completing a revolution in 14.4 hours.

Pandora and *Prometheus* work together to hold the F ring's particles and fragments in place. They orbit at about 140,000 km from Saturn and complete their revolution in about 15 hours. Pandora has a mean radius of 40.7 km and is covered by a layer of fine water ice powder. Prometheus has a mean radius of 43.1 km and a lower density than the other icy satellites, which implies a high degree of porosity. Its surface is less cratered than Pandora, Janus, or Epimetheus.

Aegaeon is Saturn's smallest known satellite to date (0.12 km in mean radius). It orbits Saturn at a distance of 167,500 km and completes its almost perfectly circular orbit in 19.4 hours. *Aegaeon* is immersed in the G ring, which was formed by debris ejected following impacts to the satellite that first produced an arc, which then spread to form the rest of the ring.

1.3.4. *Challenges for future missions in the Saturn system and Dragonfly*

Each of Saturn's satellites is distinctive. The largest have complex geological histories controlled by the heating and cooling of their interiors. But the small satellites also have unique geological histories because of their different positions in relation to Saturn, their relations with the ring system, and their interaction with the planet's magnetic field. The opportunity to observe, with Cassini, the ring-satellite interactions over a long period allowed us to understand the many evolutionary processes of material in orbit in the Saturn system: accumulation, fragmentation, renewal, aging, and capture. Their similarities and differences, which are the foundation of the Saturn system's uniqueness, show us the paths to follow to better understand the origin and overall evolution of the planetary system and invite future exploration.

The Cassini-Huygens mission discovered specifically that Titan and Enceladus possess active organic chemistries with seasonal variations, unique geological characteristics, and possibly internal liquid water oceans. It will be interesting to further explore all of these worlds now that the Cassini mission has ended in a spectacular *Grand Finale* on September 15, 2017 when it plunged into the atmosphere of Saturn. The final data collected at that time is still being analyzed, but in any case there will be many questions left to answer. For example, Saturn's deep atmosphere, Enceladus's plumes, and the two major themes of the exploration of Titan – the methane cycle as an analog of the terrestrial water cycle and the chemical transformations of complex organic molecules in the atmosphere and on the surface – justify another exploration of the Saturnian system.

Several ideas for new missions have been proposed to space agencies, and some have been or are in the process of being studied, for example, a probe to the interior of Saturn, similar to what Galileo was for Jupiter, but also orbiters, landers in the lakes, gliders, balloons, and drones dedicated to Titan or Enceladus.

In particular, *Dragonfly* is a NASA mission that will consist of sending a mobile robotic lander to Titan in order to study, among other things, the prebiotic chemistry and extraterrestrial habitability at several locations, where it will conduct vertical takeoffs and landings.[13] The mission will take advantage of the presence of a dense atmosphere and relatively low gravity: it will consist of a rotorcraft weighing about 450 kg, which will carry out multiple short flights to study the low atmosphere and the surface of Titan, using an electric battery as its energy source. This will be recharged on the ground by a radioisotope thermoelectric generator. When on the ground, it will study the composition of organic matter and ice on the surface by using a mass spectrometer and an active neutron gamma ray spectrometer. The drone will also carry instruments for meteorological and seismic studies. Dragonfly is expected to launch in 2027 and land on Titan in 2035 near the equator during the same season and at the same latitude as Huygens in 2005, although 30 years later, which is equivalent to one year on Titan.

While waiting for such an *in situ* exploration in the mid-2030s, ground-based measurements will be essential to continue exploring the satellites of Saturn.

1.4. The satellites of Uranus and Neptune

The Voyager 2 spacecraft explored the systems of Uranus and Neptune in the 1980s. Voyager 2 flew over the planet Uranus, which is characterized by a rotational axis that is almost parallel to its orbital plane, on January 24, 1986. This mission discovered a total of ten new natural satellites orbiting Uranus and studied its atmosphere and its ring system. Voyager 2 is also the only spacecraft to have flown over Neptune, on August 25, 1989, at 4,950 km from the planet's north pole and 39,790 km from Triton. As no other mission has explored these outermost planets in our Solar System to date, our knowledge of these icy giants remains limited.

1.4.1. *The satellites of Uranus*

Among the 27 known satellites of Uranus, *Oberon* and *Titania* are the largest and were the first to be discovered, by William Herschel in 1787. William Lassell

13 See also: https://dragonfly.jhuapl.edu/.

discovered the next two, *Ariel* and *Umbriel*, and Gerard Kuiper (1905–1973) discovered Miranda in 1948 (see Figure 1.13). In 1986, Voyager 2 discovered 10 additional satellites, which are only 26 to 154 km in diameter: *Juliet*, *Puck*, *Cordelia*, *Ophelia*, *Bianca*, *Desdemona*, *Portia*, *Rosalind*, *Cressida*, and *Belinda*. Since then, astronomers using the HST and ground-based telescopes have increased the total number to 27 satellites, which are for the most part tiny and dark.

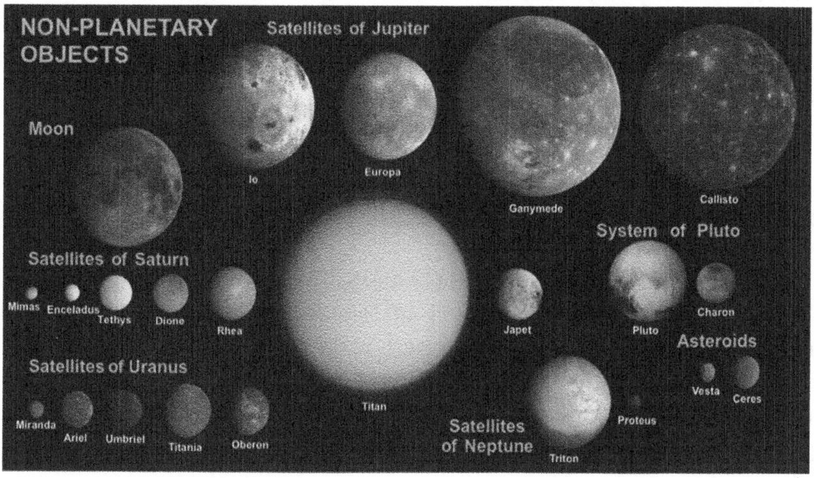

Figure 1.13. *Natural satellites of the giant planets and other non-planetary objects (source: NASA, Planetary Society). For a color version of this figure, see www.iste.co.uk/encrenaz/solar2.zip*

All of Uranus's inner satellites (those observed by Voyager 2) seem to be composed of about half water ice and half rock. The composition of the satellites outside Oberon's orbit remains unknown, but they are probably captured asteroids. Each of these satellites has unique characteristics. *Miranda*, the smallest and innermost of the five large satellites, has a surface that does not resemble that of any other satellite. It has giant canyons, up to 12 times deeper than the Grand Canyon, layers and terraces that seem very old, and others that seem much younger. Their origin on such a small object remains enigmatic. *Ariel* has the brightest and possibly the youngest surface of all of Uranus's satellites. This satellite has few large craters and many small ones, which indicates that relatively recent weak impacts have erased the large craters that would have been left by older and larger impacts. Valleys dotted with craters and crescents mark its surface.

Umbriel is old and the darkest of the five large satellites. It has many ancient large craters and harbors a mysterious bright ring on one side. *Oberon*, the outermost of the five large satellites, is old, highly cratered, and shows few signs of internal activity. An unidentified dark material appears on the floor of several of its craters. Finally, *Cordelia* and *Ophelia* are *ring shepherds* that guard Uranus's thin epsilon ring. Between them and *Miranda* are found a set of eight small satellites that do not resemble any other planetary satellite systems. This region is so crowded that astronomers fail to understand how these small satellites have avoided being smashed. They may be some of the shepherds for the planet's 10 narrow rings, and scientists think that there must be more satellites, closer than those already known, to constrain the edges of the interior rings.

1.4.2. *The satellites of Neptune*

British astronomer William Lassell (1790–1880) discovered Triton on October 10, 1846, just 17 days after the discovery of Neptune (Cruikshank 2016). Scientists using powerful telescopes and spacecraft have since discovered a total of 14 satellites in orbit around the planet. *Triton* is by far the largest of Neptune's satellites (see Figure 1.13). Gerard Kuiper discovered *Nereid* in 1949. *Proteus*, which is much larger than Nereid, and five other satellites, had to wait for Voyager 2 to be known.

Other than Triton, Neptune's satellites are some of the darkest in the Solar System. The astronomers using the most powerful ground-based telescopes discovered more satellites in 2002 and 2003. Proteus is a slightly aspherical satellite, and we think that it is at the limit of the size for an object before its gravity imparts a spherical form.

The Voyager 2 mission revealed to us the atmosphere and the surface of Triton during its flyby of Neptune in 1989. Before this date, ground-based infrared spectroscopy had shown the presence of CH_4 and N_2 ice on the surface, but its atmosphere was unknown. Voyager 2 measured a density of 2.08 g/cm^3, a surface temperature of 38 K, and a surface pressure of 14 µbar (140 Pa). The presence of ice is responsible for its high albedo (0.7). The existence of a stable atmosphere is made possible by the very low temperature, which limits gravitational escape. Triton's atmosphere, which is unique in the Solar System, is made up of a troposphere up to an altitude of 8 km, topped by a thermosphere that extends to the ionosphere at 400 km altitude. Traces of aerosols probably resulting from methane and products of photodissociation and nitrogen clouds are also present. Seasonal effects, which are notable on Triton, cause the condensation of N_2 and CH_4, which play an analogous role to CO_2 and H_2O on Mars.

Voyager 2 revealed unexpected details about Triton. Part of its surface is relatively undamaged ((Cruikshank 2016); see also Figure 1.13) while ice volcanoes elsewhere eject what is probably a mixture of liquid nitrogen, methane, and dust, which freezes instantly and falls back to the surface. An image from Voyager 2 showed a glacial plume rising 8 km above the surface and drifting for 140 km with the wind. We do not know its precise origin. Triton moves around its planet in the opposite direction from the planet's rotation (in retrograde orbit), which suggests that it may have been an independent object captured by Neptune. The disruptive effect that this may have had on the other satellites could explain why Nereid has the most eccentric orbit of all known satellites ($e = 0.75$). Neptune's gravity acts on Triton's retrograde orbit, slowing it and pulling it toward the planet. In a few million years, Triton will reach the Roche limit (see section 1.5.1) and will be destroyed by gravitational forces.

1.4.3. *Future exploration of the icy giant planets' systems*

Even if Uranus and Neptune have been recognized among the priorities of the space agencies' programs, these icy giants remain the only planets not to have been explored in detail, while even Pluto and Charon were recently visited by the New Horizons mission. The brief encounters with Voyager 2 gave us an overview of the complexity and unique character of these planets and finally raised a number of other questions. With new, large telescopes and the exploration by missions (as was the case for Jupiter and Saturn), our knowledge of the systems of Uranus and Neptune should grow. Future missions to Uranus and Neptune should provide us with important new information on the formation of these far-away systems, especially since the measurements from Voyager 2's flyby may not have been representative of their complete seasonal cycles, because of the length of these planets' orbital periods.

1.5. The rings

The first rings were discovered around the planet Saturn by Galileo on July 25, 1610. But Galileo did not understand the structure he saw: "I observed that the farthest planet has a triple form. The planet seems to have handles that appear and disappear." It was Huygens who, in 1659, understood that there was a flat disk surrounding the planet, seen from different angles depending on the date. Two hundred years later, Maxwell showed that this ring cannot be solid, and that it had to be made up of isolated particles. Galileo's instrument, Huygens's physical intuition, and Maxwell's mathematics were all needed for us to be able to grasp the idea of a planetary ring (see Figure 1.14).

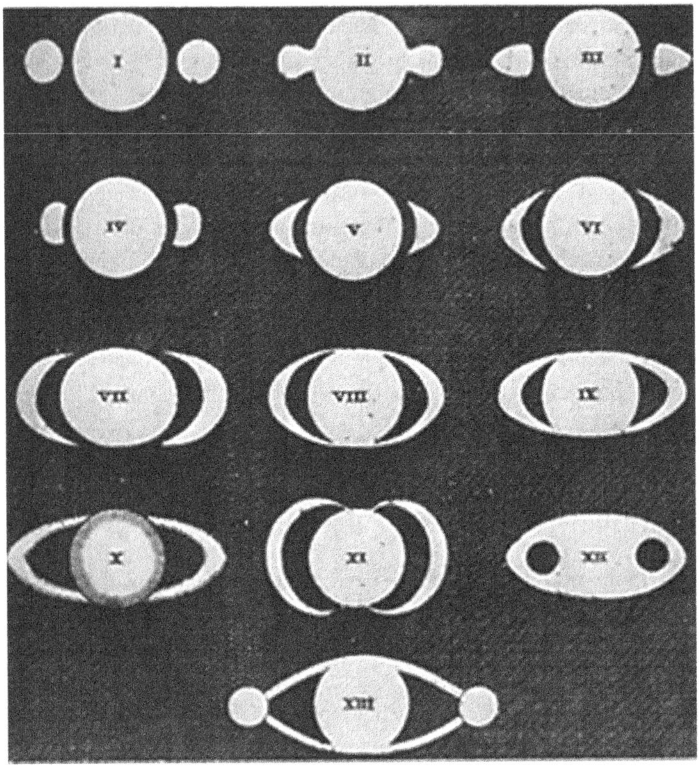

Figure 1.14. *Drawings of Saturn by different observers in the 17th century, before Huygens came to understand that it has a ring. Drawing 1 is Galileo's (source: according to Huygens, Systema Saturnium, p. 202, Library of the Paris Observatory)*

On March 10, 1977, the rings of Uranus were discovered by James Eliot's team by observing an occultation of a star by the planet. The rings appeared as a brief blocking of the star's light before and after its passage behind the planet. Jupiter's rings, which are almost invisible from Earth, were observed for the first time by the Voyager 1 probe when it approached the planet on July 9, 1979. It was not until July 22, 1984 that, as for Uranus, the observation of an occultation of a star by the planet allowed André Brahic's team to discover an arc (incomplete ring) around Neptune.

Thus, all of the giant planets have rings. These four ring systems have very different structures, narrow or wide, bright or dark, composed of particles with sizes ranging from microns to kilometers. Their commonalities are that they are in the planet's equatorial plane and they are very close to a number of small satellites (see Figure 1.15).

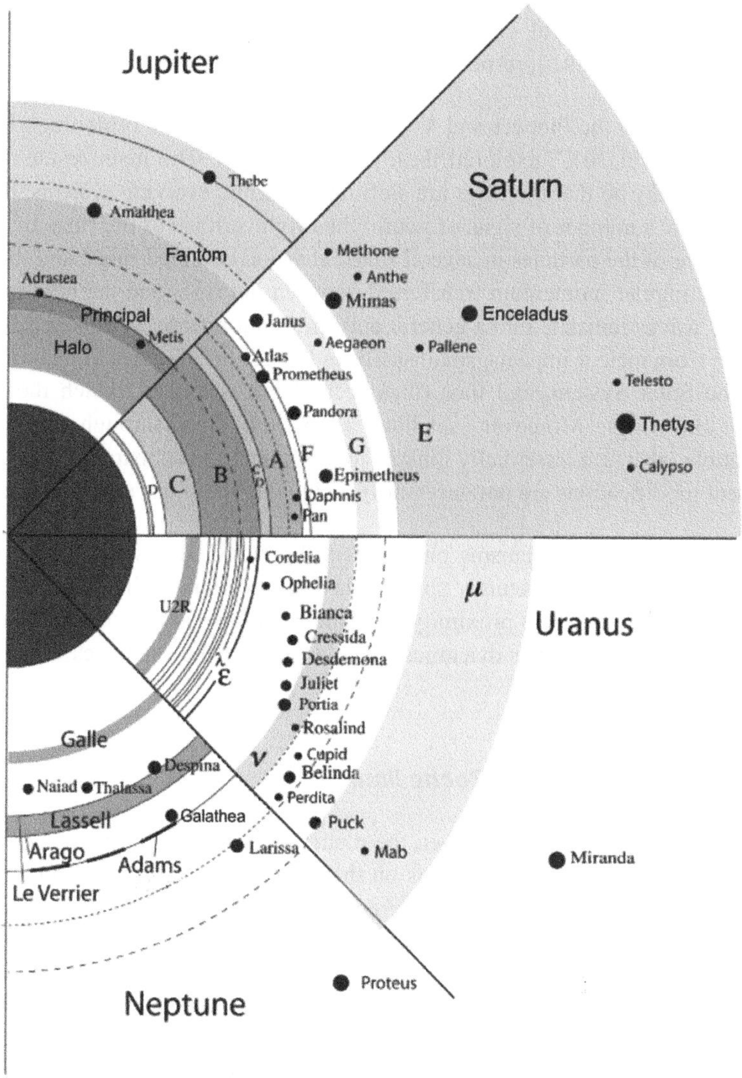

Figure 1.15. *The four ring systems of the giant planets and the near satellites. The different shades of gray indicate the optical depth of the different components. Synchronous rotation is indicated by dotted lines and the Roche limit for the density of water is indicated by dashed lines (source: from (Grun* et al. *2001))*

But rings are not the prerogative of giant planets, since stellar occultations, once again, have allowed us to discover them around the small bodies Chariklo and Haumea, in 2013 and 2017, respectively.

After flybys by the Pioneer and Voyager probes, Cassini, in orbit around Saturn between 2004 and 2017, explored these rings at length. The main result of these observations may be that the rings are shifting, dynamic, evolving systems, but also that they show a number of signs of youth: the bright surface of the rings of Saturn, the small size of the particles in several rings, arcs and confined rings, and the sharp edges. The angular momentum transferred between the rings and the near satellites by density waves translate into a general outward motion from their planet. Because of erosion from meteor impacts, small satellites have short lifetimes in relation to the age of the Solar System, and thus rings accrete into satellites, which themselves crumble into rings. Moreover, satellites shape the rings through gravitational perturbations that are essentially linked to resonances. But paradoxically, these permanent modifications are not necessarily a sign that the rings appeared recently.

Furthermore, the mechanisms observed in the rings are also at work in other flattened astronomical structures, spiral galaxies, or circumstellar disks in which planets form. The (relative) proximity of planetary rings has provided a laboratory for better understanding the dynamics of galaxies and the processes of planetary formation.[14]

1.5.1. *Tidal forces and the Roche limit*

Inside a satellite orbiting a planet, the planet's gravitational attraction is not the same between close and distant points on the satellite. This difference in attraction deforms the satellite, and when the difference between the force at the center of the satellite and at the edges is greater than the satellite's internal forces of cohesion, it will break up. The distance L_R on the planet at which these two forces cancel each other out is called the Roche limit (see section 1.2.7.2, Volume 1 and section 4.8.3 in this volume). Inside this limit, a satellite is unstable and breaks up into a ring, and outside of it, the particles in collision accrete on the satellite, and a ring cannot be maintained.

When an object approaches a planet at a distance lower than L_R, it fragments and gives rise to smaller objects. Their mutual violent collisions reduce the average size of the grains, which progressively populate a ring of matter. The collisions also

14 For a general review, see Greenberg and Brahic (1984); for a more recent review, see Tiscareno and Murray (2018).

reduce the velocity component perpendicular to the planet's equatorial plane and circularize the orbits. This evolution translates into a decrease in energy and a conservation of the particle cloud's angular momentum. The particles quickly find themselves on circular orbits in the equatorial plane. Outside the plane of the equator, particles orbit in precession, which causes collisions and brings the particles back into the equatorial plane. Once the ring has formed, weaker collisions cause it to spread out. In the absence of another disturbance, a part of the ring falls into the planet and the other part extends beyond the Roche limit, where the particles will accrete and form a satellite. Those satellites that are small enough to resist tidal effects remain near the planets and orbit between the rings. On the other hand, diffuse rings spread outside the Roche limit (see Figure 1.15). Images obtained by the Cassini probe show that this region is the site of active mechanisms of collision and accretion.

To estimate the distance from the planet of this limit, we can compare the difference in the gravitational attraction of the planet on two particles to the gravitational attraction they exert on each other. Thus, we find $L_R = 2^{4/3} R_P (\rho/\rho_P)^{1/3}$, R_P being the planet's radius, ρ_P its density, and ρ the satellite's density (see section 4.7.3 of this volume).

1.5.2. *Flattening and ring dispersion*

The study of the dynamics of a disk of colliding particles shows that it flattens rapidly within a time period of a few dozen collisions per particle. It then attains a state of quasi-equilibrium where the combined effect of differential rotation and inelastic collisions will make it spread slowly inward and outward while conserving the system's total angular momentum. In the absence of confining forces, the time to spread is the time it takes the particle to cross the ring in a random path. During collisions, some of the relative velocity due to differential rotation transforms into vertical and radial agitation. Equilibrium is established between the kinetic energy received by differential rotation and that lost through inelastic collisions. Given that the angular momentum is conserved, the energy loss of the particles moving inward is greater than the energy gained by the particles moving outward, and the disk loses a small fraction of its total energy.

1.5.3. *Jupiter's rings*

All of the ring systems include diffuse rings; those of Jupiter are the best example (see (Burns *et al.* 2004) and Figure 1.15). Their structure also shows the strong link that exists between rings and satellites. The densest ring, the *main ring*, extends inward from the orbit of the satellite Adrastea. Inside the main ring, a toroidal structure is called "the halo". Outside, two very diffuse rings, the *phantom rings*, are associated respectively with the satellites Amalthea and Thebe.

These rings are very difficult to observe from Earth because of their proximity to the planet. The solar light they scatter is extremely faint in comparison to Jupiter's luminosity. It is only in very narrow spectral domains, where the Jovian atmosphere is opaque, that ground-based observations of these rings have been carried out. Voyager 2, followed by Galileo, provided details about the properties of Jupiter's rings. The main ring is the brightest, although its optical depth is very low, on the order of $3\ 10^{-6}$. It extends over 7,000 km between a highly contrasted outer edge at 1.81 R_J (R_J being Jupiter's radius) and a more diffuse limit at 1.72 R_J. Inside, the halo, with an optical depth of 10^{-6}, extends from 1.72 R_J to Jupiter's upper atmosphere. The ghost rings, which are even less dense, surround the whole structure.

Lit from behind, the rings are twenty times brighter than in reflected light because of the size of the particles. Between 1 and 15 microns, these particles have very short lifetimes because of the dynamic effects (Poynting-Robinson drag, electromagnetic force, see section 4.9.2) and erosion (ionic and micro-meteorite bombardment, collisions). The rings of Jupiter therefore need to be constantly supplied with new dust. The particles come from the four small satellites, Thebe and Amalthea for the ghost rings and Metis and Adrastea for the main ring and the halo (see section 1.2.2).

1.5.4. *Saturn's rings*

At the end of the 17th century, the structure of Saturn's rings began to be observed. In 1675, Cassini discovered that there is not one homogeneous ring, but it is divided into two rings, A and B, separated by the division that bears his name. The inner C ring was discovered in 1850. Laplace showed in 1785 that a solid ring would not be stable under the effects of tidal forces, and this idea was verified theoretically by Maxwell in 1857 and then observationally at the end of the 19th century by measuring the rotation speed using the Doppler-Fizeau effect: the orbital speed of dust particles depends on their distance from Saturn, according to Kepler's third law. It was only during the 20th century that models managed to describe the evolution of the rings, particularly the effect of collisions between grains. The spectacular data from space probes shows that the rings are constantly evolving systems (Esposito 2010). Thinner and thinner rings have been discovered, both inside the C ring, where the existence of the D ring was confirmed, and outside the A ring (see Figures 1.7 and 1.15): the outermost E ring extends beyond the Roche limit. The F ring, on the contrary, is only 4,000 km from the A ring. The G ring is in-between. Although the optical depths of the A and B rings are close to 1, that of the E ring is only 10^{-7}. Hundreds of satellites that are a few dozen kilometers in diameter and have low density (0.5 g/cm^2), or accumulations of particles of all sizes, form and disappear, sculpt undulating edges, waves, and helically shaped forms (see Figure 1.16). Several of these satellites have *skirts*, an accumulation of matter fallen from neighboring rings (see also section 1.3.3.2).

Figure 1.16. *Daphnis and its wave on the edge of the Keeler division, inside the A ring (source: NASA/JPL-Caltech/Space Science Institute)*

At high image resolution, the rings appear in the form of thousands of very thin concentric circles that are clearly distinct. This is also the case in the Cassini division, which is not empty of grains but rather filled with a multitude of rings. At the smallest scale, some rings are twisted or contain heterogeneous regions. Radial structures, dark in reflection but bright in diffusion, change very rapidly over time. With a length that can exceed 10,000 km, they appear in a few minutes and last at most a few dozen hours. They may be submicron particles lifted up during magnetic storms.

The thickness of the rings can be locally very small, limited to the size of a few particles: the dynamic evolution of the rings leads to surprisingly effective planar confinement, if we compare the local thickness of the disk, which is less than a kilometer, to its diameter, on the order of 300,000 km. However, the rings have vertical excursions that increase their overall thickness. Not only do satellites create vertical undulations, but auto-gravity also creates temporary accumulations of particles that are relatively voluminous. These structures have been particularly highlighted in low-angle light observations by Cassini at the equinox (see Figure 1.17).

Figure 1.17. *Exterior edge of the B ring confined by 2:1 resonance with Mimas and, above, the beginning of the Cassini division. It is likely that the accumulation of matter favors the formation of large objects whose shadows we see at low angles (source: NASA/JPL-Caltech/Space Science Institute)*

The size of the particles varies widely between the diffuse D, E, F, and G rings and the main rings, where most of the particles range between a centimeter and about a dozen meters, with a power distribution of size with a factor of 3.

As in the case of Jupiter's rings, the presence of small grains with brief lifetimes in Saturn's environment implies the existence of reservoirs and constant renewal. The rings undergo constant dynamic evolution. In particular, gravitational interactions are responsible for heterogeneities of density, structural irregularities, and spiral density waves. These interactions as a whole, operating on grains that are constantly subjected to collisions, lead in the end to a stable confinement of these grains.

Crystalline water ice is the rings' primary component. A reddening of the rings' spectra greater than that on the surface of the satellites may be a sign of the presence of organic matter, iron, or carbon. The color changes under bombardment by meteors and with the interaction of the particles with the magnetosphere and the ionosphere. The interpretation of spectroscopic observations depends on complex models that take into account distributions of sizes, shapes, temperature, and particle composition. Significant variations in color appear between the rings, but we have not yet been able to connect them to the chemical composition of the grains. They indicate, on the other hand, that the motion of grains from one ring to another is not a dominant process, which confirms the stability of Saturn's ring system.

1.5.5. *Uranus's rings*

The rings of Uranus (French *et al.* 1991) were discovered on March 10, 1977, during a stellar occultation by the planet. They have been studied since then using the same technique, with the first direct photograph taken in 1984. In total, 10 very narrow rings have thus been discovered, located between 42,000 and 52,000 km from the planet's center (see Figure 1.15). They are named based on their increasing distance from Uranus, first by number: 6, 5, 4, then by Greek letters, α, β, η, γ, δ, λ, and ε. All are less than 100 km thick, and eight of them are less than 10 km thick. Their albedo is particularly low, below 5%. There are also three wide and diffuse rings, ζ, η, and μ. All of these rings are immersed in a compound, visible only through forward scattered light, which has an optical depth on the order of 0.005.

The ε ring's thickness varies from 20 to 100 km. The rings are primarily made up of grains that range from a few centimeters to a few meters wide, except for λ, which is made up of dust. The very low albedo of the ring's material may be due to the presence of organic polymers, which result from irradiation of grains of ice containing hydrocarbons by magnetospheric winds. Thirteen small satellites are associated with the rings, eleven of which were discovered by Voyager's flyby in 1986, while the two others were discovered by the Hubble Space Telescope in 2003.

1.5.6. *Neptune's rings*

With the discovery of the rings of Uranus by stellar occultation, the same technique was applied to investigate the rings around Neptune. In 1984, an occultation revealed, in Neptune's vicinity, the presence of a semi-transparent object, 15 km wide and at least 100 km long. It therefore could not be a satellite. Furthermore, no detection was recorded from the other side of the planet, where a complete ring would have interrupted the star's apparent trajectory. This was the first discovery of an incomplete ring, an arc, a structure that was previously unknown and which is unstable if isolated. Indeed, the particles of the inner edge of a ring rotate more quickly than those on the outer edge. Therefore, without a mechanism for confining them, an arc should spread quickly to form a complete ring. In 1989, Voyager 2, reprogrammed to observe these structures, observed four narrow, diffuse rings made up of micrometer-sized dust ((Porco *et al.* 1995; see also Figures 1.15 and 1.18). The narrow Adams and Le Verrier rings have radii of 62,932 and 53,200 km. The Lassell ring, 4,000 km wide, spreads inward, between the Le Verrier and Arago rings. A very faint ring may share the orbit of the satellite Galatea, 980 km inside the Adams ring. These faint rings are undetectable by occultation.

On the Adams ring, three arcs are superimposed, covering about 40 degrees of longitude and whose optical depth, on the order of 0.1, clearly explains the events seen from Earth. These three arcs are called "Liberté, Egalité, Fraternité". Since Voyager's flyby, Liberté has split into two components, Liberté 1 and Liberté 2, and a fifth arc has appeared, named "Courage".

The Adams ring is located in proximity to the 42:43 mean-motion resonance with the Galatea satellite; in other words, the ring makes exactly 42 revolutions around the planet while the satellite makes 43. The stability of Neptune's arc can be explained by resonant interactions with this satellite, which has an orbit that is both inclined and eccentric (see Figure 1.18).

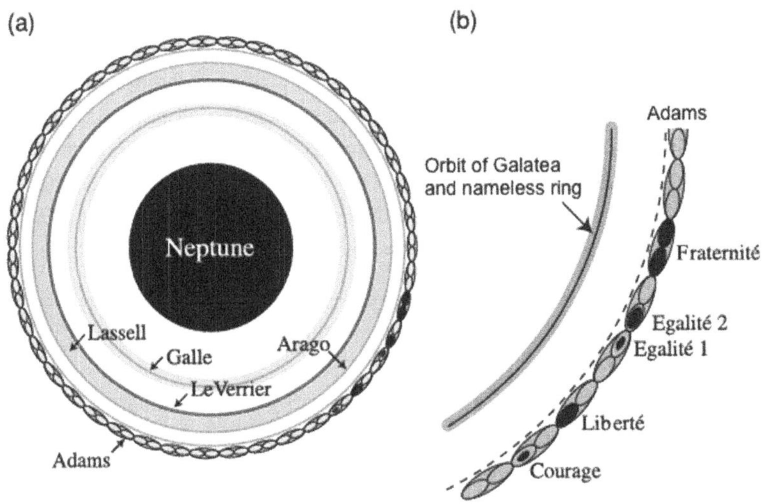

Figure 1.18. *Neptune's rings seen from the south pole (source: (Pater 2015), thanks to C. Murray)*

1.5.7. *The rings of small bodies*

1.5.7.1. *Chariklo-Haumea*

The star occultation technique revealed that the outer Solar System's small bodies can also have rings (see Figure 1.19). In a few years, ring systems have been discovered around at least two of these small bodies, Chariklo and Haumea. The discovery that small bodies a few hundred kilometers wide could be surrounded by dense rings, which we believed to be the prerogative of the giant planets, was an enormous surprise. Chariklo is a 250 km large Centaur orbiting between Saturn and Uranus. It is one of the darkest objects in the Solar System, which became captured in its orbit about 10 million years ago, a very short time compared to the age of the Solar System. The stability of Chariklo's rings requires that the particles have very low density.

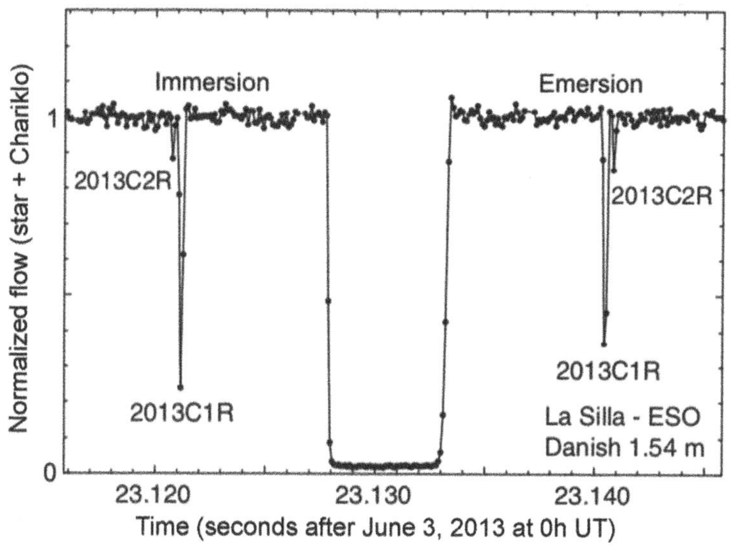

Figure 1.19. *Occultation of a star by the Chariklo system, observed on June 3, 2013: stellar flux over time (source: based on (Sicardy* et al. *2018))*

Haumea is a dwarf planet, one of the largest trans-Neptunian objects known, along with Eris and Pluto. It is an ellipsoid measuring 2,300 km by 1,705 km by 1,035 km, orbiting at 50 astronomical units from the Sun. In January 2017, a stellar occultation revealed the presence of a 70 km wide ring around this dwarf planet, probably in the same orbital plane as the satellite Hi'iaka.

Another centaur, Chiron, may also have rings, even if observations of it could rather be explained by transitory cometary activity.

In the cases of Chariklo and Haumea, the rings are narrow, with sharp edges, which requires a containment mechanism, possibly due to small satellites that have not yet been discovered.

Two orders of magnitude separate the size of planetary rings from these mini-rings, which may nevertheless have similar mechanisms. But differences exist; in particular, the gravity field of these small bodies can exhibit relatively significant irregularities. The departure from sphericity of a planet is limited to its flattening, and so a 5 km high mountain on the surface of Chariklo is equivalent to a satellite of Titan's size near the surface of Saturn. These irregularities create resonances between the rotation period of the particles and the rotation period of the central body. Because of these resonances, the particles inside the co-rotation (the place where the particles rotate at the same speed as the central body) will be slowed down and will fall to the body's surface, and the outer particles will move away from the central body. A 5 km mountain could empty the co-rotation zone in less than a million years. As for Haumea, the body's elongation has a similar effect that is even faster, within a few dozen years.

If the Roche limit is inside the resonances, the ring will quickly pass the Roche limit and be transformed into a satellite. If the Roche limit is outside the resonances, a ring can exist inside the zone between the resonances and the Roche limit. This is possible if the central body is rotating quickly enough on its axis.

Two hypotheses have been proposed to explain the origin of these mini-rings. The first is the capture of a small body that disrupts it as it enters the Roche limit, or after a collision. Another hypothesis is that the matter ejected during outgassing (due to reheating caused, for example, by being closer to the Sun) was slowed and that collisions reduced the energy of the particles to form a ring.

1.5.7.2. *Transitory rings and disks*

In the Solar System, the region between 8 and 20 au can be favorable to stable ice rings around small bodies. Indeed, in this region, ice is too cold to sublimate as it would in a collision closer to the Sun, but is fragile enough to break up in the Roche limit. Furthermore, in this region, the collisions are much less destructive than in the inner Solar System. And finally, this region is favorable to some cometary activity such as what was proposed to explain the matter detected around Chiron and Echeclus.

Mini-rings like those around Chariklo and Haumea probably played a role in the scenario of planetary formation. Crests visible on Rhea and Iapetus (see Figure 1.20 and section 1.3.3.1) may have formed by matter falling from an ancient ring onto the body's equator.

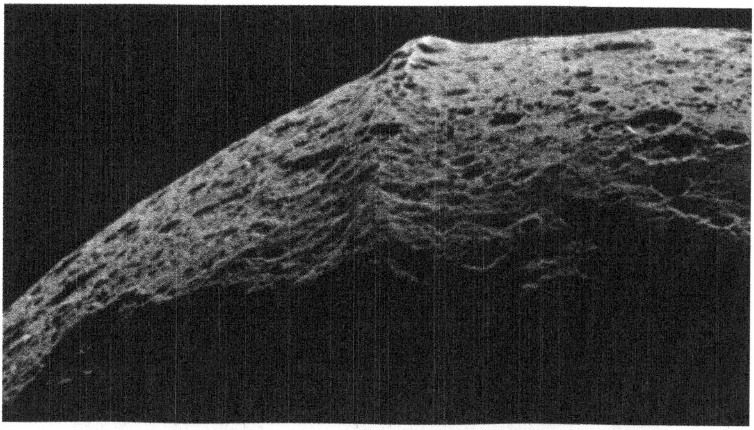

Figure 1.20. *Crest of matter on the surface of Iapetus (source: NASA/JPL-Caltech/Space Science Institute)*

The density of craters on the surface of bodies devoid of atmosphere in the Solar System shows the primordial role of such collisions.

During a collision or a gravitational encounter between two bodies, if a massive enough object stays on, the other bodies will quickly form a ring around this central body. The lifetime of this ring is highly variable depending on the environment. If the system can evolve without too much disturbance, some of the particles will fall on the central body while others will move away quickly, and, once they reach the Roche limit, accrete into a disk. This may explain the large number of double objects in the Kuiper disk. Transitory rings probably also existed in the inner Solar System. For example, the satellites of Mars are in the planet's equatorial plane. They probably formed in a circum-Martian disk that appeared during the capture of planetesimals. The rings discovered around Chariklo and Haumea may be an example of this type of transitory phenomenon, with the particularity that their lifetime must not be too short compared to the age of the Solar System since they were able to be observed.

1.5.8. *Ring dynamics*

All rings present numerous, essentially radial structures with a wide diversity in terms of size. Many are well explained, especially when they are connected to resonances with observed satellites. Some even correspond to resonances with inhomogeneities inside the planet. Many more are still misunderstood, particularly in the vastness of Saturn's rings, the extraordinary beauty of which was revealed to us by the Cassini orbiter (see Figure 1.21).[15]

Figure 1.21. *View of Saturn from the Cassini probe in natural colors (as the human eye would see it) (source: NASA/JPL-Caltech/SSI/Cornell). For a color version of this figure, see www.iste.co.uk/encrenaz/solar2.zip*

1.5.8.1. *Resonances*

When an oscillator is excited by a variable force whose frequency is almost equal to the oscillator's natural frequency, the disturbance can be significant, even for a low amplitude force. This is the case in planetary rings disturbed by satellites whose mass is very small compared to the planet.

There is resonance when the radial frequency of the particles is equal to the frequency of the satellite's horizontal component of force in a rotating frame with

15 See also: https://solarsystem.nasa.gov/missions/cassini/galleries/images/?page=0&per_page=25&order=created_at+desc&search=&tags=cassini&condition_1=1%3Ais_in_resource_list&category=51.

the particle. For these particles, the satellite is disturbed when the particles are in the same phase as its radial oscillation, whose amplitude is subsequently amplified. The particles closest to the exact location of the resonance have the greatest eccentricity. This eccentricity forced by resonance decreases as a function of distance from the resonance.

A similar effect takes place when a vertical oscillation of particles is forced by resonance with a satellite on an inclined orbit, which creates a wave. The precise location where the radial/vertical resonance manifests may be shifted by precession along the lines of apsides/regression of nodes. The effect of this shift is that the horizontal resonances will be found outside the vertical resonances (see the 5:3 resonance with Mimas on Figure 1.22).

Collisions and self-gravity between particles complicate the phenomenon somewhat. Resonant forcing translates into a secular transfer of angular momentum between the ring and its satellites, which can translate into density waves/vertical undulations, empty rings, or sharp edges. This will depend on the respective forces of the satellite and the viscous torque at the location of the resonance in the ring.

The two most important phenomena in Saturn's rings are the confinement of the outer border of the B ring by the 2:1 horizontal resonance with Mimas and the maintenance of the outer border of the A ring by the 7:6 resonance with the co-orbital satellites Janus and Epimetheus. Moderated torque of the same intensity can open empty rings in the C ring with low optical depth but can only generate waves in the A and B rings, which are denser.

Six undulations and about a hundred density waves have been detected in Saturn's rings. These waves are tightly coiled in on themselves (see Figure 1.22), with a typical wavelength of 10 km. The study of these waves has made it possible to estimate the surface density of the rings at 300–600 kg/m^2 in the A and B rings and at 10 kg/m^2 in the C ring or the Cassini division.

The torque applied by the rings on the small satellites causes them to migrate outward on a short time scale compared to the age of the Solar System. Although they are slowed by resonances with more massive outer satellites, Saturn's small satellites like Pandora, Epimetheus, and Prometheus were in the rings a few million years ago.

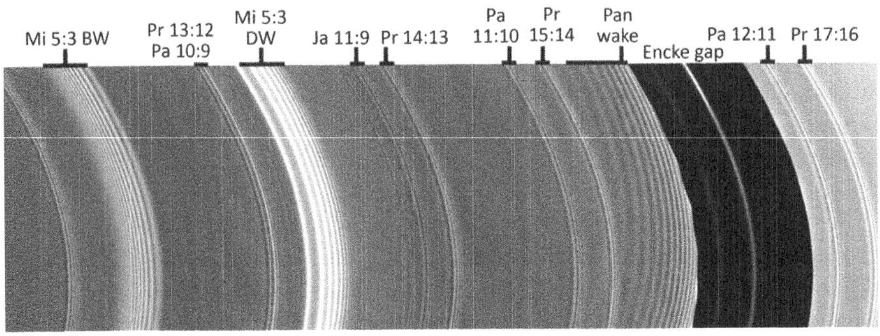

Figure 1.22. *Outer section of Saturn's A ring with a density wave (DW) and a spiral wave (BW) created by the 5:3 resonance with Mimas, as well as other waves associated with Pandora, Prometheus, and Janus (the resonances are indicated). For a color version of this figure, see www.iste.co.uk/encrenaz/solar2.zip*

COMMENTS ON FIGURE 1.22. – *The dark portion to the right is the Encke division (320 km), which contains several small rings. The wavy and spiral edges (wake) visible inside the division are caused by the recent passage of the satellite Pan (source: Lovett, NASA/JPL/CICLOPS).*

1.5.8.2. *Sharp edges*

The rings of Saturn and Uranus have particularly sharp edges corresponding to orbits in resonance with known satellites. The inner and outer edges of Uranus's ε ring are respectively confined by the 14:13 Lindblad resonance with Ophelia and the 24:25 resonance with Cordelia. The shepherd satellites of Uranus's other narrow rings have not been seen, either because they are too small to be detected or because there are other mechanisms of confinement. Saturn's F ring also has two shepherd satellites, Prometheus and Pandora, but its position does not correspond exactly to the position of equilibrium expected based on calculations. It is possible that the hypotheses made about the distribution of matter in the ring, which is included in the calculation, need to be revisited. Whatever the case, the ring and its two shepherds are migrating outward because of density waves provoked by Prometheus and Pandora in the A ring, and they are going to crash into Mimas in about 20 million years.

1.5.8.3. *Dust*

The dust rings, that is to say, the rings of Jupiter and Neptune, and the E and G rings of Saturn, have unusual dynamics. The particles in the rings are subjected to two effects:

– *Solar radiation pressure*, which pushes particles out of the Solar System. Since this is proportional to the grain's surface, while gravitational attraction is proportional to its volume, radiation pressure is particularly effective on small particles, on the order of a micrometer;

– The *Poynting-Robertson effect* (see section 4.9.2), which results from the fact that a particle in orbit around the Sun receives solar radiation along this axis but re-emits photons isotropically in its own frame. The result is preferential forward radiation (the photons finding their frequency increased because of the speed of the particle), which translates into a loss of energy for the particle. This method of slowing the pace, which constantly diminishes the eccentricity of the grain's elliptical trajectory, tends to drive it to become circular and ends up leading to a spiral trajectory. We thus calculate that a one-micrometer grain subjected to the Poynting-Robertson effect alone would fall into the Sun in a few thousand years. In practice, micrometer-sized grains are more sensitive to radiation pressure, which pushes them out of the Solar System. The Poynting-Robertson effect is therefore especially important for centimeter-sized particles.

1.5.9. *The origin of the rings*

It is very difficult to know the age and the way in which the rings formed, but we can estimate the time scale for the mechanisms that modify them. The dust that makes up Jupiter's rings is slowed by the Poynting-Robertson effect or ejected from the ring system by radiation pressure. Their lifetime is very short, on the order of a few million years. The rings are therefore constantly being resupplied by dust coming from the satellites. The arcs of Neptune are also structures that change over very short time scales, since a new arc has appeared in a few years. It is likely that they were created by collisions between large particles.

The visible arcs may be the visualization in dust of collisions between undetected bodies. The dust rings are therefore transitory structures associated with satellites. The rings of Uranus and Saturn are above all composed of large particles that are much less sensitive to non-gravitational forces. They nevertheless are subjected to the processes of erosion: radiation pressure, meteorite impacts, slowing by the exosphere of the planet. The calculation of the time scale of these phenomena for the rings of Uranus shows that particles are destroyed in a few million years. The particles of Saturn's rings are particularly bright. If they were more than 100 million years old, they would have been darkened by material from the meteorites that bombard the rings. But the particles can be constantly reformed in a ring that itself may be much older. In addition, interactions between the edge of the A ring and the satellites Prometheus and Janus translate into a distancing of the satellite on a time scale of 10 million years.

Did the rings form at the same time as the planets in a circumplanetary disk, or are they the result of more recent events such as the break-up of a satellite or the accretion of interplanetary matter? An origin of the large particles in the rings by accretion of interplanetary bodies does not explain why all of the rings rotate in the same direction. Moreover, the number of errant bodies in the Solar System has largely decreased since the planets formed. The rings may be the result of the break-up of a satellite that was destroyed upon entering the Roche limit or by a collision, then was prevented from reforming into a satellite by tidal effects.

In this scenario, it is difficult to explain how a satellite would have arrived at the Roche limit. Indeed, a mechanism can cause a satellite to fall towards the planet, but this only applies in synchronous rotation (where the satellite is orbiting at the same angular velocity as the planet rotates), but this occurs inside the Roche limit!

The rings' large particles are probably the remnants of a circumplanetary disk that has been destroyed and reformed several times.

These calculations show that the rings as we see them today have changed significantly since the formation of the Solar System 4.56 billion years ago. They may have been destroyed and reformed several times since the origin of the Solar System on time scales of several dozen million years.

A recent origin for the rings of Jupiter, Uranus, and Neptune does not create any theoretical difficulties because they have such low mass, below that of the surrounding small satellites. The problem is greater for the rings of Saturn, whose evolution is essentially controlled by resonant torques between the rings and the satellites. Understanding this phenomenon is therefore critical to understanding the origin of the rings of Saturn.

The same phenomena of the transfer of angular momentum via resonant torques and density waves plays a determining role in protoplanetary disks (see section 2.4, Volume 1). They govern, for example, the migration of young planets, whose survival we do not understand given the migration times estimated at less than a million years. The key to the formation of the planets may be found in Saturn's rings.

1.5.10. *An exo-ring*

Although thousands of exo-planets have been discovered,[16] none seem to possess rings. It is very possible that this is a result of observational bias.

Indeed, a number of exoplanets are gas giants but they are often very close to their star, which makes the detection of rings difficult and renders the rings probably very fragile. However, a phenomenon observed in 2007 may signal the discovery of exo-rings around a planet of the star J1407.

Indeed, this star's flux has shown variations that could be interpreted as its transit behind an immense ring, 200 times bigger than Saturn's, that surrounds the planet (Kenworthy and Mamajek 2015). As for the planet itself, it has not transited in front of the star (see Figure 1.23). Models show that this type of ring system would not be stable if it is prograde. But if they are retrograde, these rings have a lifetime of more than 10,000 years.

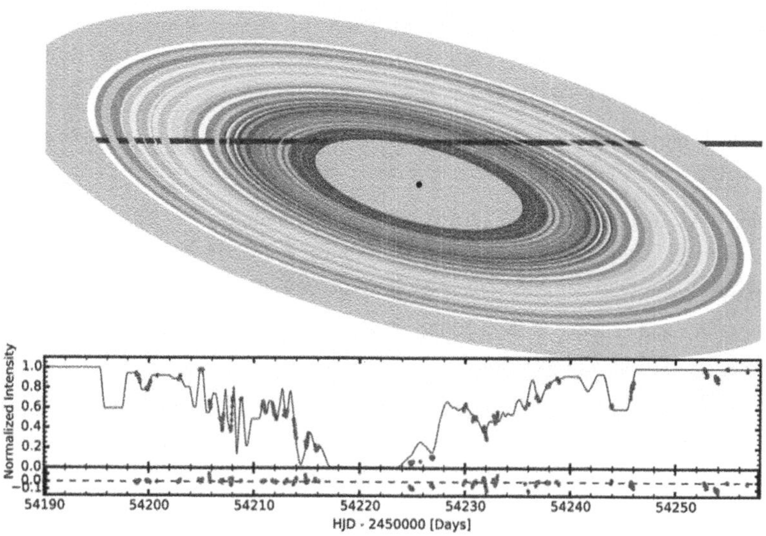

Figure 1.23. *Possible detection by transit of an exo-ring around a planet of the star J1407. Above, a model of the ring and the trajectory of the star at the rear; the planet is represented by a black dot. Below, variations in the star's flux, superimposed on the ring model (source: based on (Kenworthy and Mamajek 2015)). For a color version of this figure, see www.iste.co.uk/encrenaz/solar2.zip*

16 See also: http://exoplanet.eu/catalog/.

1.6. References

Atreya, S.K., Pollack, J.B., Matthews, M.S. (1989). *Origin and Evolution of Planetary and Satellite Atmospheres*. University of Arizona Press, Tucson.

Burns, J.A., Simonelli, D.P., Showalter, M.R., Hamilton, D.P., Porco, C.C., Throop, H., Esposito, L.W. (2004). Jupiter's ring-moon system. In *Jupiter: The Planet, Satellites and Magnetosphere*, Bagenal, F., Dowling, T., McKinnon, W. (eds). Cambridge University Press, Cambridge.

Catling, D.C. and Kasting, J.F. (2017). *Atmospheric Evolution on Inhabited and Lifeless Worlds*. Cambridge University Press, Cambridge.

Coustenis, A. (2014). Titan. In *Encyclopedia of the Solar System*, 3rd edition, Spohn, T., Breuer, D., Johnson, T.V. (eds). Elsevier, Saint-Louis.

Coustenis, A. (2015). The Cassini-Huygens mission. In *Encyclopedia of Astrobiology*, 2nd edition, Gargaud, M., Amils, R., Cernicharo, J., Cleaves II, H.J., Kobayashi, K., Pinti, D., Viso, M. (eds). Springer, Berlin.

Coustenis, A. (2021). The atmosphere of Titan. In *Oxford Research Encyclopedia of Planetary Science*, Read, P. (ed.). Oxford University Press, Oxford. DOI: 10.1093/acrefore/9780190647926.013.120.

Coustenis, A. and Encrenaz, T. (2013). *Life Beyond Earth: The Search for Habitable Worlds in the Universe*. Cambridge University Press, Cambridge.

Coustenis, A. and Raulin, F. (2015). Titan astrobiology. In *The Encyclopedia of Astrobiology*, 2nd edition, Gargaud, M., Amils, R., Cernicharo, J., Cleaves II, H.J., Kobayashi, K., Pinti, D., Viso, M. (eds). Springer, Berlin.

Coustenis, A. and Taylor, F.W. (2008). *Titan: Exploring an Earth-like World*. World Scientific Publishing, Singapore.

Coustenis, A., Bampasidis, G., Achterbergh, R.K., Lavvas, P., Jennings, D.E., Nixon, C.A., Teanby, N.A., Vinatier, S., Flasar, F.M., Carlson, R.C. (2013). Evolution of the stratospheric temperature and chemical composition over one Titanian year. *Astrophysical Journal*, 799(177).

Coustenis, A., Taylor, F.W., Plainaki, C. (2018). Climate issues from the planetary perspective and insights for the Earth. In *Global Change and Future Earth: The Geoscience Perspective*. Beer, T., Li, J., Alverson, K. (eds). Cambridge University Press, Cambridge.

Coustenis, A., Jennings, D., Achterberg, R., Lavvas, P., Bampasidise, G., Nixon, C.A., Flasar, F.M. (2020). Titan's neutral atmosphere seasonal variations up to the end of the Cassini mission. *Icarus*, 344, 113413.

Cruikshank, D.P. (2016). *Neptune and Triton.* University of Arizona Press, Tucson.

Dougherty, M., Coustenis, A., Lorenz, R. (2010). Titan beyond Cassini-Huygens. In *Titan from Cassini-Huygens*, Brown, R.H., Lebreton, J.-P., Waite, H. (eds). Springer-Verlag, New York.

Encrenaz, T. and Coustenis, A. (2017). Atmospheres of terrestrial planets: Mars, Venus and Titan. In *Handbook of Exoplanets*, Deeg, H.J., Belmonte, J.A. (eds). Springer, Berlin.

Esposito, L. (2010). Composition, structure, dynamics and evolution of Saturn's rings. *Annual Review of Earth and Planetary Sciences*, 38, 383–410.

French, R., Nicholson, P., Porco, C., Marouf, E. (1991). Dynamics and structure of the uranian rings. In *Uranus*, Bergstralh, J. (ed.). University of Arizona Press, Tucson.

Greenberg, R. and Brahic, A. (1984). *Planetary Rings.* University of Arizona Press, Tucson.

Grün, E., Gustavson, B.A.S., Dremott, S.F., Fechtig, H. (2001). *Interplanetary Dust.* Springer-Verlag, Berlin.

Hand, K.P., Sotin, C., Hayes, A., Coustenis, A. (2020). On the habitability and future exploration of ocean worlds. *Space Science Reviews*, 216, 95.

Horst, S.M. (2017). Titan's atmosphere and climate. *J. Geophys. Res. Planets*, 122, 432–482. DOI:10.1002/2016JE005240.

Kenworthy, M.A. and Mamajek, E.E. (2015). Modeling giant extrasolar ring systems in eclipse and the case of J1407b: Sculpting by exomoons? *Astrophysical Journal*, 800, 126.

Lang, K.R. (2011). *The Cambridge Guide to the Solar System.* Cambridge University Press, Cambridge.

Lissauer, J.J. and de Pater, I. (2019). *Fundamental Planetary Science: Physics, Chemistry and Habitability.* Cambridge University Press, Cambridge.

Lorenz, R.D. (2020). *Saturn's Moon Titan: From 4.5 Billion Years Ago to the Present – An Insight into the Workings and Exploration of the Most Earth-like World in the Outer Solar System (Owners' Workshop Manual).* Haynes Publishing UK, Somerset.

Lorenz, R.D. and Mitton, J. (2008). *Titan Unveiled.* Cambridge University Press, Cambridge.

Pappalardo, R.T., McKinnon, W.B., Khurana, K. (2009). *Europa.* University of Arizona Press, Tucson.

de Pater, I. and Lissauer, J.J. (2015). *Planetary Sciences.* Cambridge University Press, Cambridge.

Porco, C.C., Nicholson, P.D., Cuzzi, J.N. (1995). Neptune's ring system. In *Neptune and Triton*, Cruikshank, D. (ed.). University of Arizona Press, Tucson.

Schenk, P.M. (2010). *Atlas of the Galilean Satellites*. Cambridge University Press, Cambridge.

Schenk, P.M., Clark, R.N., Howett, C.J.A., Verbischer, A., Waite, J.H. (2018). *Enceladus and the Icy Moons of Saturn*. University of Arizona Press, Tucson.

Sicardy, B., El Moutamid, M., Quillen, A.C. (2018). Rings beyond the giant planets. In *Planetary Ring Systems, Properties, Structure and Evolution*. Tiscareno, M.S. and Murray, C.D. (eds). Cambridge University Press, Cambridge.

Teanby, N.A., Sylvestre, M., Sharkey, L., Nixon, C.A., Vinatier, S., Irwin, P.G.J. (2019). Seasonal evolution of Titan's stratosphere during the Cassini mission. *Geophysical Research Letters*, 46, 3079–3089.

Tiscareno, M.S. and Murray, C.D. (2018). *Planetary Ring Systems, Properties, Structure, and Evolution*. Cambridge University Press, Cambridge.

2

Comets, Asteroids, and Dwarf Planets

Jacques CROVISIER and Marcello FULCHIGNONI
LESIA, Paris Observatory, PSL University, Paris, France

This chapter concerns comets (section 2.1), asteroids, including the trans-Neptunian objects (sections 2.2 and 2.3) and the dwarf planets (section 2.4), which include other small bodies that were previously considered to be asteroids, as well as Pluto, which was classified as a planet until recently. As we will see, the distinction between these various categories is fuzzy, with the discovery of a range of small bodies intermediate between comets and asteroids, hence the difficulty of reaching consensus on the distinction between planetoids and large asteroids. All of these small bodies are left over from the formation of our Solar System 4.6 billion years ago. While planets and satellites have changed over the millennia, many of these small pieces of ice, rock, and metal have not, such that they are now to some extent fossils representing the first phases of planetary evolution.

2.1. Comets

Comets no long frighten us, but they still surprise us...

D.K. Yeomans (1991)

The comets – sometimes spectacular, often unexpected, and long unexplained phenomena – have intrigued us and even frightened us since the beginning of time. Some of their mystery faded beginning in the 17th century with the determination of their orbits, but the fear of a collision with Earth arose instead. Knowledge of their

The Solar System 2,
coordinated by Thérèse ENCRENAZ and James LEQUEUX. © ISTE Ltd 2021.

physical nature and the mechanism of their activities only came in the 19th century with improvements in observation methods and especially the development of spectroscopy. But it was only at the end of the 20th century, with space exploration, that we were finally able to obtain images of their cores and to analyze cometary matter *in situ* (see Figures 2.1, 2.2, and (Yeomans 1991; Festou *et al.* 2004; Meierhenrich 2014; Rickman 2018)).

Figure 2.1. *An active comet seen from a distance. The comet C/1995 O1 (Hale-Bopp) observed from Earth at a distance of 1.4 au on April 6, 1997. The dust and ionized gas tails extend over dozens of millions of kilometers (source: Nicolas Biver). For a color version of this figure, see www.iste.co.uk/encrenaz/solar2.zip*

2.1.1. *Definition and nomenclature*

The definition of a comet is a delicate affair. The International Astronomical Union, which specified the difference between planets and asteroids in 2006, has declined to address the subject. A comet is, in effect, both an icy small body and the phenomenon that occurs when this body approaches the Sun and is heated: the ice sublimates, releasing an atmosphere and trailing dust – this is the coma. Under the Sun's radiation pressure, a wide, curved dust tail develops. And the interaction with solar wind produces another fine, straight tail, composed of ions (see Figure 2.1). The comet's nucleus itself is too small to be able to be seen from Earth, and so direct study requires space exploration (see Figure 2.2). Since there is a wide range of intermediate small bodies between asteroids and comets, real ambiguity lingers over their classification. Distant icy objects such as those in the Kuiper belt are still potential comets if they happen to approach the Sun.

Figure 2.2. *An active comet seen from close up. The comet 67P/Churyumov-Gerasimenko observed from a distance of 198 km by the Rosetta space probe on February 18, 2015. The nucleus is about 4 km wide. The escaping dust jets were intensified using appropriate image processing (source: ESA/Rosetta/NAVCAM)*

Under the aegis of the International Astronomical Union, the *Central Bureau for Astronomical Telegrams* and the *Minor Planet Center* are responsible for taking a census of new comets and their orbits, and for attributing an official name to them. Following the rule established in 1995,[1] each new comet that is discovered is designated by the prefix "C/" followed by the year, a letter indicating the fortnight, then an order number.

Thus, C/2012 F6 is the sixth comet discovered during the second fortnight of March 2012. If the comet's orbital period has been determined and is less than 200 years, the prefix becomes "P/" (for "periodic"). When a periodic comet has been observed several times, the registration number simply becomes an order number preceding the "P". The first, 1P, is Halley's comet. On December 18, 2020, there were 409 periodic comets numbered in this way. The prefix is changed to "D/" (for "disappeared" or "defunct") when the comet cannot be found; the most famous, 3D/Biela, which it was feared would encounter Earth, was expected for most of the 19th century. Sometimes we use the prefix "X/" for comets that have appeared, but whose orbit could not be established precisely, and the prefix "A/" for small bodies incorrectly classified as comets that have turned out to be asteroids (see Tables 2.1 and 2.2).

1 See also: https://www.minorplanetcenter.net/iau/lists/CometResolution.html.

Comet	T	q (au)	e	i (°)	P (years)	Comments
C/1743 X1 (Chéseaux's Comet)	3/1/1744	0.222	1.00			
D/1770 L1 (Lexell)	8/14/1770	0.674	0.786	1.6	5.6	Passage near Earth
C/1811 F1 (Great Comet)	9/12/1811	1.035	0.995	106.9	395	
C/1819 N1 (Great Comet)	6/28/1819	0.341	1.00	80.6		
C/1858 L1 (Donati)	9/30/1858	0.578	0.996	117.0	2,000	
C/1864 N1 (Donati)	8/16/1864	0.909	0.996	178.1	4,000	First cometary spectrum
C/1910 A1	1/17/1910	0.129	0.9999	138.8		
C/1957 P1 (Mrkos)	8/1/1957	0.355	0.9994	93.9		
C/1965 S1 (Ikeya-Seki)	10/21/1965	0.0078	1.00	141.9		Sungrazing comet
C/1969 Y1 (Bennett)	3/20/1970	0.538	0.9962	190.		
C/1973 E1 (Kohoutek)	12/28/1973	0.142	1.00	14.3		Observation campaign
C/1975 V1 (West)	9/25/1976	0.197	1.00	43.1		Rupture of the nucleus
C/1981 E1 (Bowell)	3/12/1982	3.364	1.057	1.7		Hyperbolic comet
C/1983 H1 (IRAS-Aracki-Alcock)	5/21/1983	0.991	0.990			Passage near Earth
C/1988 Y1 (Yanaka)	11/12/1988	0.488	1.00	71.0		Atypical composition
D/1993 F2 (SL 9)						Collision with Jupiter
C/1996 B2 (Hyakutake)	5/1/1996	0.230	0.9998	124.9	9,000	Passage near Earth
C/1995 O1 (Hale-Bopp)	4/1/1997	0.914	0.995	89.4	4,000	Observation campaign
C/2006 P1 (McNaught)	1/12/2007	0.171	1.00	77.8		
C/2012 S1 (ISON)	11/28/2013	0.0124	1.00	62.4		Sungrazing comet
C/2014 Q2 (Lovejoy)	1/30/2015	1.290	0.998	80.3	13,900	
C/2016 R2 (PanSTARRS)	5/9/2018	2.602	0.996	58.2	19,500	Atypical composition
2I/2019 Q4 (Borisov)	12/8/2019	2.006	3.35	44.1		Interstellar comet

Table 2.1. *Some notable unnumbered comets*

The practice in the past was to designate comets with the name of their discoverer. That is how Charles Messier (1730–1817), Caroline Herschel (1750–1848), Jean-Louis Pons (1761–1831), and others became immortalized (limiting ourselves to the great comet hunters of the 18th and 19th centuries). This tradition was maintained in 1995 by keeping the name of the discoverer(s) in the registration of comets. (This practice is therefore different from the one used for asteroids, where the discoverer has the privilege of naming the object). Two names (previously up to three) can be associated with simultaneous discoveries. This name can also be that of the observatory or the station where the discovery occurred (such as for Soho, Linear, Panstarrs, and Ison). Thus, C/2012 F6, discovered at the Mount Lemmon Observatory in Arizona, became C/2012 F6 (Lemmon). But it is important to be careful not to leave out the registration (as is often the case in popular writing and even in some professional publications) since a single observer is often the discoverer of several comets.

Comet	T	q (au)	e	i (°)	P (years)	Comments
1P/Halley	2/9/1986	0.587	0.967	162.2	76.0	Space exploration
2P/Encke	6/27/2015	0.336	0.848	11.8	3.30	
3D/Biela	11/26/1832	0.879	0.751	13.2	6.65	Rupture of the nucleus
9P/Tempel 1	7/5/2005	1.505	0.518	10.5	5.51	Space exploration
19P/Borrelly	9/14/2001	1.358	0.624	30.3	6.86	Space exploration
21P/Giacobini-Zinner	9/5/1985	1.028	0.707	31.9	6.59	Space exploration
28P/Neujmin 1	3/6/2006	1.551	0.776	14.3	18.2	Extinct comet
29P/Schwassmann-Wachmann 1	9/4/2019	5.761	0.041	9.4	14.7	Atypical comet
43P/Wolf-Harrington	8/19/2016	1.357	0.595	16.0	6.13	Atypical composition
46P/Wirtanen	12/12/2018	1.055	0.658	11.7	5.43	Passage near Earth
49P/Arend-Rigaux	2/10/2012	1.423	0.600	19.0	6.72	Extinct comet
67P/Churyumov-Gerasimenko	8/13/2015	1.243	0.641	7.0	6.44	Space exploration
73P/Schwassmann-Wachmann 3	6/6/2006	0.939	0.693	11.4	5.36	Rupture, passage near Earth
81P/Wild 2	9/25/2003	1.594	0.538	3.2	6.41	Space exploration
95P/Chiron	2/17/1996	8.421	0.382	6.9	50.4	Centaur
103P/Hartley 2	10/28/2010	1.059	0.695	13.6	6.47	Space exploration
109P/Swift-Tuttle	12/12/1992	0.958	0.964	113.4	133	Perseids
133P/Elst-Pizarro	2/9/2013	2.650	0.162	1.4	5.62	Main belt
153P/Ikeya-Zhang	3/10/2002	0.507	0.990	28.1	366	Halley family

Table 2.2. *Some notable short-period comets (among the numbered comets). The return to be referenced had to be selected (date of passage at perihelion T): it may indicate the date of space exploration, of a passage near Earth, or other*

These rules of nomenclature are crawling with exceptions, usually for historical reasons. Thus, some comets, such as 1P/Halley or 2P/Encke, bear the name of the astronomer who identified the orbit rather than one of the discoverers. Asteroids that turn out to have cometary activity keep their asteroid name preceded by the prefix "P/". However, some small bodies have a dual classification: this is the case for the Centaurs (2060) Chiron = 95P/Chiron and (60 558) Echeclus = 95P/Echeclus.

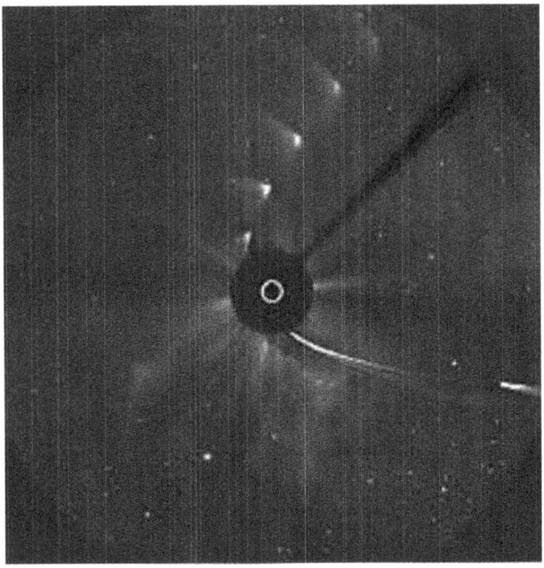

Figure 2.3. *The sungrazing comet C/2012 S1 (ISON) as seen by the Soho space coronagraph; montage of several images taken during its passage near the Sun; the white circle represents the solar disk (source: NASA). For a color version of this figure, see www.iste.co.uk/encrenaz/solar2.zip*

The 17th edition of the *Catalogue of Cometary Orbits* (2008) edited by the *Central Bureau* lists 3,700 appearances of 2,844 distinct comets. The *Central Bureau* decided to stop cataloging SOHO comets discovered by the Solar and Heliospheric Observatory, which are essentially sungrazing comets that are basically debris a few meters in size, for the most part probably lacking in ice, detected because of their passage very close to the Sun (see Figure 2.3 and (Jones *et al.* 2018)). The role of amateur astronomers in the study of comets should be highlighted. They are still involved in the discovery of new comets, despite competition from automatic stations that scan the sky constantly. Moreover, their diligence remains crucial for monitoring how comets change, which can be done with modest instruments based on visual observations.

2.1.2. *The orbits and families of the comets*

Cometary orbits are updated and distributed by the *Central Bureau for Astronomical Telegrams*, which was succeeded in this task by the *Minor Planet Center,*[2] under the auspices of the International Astronomical Union, and independently by the Horizons team at the Jet Propulsion Laboratory (JPL, NASA)[3] and the *Institut de mécanique céleste et de calcul des éphémérides* (IMCCE)[4] in Paris.

As for the other objects in the Solar System, cometary orbits follow Kepler's laws (see section 1.2.2, Volume 1) and are described using six parameters, presented in Box 2.1.

– The time T of the passage at perihelion.

– The distance q of the comet at perihelion.

– The argument ω of perihelion.

– The longitude Ω of the ascending node.

– The inclination i of the orbital plane.

– The eccentricity e of the conic.

For elliptical orbits ($e < 1$), we determine the semi-major axis a as follows: $q = a(1-e)$, and the distance at aphelion as follows: $Q = a(1+e)$. The orbital period P is linked to a according to Kepler's third law $a^3 = P^2$ (the distances q, Q, and a being expressed in astronomical units and the period P in years).

Box 2.1. *The orbital elements*

The discovery of the periodicity of the return of comets is attributed to Edmond Halley (1656–1742), who established the first catalog of cometary orbits by using Newton's methods. He suggested that the comets that appeared in 1531, 1607, and 1682, which had similar orbital elements, could be one and the same object. The confirmation came in 1759 with the return of the comet that now bears his name.

2 See also: https://www.minorplanetcenter.net/iau/Ephemerides/Comets/index.html.

3 See also: https://ssd.jpl.nasa.gov/?horizons.

4 See also: https://www.imcce.fr/services/ephemerides/.

But the comets do not exactly follow Kepler's laws, and their orbits are subjected to several perturbations. First, gravitational perturbations by the giant planets – primarily Jupiter. This must be taken into account to understand the irregularities of Halley's comet and the appearance and subsequent disappearance of the comet D/1770 H1 (Lexell). Then, nongravitational perturbations caused by the very nature of the comets, provoked by a non-isotropic ejection of matter from the nucleus, can induce a rocket effect. It is this effect that explains the variations in the orbit of comet 2P/Encke and that led Fred Whipple (1906–2004) to propose his "dirty snowball" model.

Consulting catalogs of cometary orbits immediately indicates that there are two main families:

– short-period comets with low inclinations to the ecliptic. Most have orbital periods close to 6 or 12 years. The evolution of their orbits is dominated by strong gravitational interactions with Jupiter. We call them "Jupiter-family comets" or "ecliptic comets". We believe they come from the Kuiper belt. These comets have made several returns near the Sun. They are exhausted and have weak activity;

– comets with any inclination whatsoever on the ecliptic, known as "*nearly isotropic comets*". The most spectacular comets belong to this category. They can have long orbital periods: these are "dynamically new" comets with quasi-parabolic orbits. They also can have a short period, such as Halley's comet: these are the "Halley family comets". It was the constant arrival of these new comets that led, following the proposition of Ernst Öpik (1893–1985) and Jan Oort (1900–1992), to the hypothesis of a distant, isotropic reservoir of comets, the Oort cloud. For this reason, these comets are often called "Oort cloud comets".

To distinguish Jupiter-family comets, a convenient and precise indicator is Tisserand's parameter relative to Jupiter:

$$T_J = \frac{a_J}{a} + 2\sqrt{\frac{a}{a_J}(1 - e^2)}\cos i \qquad [2.1]$$

where a_J is the semi-major axis of Jupiter's orbit (5.2 au), a that of the comet, e and i its eccentricity and its inclination. This parameter has the distinctive characteristic of being invariant during gravitational interaction with Jupiter. For Jupiter-family comets $2 < T_J < 3$, while for long-period or Halley family comets, $T_J < 2$ and for the asteroids, $T_J > 3$.

In the *Catalogue of Cometary Orbits*, we find comets with hyperbolic orbits. Their eccentricities are only slightly higher than 1, with the most eccentric being the comet C/1981 E1 (Bowell) with e=1.057. Each time we have been able to make this calculation, we have realized that the orbit was initially elliptical and became hyperbolic after a gravitational disturbance, generally a passage near Jupiter. Therefore, these are not objects from outside the Solar System. The first exception occurred in October 2017 with the discovery of the object now called 1I/'Oumuamua. It had an eccentricity of 1.20, which caused it to be immediately cataloged as a comet, even though it did not show obvious cometary activity. Once the object's interstellar origin had been confirmed, a class "I" (for *interstellar*) was specially created for it. But a careful analysis of its orbit seemed to draw attention to nongravitational forces, which could be due to weak comet-type activity (Oumuamua 2019). Then, a veritable interstellar comet was discovered in August 2019. Designated C/2019 Q4 (Borisov) upon discovery, in the end it was named 2I/2019 Q4 (Borisov) after its interstellar origin was confirmed, as shown by its very strong eccentricity (3.3).

Another exotic object, D/1993 F2 (Shoemaker-Levy 9, "SL 9"), crashed into Jupiter in July 1994 (see Figure 2.4; Box 5.1, Volume 1; (Noll *et al.* 1996)). Before its discovery and its orbit around Jupiter, it was unknown and probably inactive. But was it a comet or a simple asteroid (from the Hilda family) broken up by tidal forces during its first passage near Jupiter, with the tails accompanying the fragments being dust that was ejected during fragmentation? No trace of gas has been detected around these fragments. Water vapor was observed in the impact scars after the collision with Jupiter, and this water could have come from the object itself.

Figure 2.4. *D/1973 F2 (Shoemaker-Levy 9), the comet (if it is one) that was destroyed by tidal forces after a passage near Jupiter, before it crashed into the planet (source: NASA). For a color version of this figure, see www.iste.co.uk/encrenaz/solar2.zip*

It now seems that the difference between comets and asteroids is not as cut and dried as we thought, and that there is a *continuum* between comets and asteroids. We find active asteroids (archetype: (3200) Phaethon); the Centaurs (archetype: Chiron); comets in the main belt (archetype: 133P/Elst-Pizarro); and depleted comets that have become practically inactive (archetypes: 28P/Neujmin 1 and 49P/Arend-Rigaux; Jewitt *et al.* 2015; Snodgrass *et al.* 2017); see also section 2.2, on asteroids.) The dwarf planet (1) Ceres (see section 2.4.1) and the satellite Enceladus also exhibit pseudo-cometary activity (see section 1.3.2).

The relatively late discovery of the comets in the main belt may be a result of their weak activity: the first to be discovered, 133P/Elst-Pizarro, was only found in 1996. Their activity is probably due to ice present in the interior. But the direct observation of water or other volatile molecules is still to come; their weak production rate is still below the detection threshold of instruments such as those on the Herschel satellite. Activity due to impacts rather than to ice sublimation has however proven true for some objects, such as P/2010 A2 (Linear) and P/2016 G1 (Panstarrs).

Sungrazing comets are short-period comets at perihelion, typically less than two Solar radii (0.01 au). Some can become very spectacular, such as C/1965 S1 (Ikeya-Seki). Once raised to very high temperatures, their nuclei rarely survive their passage near the Sun. Spatial coronagraphs, like Solwind, Soho/Lasco, and Stereo/Secchi have made it possible to observe many of them, with most being only debris a few meters in size. Heinrich K. Kreutz (1854–1907), remarking that many of these comets have similar orbits, suggested that they were the result of fall-out from a single body: this is called the Kreutz group. C/1965 S1 (Ikeya-Seki) is one of them ((Jones *et al.* 2018); see also Figure 2.3).

2.1.3. *Cometary magnitude*

When a comet is inactive (which can be the case far from the Sun), its magnitude is equivalent to that of its nucleus, similar to an asteroid:

$$m_v = H_n + 5 \log \Delta + 5 \log r_h$$

where Δ is the distance from the observer and r_h is the heliocentric distance.

When a comet is active, it is most often the brightness of its coma that dominates:

$$m_v = H_0 + 2{,}5\, k \log \Delta + 2{,}5\, n \log r_h$$

Usually, $k = 2$, and n is typically between 3 and 4.

Box 2.2. *Cometary magnitude*

For comets, we can consider two types of magnitudes: total magnitudes (including the brilliance of the coma, which develops during passage at perihelion) and nuclear magnitudes, which only concern the nucleus. Far from the Sun, in the absence of outgassing, the total magnitude of a comet is reduced to its nuclear magnitude.

If the nuclear magnitude of a comet is known (which depends on the size and albedo of the nucleus) in a given position, its nuclear magnitude can be calculated for any other position: the flux of the nucleus observed from Earth varies as r^{-2} and Δ^{-2}, r being its heliocentric distance and Δ its geocentric distance. We therefore have the following equation:

$$m_v = H_n + 5 \log \Delta + 5 \log r_h \quad [2.2]$$

where Δ and r are expressed in au; H_n is the magnitude that the nucleus would have at 1 au from Earth and 1 au from the Sun, in the absence of a coma. This relation is valid for any filter, and even for the flux integrated over the whole visible range. For example, in the case of Halley's comet, we have, for the entire visible range, $H_0 = 13.8$.

The variation of the total magnitude of a comet is more difficult to calculate precisely. When the comet becomes active, the observed flux has two components: the emission from the gas by fluorescence, and emissions due to dust, which is the diffused solar flux. To describe the variation of the total magnitude of a comet as a function of r and Δ, we can use empirical formulas deduced from observations and whose coefficients therefore vary from one comet to another.

Asymmetry in the magnitude curve (in relation to passage at perihelion) is common, as are gaps observed in relation to the simplistic empirical law described above. These variations illustrate the complexity of the phenomena involved in cometary activity. The comet nucleus is heterogeneous, and its abrasion can lead to the exposure of deep layers rich in ice, or, conversely, to the surface being covered by a protective layer of dust that prevents ice from sublimating. The inclination of the nucleus's axis of rotation on the orbital plane can lead to seasonal effects by periodically exposing or hiding active regions to sunlight. Finally, bursts of activity can occur unpredictably after partial or total ruptures of the nucleus.

Databases of cometary magnitudes have been compiled.[5] The contribution of amateur astronomers who provide this information is essential.

5 See also: http://www.icq.eps.harvard.edu/CometMags.html; or http://www.aerith.net/index.html; or https://cobs.si/.

2.1.4. *Space exploration of the comets*

Observations from artificial satellites have taught us much about the comets by extending our access to new spectral ranges. We can specifically mention IUE for ultraviolet observations, IRAS, ISO (comet C/1995 O1 (Hale-Bopp)) and Spitzer for the infrared, Swas, Odin, and Herschel for sub-millimeter observations. And we can't forget the contributions of the Hubble Space Telescope (the collision of SL9 on Jupiter and observations of many comets). But the decisive advances were contributed by *in situ* space exploration of several comets. Eight comets (counting 21P/Giacobini-Zinner and 26P/Grigg-Skjellerup, which were the target of flybys by the ICE and Giotto probes, without functioning cameras) have been explored to date by space probes (see Table 2.3).

Comet	Probe	Agency	Date	D (km)	V (km s^{-1})	
21P/Giacobini-Zinner	ICE	NASA	9/11/1985	7,800	21	(a)
1P/Halley	VEGA 1	IKI	3/6/1986	8,890	79	
1P/Halley	VEGA 2	IKI	3/9/1986	8,030	77	
1P/Halley	Giotto	ESA	3/14/1986	596	68	
1P/Halley	Sakigake	ISAS	3/11/1986			(a)
1P/Halley	Suisei	ISAS	3/18/1986			(a)
26P/Grigg-Skjellerup	Giotto	ESA	7/10/1992	200	14	(a)
19P/Borrelly	Deep Space 1	NASA	9/22/2001	2,170	17	
81P/Wild 2	Stardust	NASA	1/2/2004	237	6	Return
9P/Tempel 1	Deep Impact	NASA	7/4/2005	500	10	Impact
103P/Hartley 2	Deep Impact (EPOXI)	NASA	11/4/2010	700	12	
9P/Tempel 1	Stardust (Stardust-NExT)	NASA	2/14/2011	178	11	
67P/Churyumov-Gerasimenko	Rosetta	ESA	2014–2016	0	0	

Table 2.3. *Space exploration of the comets; D and V are distance and speed relative to the moment of flyby; (a) indicates no imaging possible*

2.1.4.1. *Before Rosetta*

After an approach in 1985 of the comet 21P/Giacobini-Zinner by the ICE probe (International Comet Explorer, a probe initially dedicated to the study of interplanetary plasma that lacked imaging equipment), space exploration of the comets really began in 1986 with the Halley's comet flyby. Two similar Soviet probes, Vega 1 and 2 (Vega is the contraction of Ve for Venus, the probes' first target, and Ga for Halley in Russian), and an ESA probe, Giotto, photographed its nucleus and conducted *in situ* observations during a flyby. Two Japanese probes (Sakigake [pioneer] and Suisei [comet]) also joined this flotilla, from farther away. These space-based studies were supplemented by a major ground-based observation initiative, taking advantage of the International Halley Watch organization. Finally, a comet nucleus could be observed directly, and its size and shape were determined (see Figure 2.5).

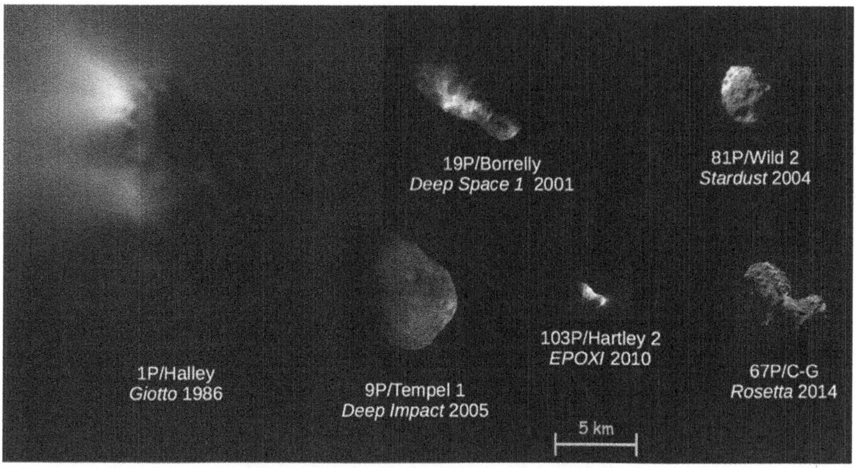

Figure 2.5. *The nuclei of the comets observed by space missions, represented at the same scale (source: ESA, NASA, MPIfA)*

Next was Deep Space 1, a technology mission from NASA destined to test an ion-propulsion engine. It completed a successful flyby of comet 19P/Borrelly in 2001.

The main objective of the Stardust mission, launched by NASA on February 7, 1999, was to retrieve samples taken from the coma of 81P/Wild 2. The comet was encountered on January 2, 2004, with a passage at 237 km from the nucleus. The cometary grains, collected at a speed of 6 km s^{-1}, were trapped in low-density silica gel, an "aerogel", which made it possible to slow them down without destroying them. The samples, contained in a capsule that separated from the probe, returned to Earth on January 15, 2006. These thousands of grains can now be studied in the laboratory, as well as the small impact craters on the collector's casing, equally rich in information. However, the collection process did not make it possible to preserve the ice and other volatile compounds from the cometary material (Brownlee *et al.* 2006).

The Deep Impact mission, launched by NASA on January 12, 2005, conducted an original active experiment on the comet 9P/Tempel 1. On July 4, 2005, a 370 kg "impacter" was detached from the probe and fired at the nucleus at a speed of 10.2 km s^{-1}, equivalent to 19 GJ of energy or 5 tons of TNT. The objective was to force out the nucleus' primitive matter, unaltered by the aging that affects the surface layer. The impact and the development of a plume were observed by the probe, which flew by at 500 km (see Figure 2.6). The ballistic trajectory of the ejected particles made it possible to estimate the nucleus's density (350 kg m^{-3}; see below). The crater created by the impact was observed later by the Stardust probe, renamed Stardust-NExT (New Exploration of Tempel 1) and redirected for a flyby at 178 km from the comet nucleus in February 2011. Renamed Epoxi (Extrasolar Planet Observation and Deep Impact Extended Investigation), Deep Impact was later redirected to comet 103P/Hartley 2, which it flew by in February 2010 (A'Hearn and Johnson 2015).

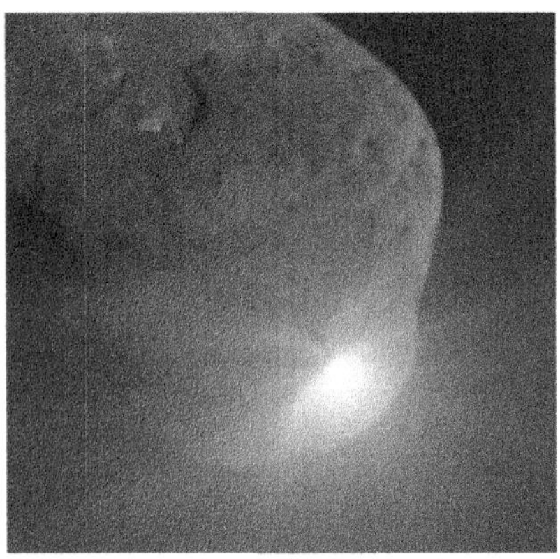

Figure 2.6. *The impact of the Deep Impact probe on the nucleus of 9P/Tempel 1 on July 4, 2005; the nucleus exhibits, in addition to craters, clear spots corresponding to exposed ice (source: NASA/JPL/UMD). For a color version of this figure, see www.iste.co.uk/encrenaz/solar2.zip*

2.1.4.2. *Rosetta*

Our knowledge of the comets was recently increased by the Rosetta mission. While the preceding cometary missions consisted of flybys of these objects, the Rosetta probe entered orbit around a comet and followed it for more than two years. This probe observed at a close distance (from a few kilometers to a few thousand kilometers) the nucleus and its environment with spectrographic cameras. It directly studied the gas, dust, and plasma in the coma in which it flew with its mass spectrometers, microscopes, and plasma probes. A lander, Philae, was sent to the comet's surface (Schulz *et al.* 2009).

The Rosetta probe was launched by ESA on March 2, 2004. Its trajectory was complex and included two gravity-assisted maneuvers around Earth, one around Mars, and the flyby of asteroids (2867) Steins and (21) Lutetia (see section 2.2.4.1 and Figure 2.33). It only arrived at the comet 67P/Churyumov-Gerasimenko in 2014. Then it began a succession of complex orbits in the comet's weak gravitational field. From Rosetta's saga, we should remember a few key moments: its awakening on January 24, 2014 after a two-and-a-half-year hibernation; the surprise caused by the shape of the comet's nucleus, revealed in July 2014 (see Figures 2.2, 2.7, and 2.9); the problems with the Philae module's landing and its destiny; the probe's disorientation after finding itself in a particle cloud that threw off its stellar sensors on March 28, 2015; the voluntary end of the mission on September 30, 2016.

Figure 2.7. *The "neck" of comet 67P/Churyumov-Gerasimenko, in the region connecting the two lobes, is the site of intense outgassing activity (source: ESA/Rosetta/MPS)*

Thus, on November 12, 2014, Philae's landing was watched live. After the first moments of euphoria, the truth had to be accepted: the Philae module, after having ricocheted, landed sideways in an unexpected location. Unable to recharge its batteries as expected, to communicate easily, or to take samples of the nucleus via drilling, Philae was nevertheless able to fulfill part of its mission. Its precise landing location was not known until the end of the mission.

Rosetta was able to observe comet 67P without interruption from June 2014 to the end of September 2016. Its distance from the Sun decreased from 3.95 au (when the comet was observed after the probe's awakening) to 1.24 au (at the moment of the comet's passage at perihelion on August 13, 2015), then increased to 3.83 au on September 30, 2016 (end of mission). Its activity, which we can quantify using the production of water coming from sublimation of the nucleus's ice, varied by more than a factor of a thousand. Sporadic bursts of activity were also observed, not globally as was the case of observations from Earth, but locally, by identifying the discrete sources on the nucleus's surface that were responsible (see Figure 2.7). Rosetta was able to study in detail the different cycles regulating cometary activity. Rosetta was also able to observe the morphological evolution of the nucleus's surface: the appearance of fissures and fractures, the collapse of cliffs, the displacement of pebbles (see Figure 2.8).

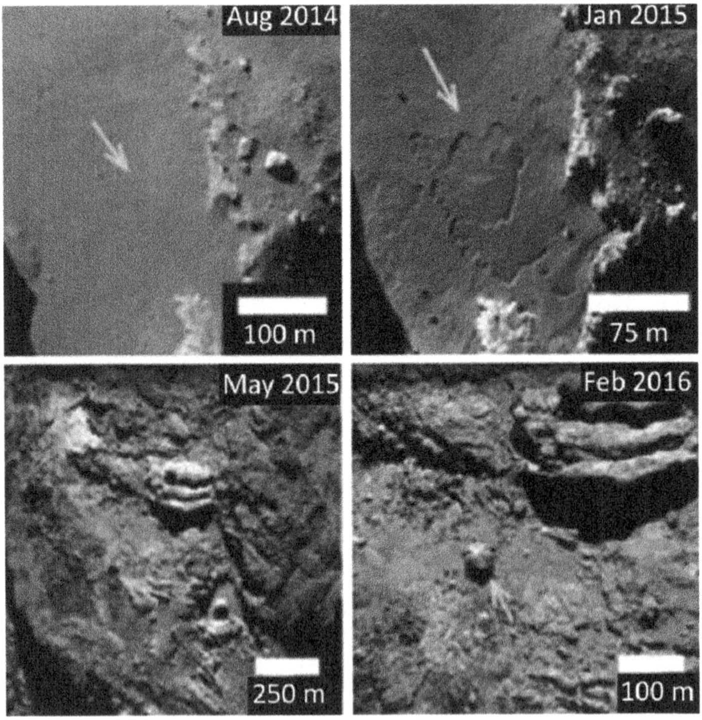

Figure 2.8. *Changes on the surface of 67P/Churyumov-Gerasimenko detected by Rosetta. Above, we see the appearance of strange escarpments on a surface that was previously smooth. Below, we see the displacement over 140 m of a rock 30 m in size (source: ESA/Rosetta/MPS). For a color version of this figure, see www.iste.co.uk/encrenaz/solar2.zip*

2.1.4.3. *After Rosetta*

67P is a short-period comet in the Jupiter family. A mission to a really primitive, long-period comet would be desirable. A high-speed flyby of such a comet, like the one for Halley's comet in 1986 (at a speed of 68 km S^{-1}), is possible with current techniques. But such a rendezvous poses a real challenge, because it is much more demanding in terms of energy. And the often belated discovery of a new comet only leaves a short window for organizing such a mission.

This challenge was accepted by Comet Interceptor, a mission undertaken by ESA for launch in 2028 in the framework of its Cosmic Vision program. Three probes, stationed at the L2 Lagrange Point, are supposed to swoop down on a dynamically new comet, or possibly an interstellar comet. Another project underway is the Chinese ZhengHe mission (launch in 2022) whose second target (after the asteroid 2016 HO_3, where it is supposed to retrieve a sample) is the main-belt comet 133P/Elst-Pizarro.

Finally, the return of a cometary sample preserved in a refrigerated container should logically be the next step for cometary exploration. Several missions have been successively proposed: CESR (Comet Exploration Sample Return, which was the primitive version of Rosetta in 1986), Triple-F (Fresh From Fridge in 2009), Caesar (Comet Astrobiology Exploration Sample Return in 2017), which have since been ruled out. In 2019, the Ambition project, proposed in the ESA's Voyager 2050 long-term project framework, entered the competition.

But space exploration has its limitations. We know that the cometary population presents a wide diversity of orbits, sizes, shapes, and compositions, probably linked to the various locations in which they formed and their diverse evolutions (see section 2.1.8). The cost of space exploration only allows us to explore a small number of these objects. These explorations must therefore be complemented by studies conducted using more traditional means, relating to the entire cometary population; studies that are likely to be both more frustrating and systematic. As in many other scientific disciplines, progress in astronomy cannot come only from spectacular, one-time results at the mercy of technological breakthroughs and irregular financing, but must also proceed by the patient accumulation, austere and lacking in short-term gratification, of observations that will fill databases and where participatory science led by amateur astronomers also has its place.

2.1.5. *The nucleus*

2.1.5.1. *Size, albedo, and rotation*

The determination of the size and shape of cometary nuclei has only been done reliably for a few objects that have benefited from space exploration-based imaging (see Figure 2.5). Another possibility is provided by radar surveys. However, this is only possible for comets that make close approaches to Earth, since the signal received is proportional to Δ^{-4}. For active comets, it is also necessary to distinguish the signal from the nucleus and the one from the surrounding dust (Harmon *et al.* 2004).

When it is possible to separate the brightness of the nucleus from that of the coma, the nuclear magnitude makes it possible to evaluate the size of the nucleus, provided that the albedo is known, just as for asteroids (Lamy *et al.* 2004).

The largest known comet is C/1995 O1 (Hale-Bopp), which has a radius on the order of 30 km. The atypical comet 19P/Schwassmann-Wachmann 1 has a radius of 15 to 20 km, and Chiron, if we consider it to be a comet, is 50 km. The large majority of cometary nuclei have radii of a few kilometers, but rarely less than one kilometer, which suggests that the smallest comets do not survive – or that they are too weak to be detected. Most sungrazing comets detected by Soho are debris of a few meters. But should these be considered comets?

The geometric albedos of cometary nuclei are very weak, in the range of 0.02–0.06, which is comparable to those of primitive asteroids. These bodies are therefore among the darkest natural materials. There is no, or very little, exposed ice on the surface of nuclei, which has been confirmed by the images (see Figure 2.6) and spectra obtained by space probes. The snow in the "dirty snowball" model is really very dirty.

For comets that haven't had the chance to be visited by space probes, the rotation of the nucleus is studied based on the light curve for mostly inactive comets, or using radar observations for comets that approach Earth, or even based on the configuration of the coma, where spiral jets can reveal successive illuminations of active regions. The rotation of nuclei is not always a simple rotation around an axis, but sometimes is a complex motion referred to as the "Euler-Poinsot case". The state of rotation can modify itself under the influence of nongravitational forces that, depending on the situation, accelerate or slow the rotation. The rotation period gives us another constraint on mass and density. A critical value for the rotation period is indeed $P = 3.3\ \rho^{-1/2}$, where the rotation period P is expressed in hours and the density ρ in g cm^{-3}. For shorter periods, the regolith cannot survive in the equatorial region, and in the absence of forces of cohesion, the nucleus will break apart. The shortest periods observed for cometary nuclei are about 6 hours (Samarasinha *et al.* 2004).

2.1.5.2. *Mass, density, and structure*

Our knowledge of the density of cometary nuclei was based first on an estimate of their size, which was itself based on their magnitude and their albedo and also on an estimate of their mass based on a model of the perturbation of their orbits by nongravitational forces. These estimates indicated a low density, but the uncertainty level was very high.

Then, with Deep Impact, it was possible to measure the density of comet 9P/Tempel 1 at 350 kg m^{-3}, based on the trajectory of the particles ejected during impact. Next, Rosetta brought us a precise determination of the mass of comet 67P by orbiting the nucleus. The resulting density was assessed to be 530 kg m^{-3}. These values are much lower than for the least dense meteorites (1,580 kg m^{-3} for Orgueil) or the known asteroids (1,200 kg m^{-3} for (101955) Bennu and (162173) Ryugu).

Given that their density is lower than that of compact ice and much lower than that of rocky grains, cometary nuclei are therefore porous. They are also fragile, which is shown by the numerous observations of nucleic ruptures (3D/Biela in 1846, C/1975 V1 (West) in 1976, 73P/Schwassmann-Wachmann 3 in 1995, and many others).

Figure 2.9. *Studies of the nucleus of the comet 67P/Churyumov-Gerasimenko: geometric model of its shape and decomposition into geological terrains (source: ESA/Rosetta/CNES/OSIRIS/LAM/Latmos). For a color version of this figure, see www.iste.co.uk/encrenaz/solar2.zip*

While some nuclei present a globular, almost spherical shape (81P/Wild 2, 9P/Tempel 1), others are elongated (1P/Halley, 19P/Borrelly), and some are even "bi-lobed," suggesting that two nuclei collided and stuck together (103P/Hartley 2, 67P/Churyumov/Gerasimenko; see Figures 2.5 and 2.9). These shapes are also seen in some asteroids, such as (486958) Arrokoth (see section 2.2.4).

2.1.6. *The atmosphere*

2.1.6.1. *The sublimation of ice*

The sublimation of ice in the nucleus is the engine of cometary activity, leading to the development of an atmosphere. When a comet nucleus, slowly rotating on its axis, slowly approaches the Sun, the temperature on its surface slowly rises. If this surface is coated with volatile matter, the state of equilibrium is represented by the following relation.

$$F(1 - A) / r_h^2 = \eta\,[\sigma T^4 + L Z]$$

where:

F = solar constant = 1,360 W m^{-2} at 1 au;

A = albedo;

$\eta = 1$ (surface perpendicularly exposed to the Sun) to 4 (rapidly rotating body);

σ = Stephan's constant: = 5.67×10^{-8} W m² K^{-4};

L = latent heat of sublimation = 2,500 kJ kg^{-1} for water ice;

Z = sublimation rate in kg s^{-1} m^{-2}.

Box 2.3. *Thermal balance and the sublimation of ice*

The first term in the equation expresses the solar flux absorbed by the nucleus. A is the nucleus's albedo, F the solar flux at a distance of 1 au, and r_h is the heliocentric distance. The second term expresses the energy emitted by the surface: on one hand, thermal emission (σ being Stephan's constant and T temperature); on the other hand, the result of the latent heat of sublimation $L(T)$ by the gas production rate $Z(T)$.

While the sublimation of water on the nucleus's surface can explain cometary activity up to about 4 au from the Sun, at greater distances, the surface temperature falls below 140 K, preventing the effective sublimation of water ice.

Nevertheless, there are many examples of active comets at greater distances: C/1995 O1 (Hale-Bopp) was already active when it was discovered at 7 au (see Figure 2.11); 29P/Schwassmann-Wachmann 1, on a quasi-circular orbit at 6 au, has unexpected and abrupt bursts of activity; the Centaur 95P/Chiron at 10 au; 1P/Halley at 15 au, etc. This distant activity can be explained by the sublimation of more volatile ices: CO_2, CO, and possible N_2 (see Figure 2.10). These are the "hyper-volatiles". Small grains heated by the Sun can also sublimate the water ice they contain much more efficiently that the surface of the nucleus. In this way, we can also explain the hyperactivity of some less distant comets whose production of water significantly exceeds what is indicated by models for the sublimation of ice in nuclei: a major additional source may come from the sublimation of frozen grains in the coma.

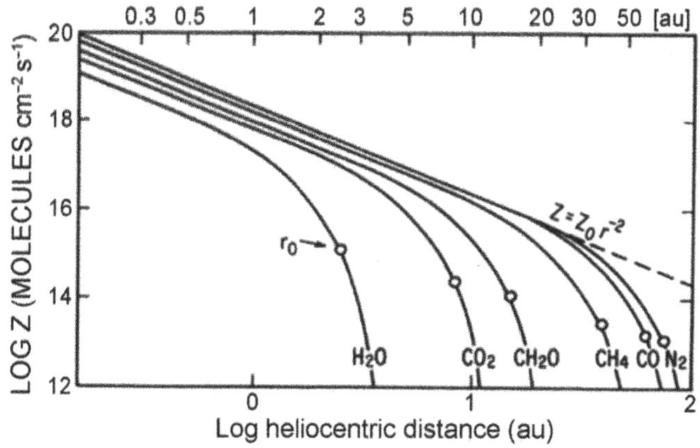

Figure 2.10. *Sublimation rate for some molecules of cometary ices, as a function of the distance from the Sun (source: based on (Delsemme 1982))*

Comet	D km	Q_{gas} 10^{28} molec. s^{-1}	Q_{gas} kg s^{-1}	Q_{dust} kg s^{-1}
67P/Churyumov-Gerasimenko	3	1	300	500
21P/Giacobini-Zinner	6	5	1,500	400
1P/Halley	10	100	30,000	20,000
C/1995 O1 (Hale-Bopp)	60	1,000	300,000	600,000

Table 2.4. *Orders of magnitude for the size and activity of several typical comets: diameter equivalent to the nucleus, dust and gas production rate (H_2O) near r = 1 au*

2.1.6.2. *The hydrodynamics of cometary atmospheres*

The gravity of comet nuclei is too low to retain a captive atmosphere. Cometary atmospheres are in quasi-adiabatic expansion and resemble supersonic jets that can be reproduced using nozzles in the laboratory. As one moves away from the nucleus, the temperature drops quickly. The density does as well, collisions become rare, and the jet becomes a free molecular jet.

Based on the theory of sublimation described in the previous section, we can estimate the velocity of sublimated molecules at the moment they leave the nucleus: they have a velocity V_o between 1/2 and 2/3 the average velocity V of the molecules at temperature T:

$$V = (8\ \mathrm{k}T/m)^{1/2} \qquad [2.3]$$

m being the mass of the molecule and k the Boltzmann constant. Let $V = 0.40$ km s^{-1} for water molecules leaving the surface at a temperature of 140 K.

The molecules are then accelerated radially by the interplay of collisions, until they attain the sound velocity:

$$V_s = (\gamma \mathrm{k}T/m)^{1/2} \qquad [2.4]$$

γ being the ratio between the specific heats C_p and C_v. In the case of water, $V_s = 0.71\ V$.

The gas then continues to propagate adiabatically until it attains a limit speed that, in the case of water vapor, is on the order of 0.8 km s^{-1}. However, the expansion of the atmosphere is not completely adiabatic. When acted upon by solar radiation, other processes occur in the coma: molecular photo-excitation, photo-dissociation, and photo-ionization. These processes have the effect of heating the atmosphere (with a rate in r^{-2}), by injecting atoms and radicals at high speed, which, if they are able to be heated by collisions, augment the temperature and the speed of the expansion. On the other hand, the atmosphere cools through radiation, essentially the emission of rotational lines of H_2O in the far infrared, with a rate in T^2. This mechanism is limited by the high level of opacity of these lines. Modeling these phenomena is delicate, because it requires taking into account the interaction between the thermodynamics of the internal coma, the excitation state of the water molecule, and the radiative transfer of these lines. This is also the case with the geometry of the problem: observations have shown us that it can be very different from the isotropic spherical model we used previously. An example of a model in a simple case is shown in Figure 2.12 (Combi *et al.* 2004).

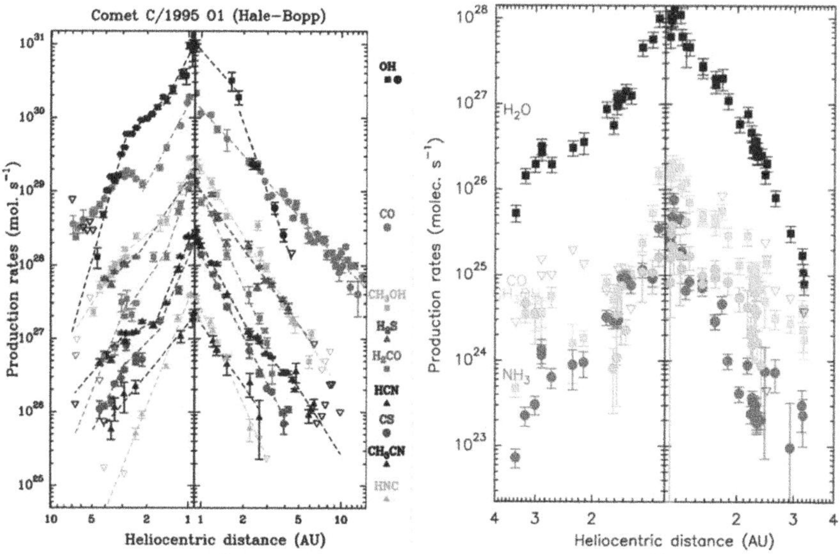

a) b)

Figure 2.11. *The evolution of production rates for different molecules observed in a comet as a function of its distance from the Sun, before and after perihelion. For a color version of this figure, see www.iste.co.uk/encrenaz/solar2.zip*

COMMENTS ON FIGURE 2.11.– *(a) The comet C/1995 O1 (Hale-Bopp) (source: based on (Biver* et al. *2002)). The data for the OH radical (black squares), which traces the production of water, comes from observations made by the Nançay radio telescope (Cher, France), while those of the other molecules were obtained by the Iram 30 m radio telescope (in Spain) or by the Sest 15 m radio telescope (in Chili). (b) The comet 67P/Churyumov-Gerasimenko observed in situ by the Rosetta's small sub-millimeter radio telescope/Miro (source: based on (Biver 2019)).*

Radio spectroscopy, with its very high spectral resolution, is an ideal tool for studying the kinematics of cometary atmospheres (see Figure 2.13). Using the Doppler effect, the profile of the rotational lines of cometary molecules accurately traces the distribution of velocities on the line of sight. The thermal agitation of the molecules contributes only slightly to the widening of these lines at the low temperatures of cometary gas. The measurement of velocities *in situ* was also made possible by Giotto and with the Rosetta's Rosina instrument.

The velocities thus observed ranged from less than 0.5 km s^{-1} up to about 2 km s^{-1}. Conforming to hydrodynamic models (see Figure 2.12), they increase when the comet approaches the Sun because of the increase in heating. They also increase with cometary activity because collisions, which determine the efficiency of the heating, are more frequent.

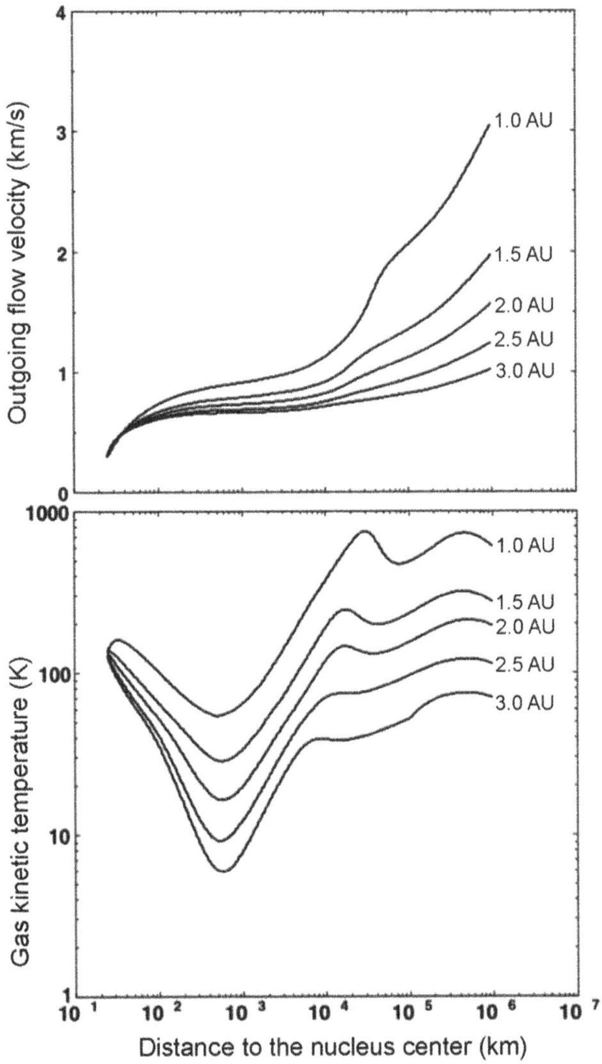

Figure 2.12. *Velocity of expansion and temperature of the atmosphere for a comet as a function of the distance from the nucleus and the heliocentric distance. The production rate for water is based on that of the comet Hale-Bopp and varies from* 4×10^{29} *molecules* s^{-1} *at 3 au to* 8×10^{30} *at 1 au (source: based on the hydrodynamic models in (Combi* et al. *1999))*

The temperature of the gas is measured based on the distribution of molecules among their different rotational levels. When the rate of collisions is sufficient to guarantee local thermal equilibrium (LTE), the molecules are distributed according to Boltzmann's law, which only depends on temperature:

$$\frac{\langle N_u \rangle}{g_u} = \frac{\langle N_{tot} \rangle}{Z(T_{rot})} \exp\left(-\frac{E_u}{kT_{rot}}\right) \qquad [2.5]$$

where $\langle N_u \rangle$ is the population density of the state of rotation u, E_u its energy, and g_u its statistical weight. T_{rot} is the rotation temperature, equal to the kinematic temperature if it is in equilibrium. Z is the partition function. Temperatures are thus estimated from the relative intensities of the rotational lines for a single molecule (for example, radio observation of methanol lines).

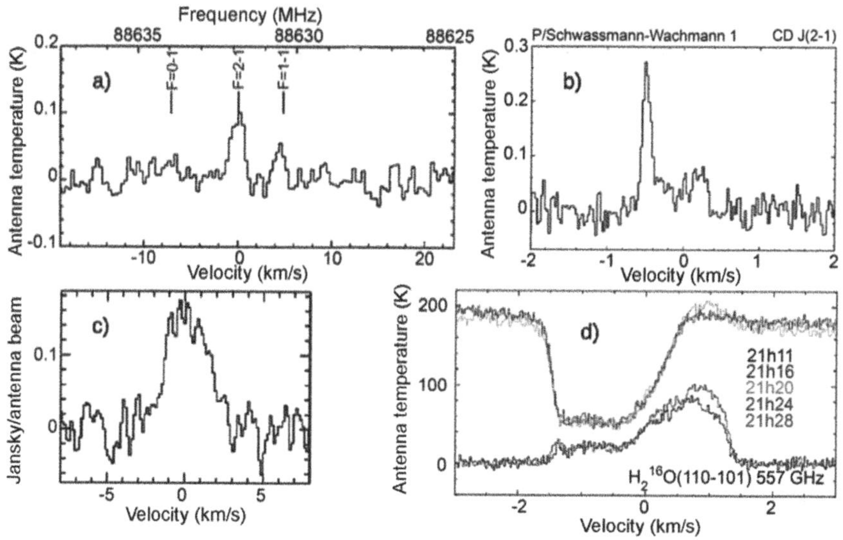

Figure 2.13. *Examples of the profiles of molecular lines observed by radio. For a color version of this figure, see www.iste.co.uk/encrenaz/solar2.zip*

COMMENTS ON FIGURE 2.13.– *(a) The line J(1-0) for HCN and its three hyperfine compounds observed by Iram in Halley's comet (source: based on (Despois* et al. *1986)). (b) The line J(2-1) of CO and its "cold" compound in the comet 29P/Schwassmann-Wachmann 1 observed by Iram (source: based on (Crovisier* et al. *1995)). (c) The line of OH at 18 cm observed by the Nançay radio telescope in Halley's comet (source: based on (Gérard* et al. *1987)). (d) The saturated line of* H_2O *at 557 GHz observed in absorption on the continuum of the nucleus of*

67P/Churyumov-Gerasimenko (in red) and in limb emissions (in blue) by Rosetta/MIRO (source: based on (Biver et al. *2019)).*

We thus observe temperatures ranging from a few dozen to more than a hundred kelvins, depending on the intensity of the cometary activity and its distance from the Sun. For comet 29P/Schwassmann-Wachmann 1, the radio line of CO has a very narrow component that indicates a very low gas temperature, on the order of 5 K (see Figure 2.13b).

2.1.6.3. *The photochemistry of cometary atmospheres*

The chemistry of cometary atmospheres is dominated by photolytic processes (see Figure 2.14). Molecules from cometary ices directly sublimated under the influence of solar radiation are progressively photodissociated into atoms and radicals. These are therefore reactions of the type: AB + hν → A + B. A compilation of these reactions and their rates is available in (Huebner and Mukherjee 2015). Two-body reactions of the type: A + B → C + D are not important because of their low density and the low reactivity of the molecules that dominate cometary atmospheres. Through the influence of solar wind (see section 2.1.9), these molecules or the products of their dissociation can be ionized. Ions are much more reactive, but their abundance remains low in the internal coma.

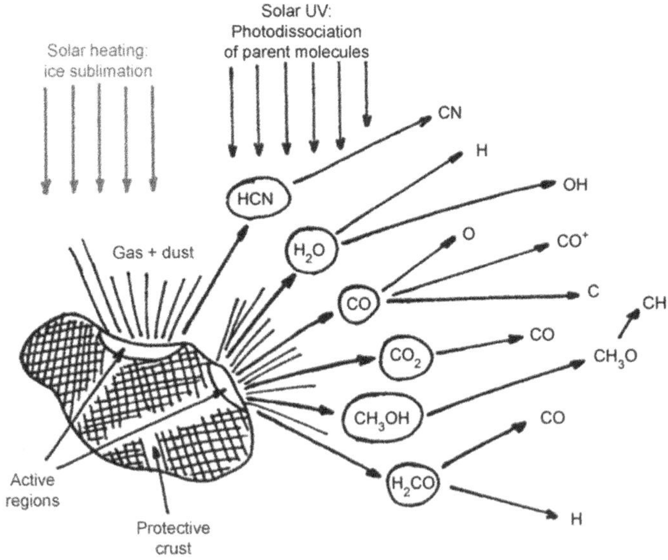

Figure 2.14. *Photolytic process in a cometary atmosphere*

Customarily, we call the sublimated molecules of cometary ice *parent molecules* and their products of dissociation *daughter molecules* (which in fact can also be atoms or radicals). A simple model postulating spherical symmetry and radial expansion at a constant speed summarizes the spatial distribution of parent molecules and daughter molecules. This is the Haser model, after the astronomer Leo Haser (1928–2016), who proposed it.

We are situated in the simplified case of a spherical symmetry. For a parent molecule with a scale length of l_m:

$$n_m(r) = \frac{Q}{4\pi r^2 v_{exp}} \exp\left(-\frac{r}{l_m}\right)$$

and for its daughter molecule with a scale length of l_f:

$$n_f(r) = \frac{Q}{4\pi r^2 v_{exp}} \frac{l_f}{l_m - l_f} \left[\exp\left(-\frac{r}{l_m}\right) - \exp\left(-\frac{r}{l_f}\right)\right]$$

Box 2.4. *The Haser formulas*

For a constant expansion speed v_{exp}, the Haser formula describe the density $n(r)$ at distance r from the nucleus, Q being the production rate in molecules/s (for a very active comet, the gas production rate can be on the order of 10^{30} molecules/s at a heliocentric distance of 1 au; see Table 2.4). After a characteristic time t_m, the lifetime, the parent molecule is dissociated, and its density beyond a sphere of radius: $l_m = v_{exp}\, t_m$ – the scale length – decreases exponentially. The daughter molecule is itself dissociated after an average time t_f and has a scale length of: $l_f = v_{exp}\, t_f$.

2.1.6.4. *Parent molecules*

Since the first spectroscopic observation of a comet, C/1864 N1 (Tempel), by the Italian Giovanni Donati (1826–1873), cometary atmospheres have become one of the preferred laboratories for molecular spectroscopists. Because of the unusual physical conditions in cometary atmospheres – low density and low temperature – they were able to observe the lines of molecules, radicals, and ions often long before being able to do so in the laboratory.

Molecule	X	T_{subl} (K)	t_{life} (s)	Secondary products
Water H_2O	100	152	80,000	OH, H, O, H_2O^+
Dioxygen O_2	2	25	200,000	O
Dinitrogen N_2	0.1	22	800,000	N
Carbon monoxide CO	1–30	24	1,300,000	C, CO^+
Carbon dioxide CO_2	5	72	500,000	C, CO, CO^+
Methane CH_4	1	31	125,000	CH
Acetylene C_xH_2	0.1	54	70,000	CH, C_2
Ethane C_xH_6	0.5	44	90,000	CH, C_2
Formaldehyde H_2CO	1	64	5,000	CO
Methanol CH_3OH	2	99	75,000	CO
Ammonia NH_3	0.5	78	7,000	NH, NH_2
Hydrogen cyanide HCN	0.2	95	70,000	CN
Hydrogen sulfide H_2S	1	57	4,000	SH
Sulfur dioxide SO_2	0.2	83	5,000	SO

Table 2.5. *Properties of some molecules outgassed from cometary ice. X is the abundance (relative production rate) relative to water, which can vary from one comet to another. T_{subl} is the sublimation temperature for the ice in the physical conditions on the surface of the nucleus. t_{life} is the lifetime at 1 au from the Sun*

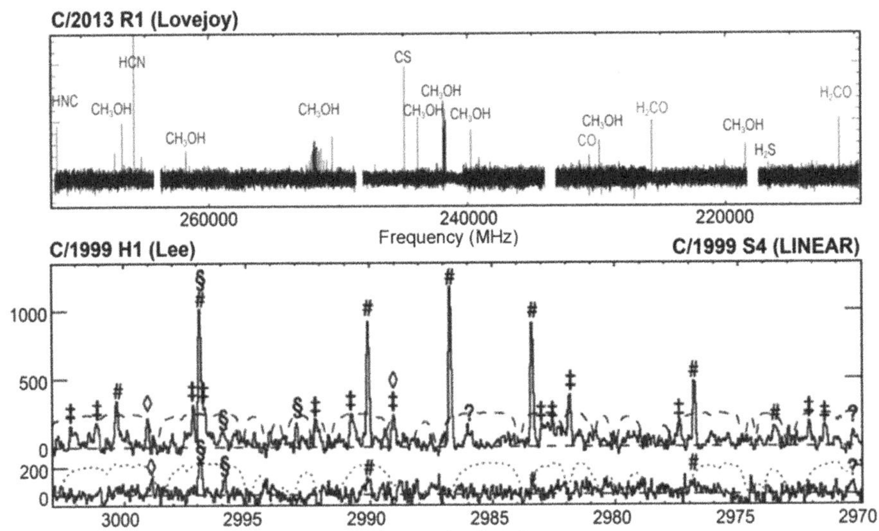

Figure 2.15. *Examples of cometary spectra showing the lines of several molecules*

COMMENTS ON FIGURE 2.15.– *(a) Part of the radio spectrum near a wavelength of 1.2 mm from the comet C/2013 R1 (Lovejoy) obtained with the Iram 30 m radio telescope (source: adapted from (Biver* et al. *2016)). (b) Infrared spectrum near a wavelength of 3.3 µm for the comets C/1999 H1 (Lee) and C/1999 S4 (LINEAR) obtained with the Keck telescope; we see there the lines of* C_2H_6 *[#],* CH_4 *[◊],* CH_3OH *[ǂ], OH [$], and unidentified lines [?] (source: adapted from (Mumma* et al. *2001)).*

The discovery and establishment of the composition of parent molecules are relatively recent and result from the development of spectroscopy in the radio and infrared spectra on one hand, and of the implementation of mass spectrometry during *in situ* exploration on the other hand.

Directly produced by the sublimation of ice from the nucleus, parent molecules supply one of the keys to the chemical composition of comets. Their observable spectral signatures correspond to the rotational-vibrational and rotational lines of molecules, in the infrared and millimeter-wave spectra, respectively (see Figures 2.15 and 2.16). We should recall that spectroscopy in the visible and UV ranges, which is so effective for the study of radicals, does not lend itself to the observation of parent molecules, since the excitation of the electronic band of these molecules leads to their dissociation rather than to emission by fluorescence. Only carbon monoxide CO could be detected in the UV spectrum as early as 1975, with the observation of the comet C/1975 V1 (West).

Since the most abundant parent molecule is water, the abundances of other molecules are compared to it. It is important to note that the abundances indicated in Table 2.5 and Figure 2.17 correspond to a typical heliocentric distance of 1 au; gases more volatile than water (such as CO) are present in relatively higher abundances at a greater heliocentric distance.

The first parent molecules H_2O, CO_2, and HCN were detected on Halley's comet. The list of cometary molecules was then considerably lengthened, particularly thanks to the observation of comets C/1996 B2 (Hyakutake) and C/1995 O1 (Hale-Bopp). Thus, through remote observation, we know of about thirty parent molecules (see Figure 2.17). The chemical composition of comets in the Jupiter family, which are less bright, is more difficult to determine. About half as many molecules have however been measured recently in comet 46P/Wirtanen by taking advantage of its passage near Earth. The number of identified cometary molecules (Bockelée-Morvan *et al.* 2004; Bockelée-Morvan and Biver 2017) is still far from the two hundred or so molecules known in the interstellar medium.

Observations from *Rosetta*'s mass spectrometer allowed us to double the number of known molecules from comet 67P. A detailed list is given in (Altwegg *et al.* 2019). This is due to the excellent resolution of its instruments (see Figure 2.18), whereas during a flyby of Halley's comet, the resolution of *Giotto*'s spectrometer was not sufficient to separate molecules of neighboring masses such as CO and N_2. But mass spectrometry cannot distinguish isomers such as HCN and HNC, as well as ethanol CH_3CH_2OH and dimethyl ether CH_3OCH_3.

One of *Rosetta*'s contributions was to highlight the hyper-volatile molecules that had not been seen by traditional spectroscopic observations because they do not have observable lines. The first one was dinitrogen N_2, which Giotto's mass spectrometer could not distinguish from the CO molecule in Halley's comet. The abundance of this molecule, supplemented with that in the already-known ammonia and nitriles, does not reach the level of its cosmological abundance.

Next, we should note the surprising discovery of dioxygen O_2, whose important abundance (several %) makes it one of the major constituents of cometary ice. O_2 is an equally abundant molecule in the interstellar medium, but it is very difficult to observe, since its lines, which are forbidden, are extremely weak. The origin of O_2 is still poorly understood. It may come from reactions between the gas and the grains in the pre-solar nebula.

The rare gases argon, krypton, and xenon are also present, with the most volatile, argon, showing a deficiency. How could these rare gases and these hyper-volatile molecules (to which we must add CO and CH_4) have been preserved in cometary ice? One possibility is that they were trapped in crystalline water ice lattice that contains them in a cage: this is the clathrate hypothesis.

High-resolution measurements of water in the infrared spectrum (initiated by the Kuiper Airborne Observatory for Halley's comet and the ISO satellite for comet Hale-Bopp, then followed up by ground-based observations of other periodic comets) supplied precious information about the physical conditions inside the coma. Based on the relative intensities of H_2O in the fundamental vibrational bands ν_1 and ν_3 at 2.6 µm, it was possible to determine the rotational temperature of H_2O (about 25 K), which gave an estimate of the medium's kinematic temperature. The comparison of the intensity of H_2O lines in the ortho and para modes of the molecule have made possible the measurement of the spin temperature of H_2O (from 25 to 35 K), which could be linked to the conditions of comet formation (see Box 2.5).

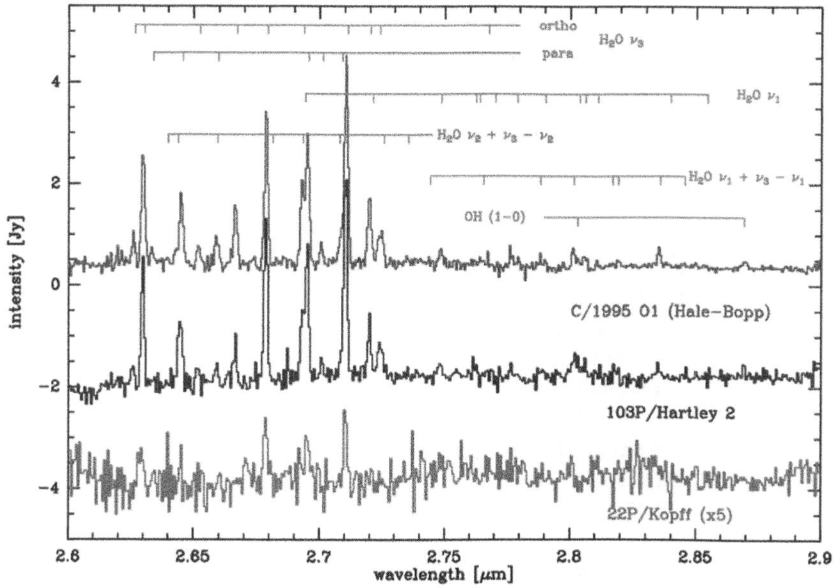

Figure 2.16. *Water vibration bands observed in infrared by the ISO satellite in different comets. The lines corresponding to the ortho and para states are identified, and their intensity ratio corresponds to a temperature of 30 K (see Box 2.5; source: based on Crovisier* et al. *in (Müller* et al. *2005)). For a color version of this figure, see www.iste.co.uk/encrenaz/solar2.zip*

Water molecules exist in two forms: the ortho form, where the two hydrogen atoms have parallel spins, and the para form, where they are antiparallel. The ortho/para ratio of the two species' abundance depends on temperature: roughly equal to 3 for temperatures above 50 K, it tends toward zero when the temperature drops. But this is the kinematic temperature of the gas. Indeed, transitions between the ortho and para states are prevented, both during collisions and during radiative transitions (when the water molecule passes from one spin level to another with the emission or absorption of a photon). Only chemical reactions can make water shift from one state to another, and it is as if we are dealing with two distinct chemical species.

The ortho-to-para ratios observed for cometary water from the infrared spectra (like those in Figure 2.16) correspond to temperatures between 25 and 35 K. This may have been the prevailing temperature during its formation, although a re-equilibration was nevertheless possible during the long periods of storage in cometary nuclei.

Box 2.5. *Ortho and para H_2O*

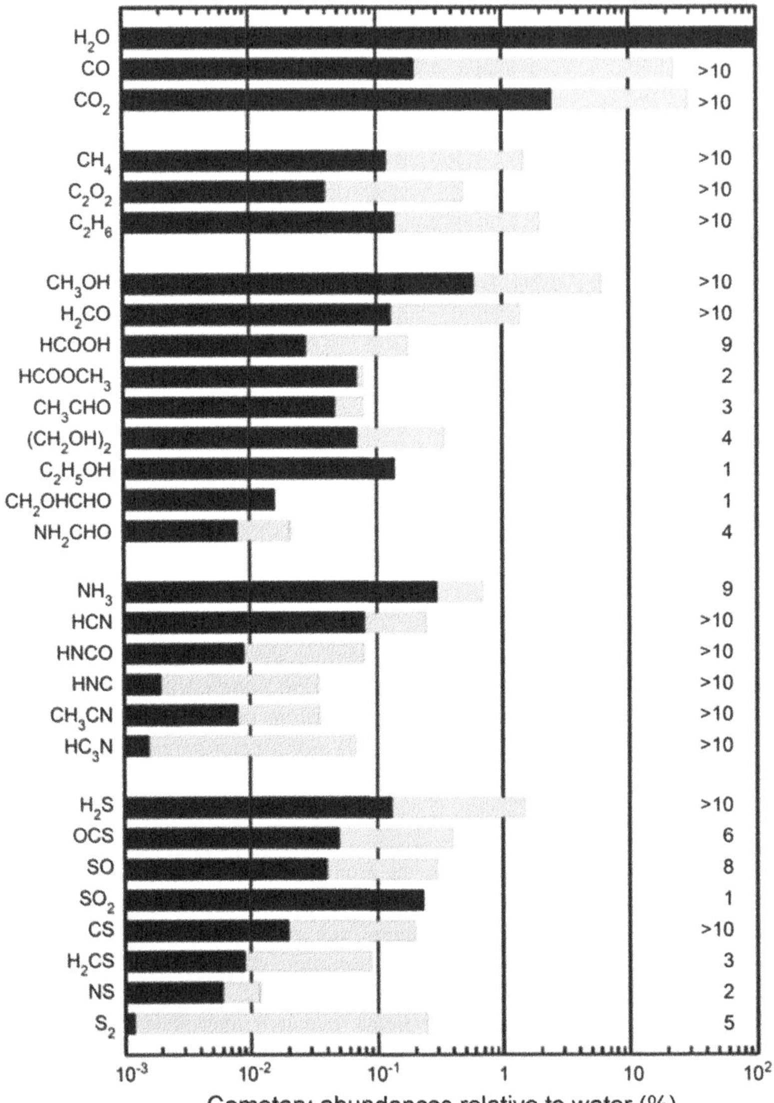

Figure 2.17. *Cometary molecules observed remotely using spectroscopy and their abundances compared to water. The bars in blue show the observed variations from one comet to another. The number of comets where these molecules have been observed (up to 2017) is indicated on the right (source: based on (Bockelée-Morvan* et al. *2004; Bockelée-Morvan and Biver 2017)). For a color version of this figure, see www.iste.co.uk/encrenaz/solar2.zip*

Figure 2.17 shows a panorama of molecules remotely observed in comets. We note that the abundance of CH_4 and NH_3 is low compared to that of CO, CO_2, and CH_3OH, which can exceed 5% in some comets. The origin of CO, like that of H_2O, seems to be both primary (coming directly from the nucleus) and secondary (produced by the dissociation of other parent molecules, particularly CO_2). Finally, we should point out the presence of hydrocarbon and complex "CHO" type molecules, detectable in the infrared bands near 3 μm, also observed in the interstellar medium, attributed to saturated or unsaturated hydrocarbon compounds. "Complex organic molecules" or COMs, are, in the lingo of cosmochemists, carbon compounds with at least six atoms. Using their millimeter-wave radio lines, we have thus identified acetaldehyde CH_3CHO, ethanol CH_3CH_2OH, glycolaldehyde CH_2OHCHO, methyl formate $HCOOCH_3$, etc., with the most complex molecule (10 atoms) being ethylene glycol $(CH_2OH)_2$. Compared to water, these compounds have abundances of 0.02 to 0.2% (Biver and Bockelée-Morvan 2019).

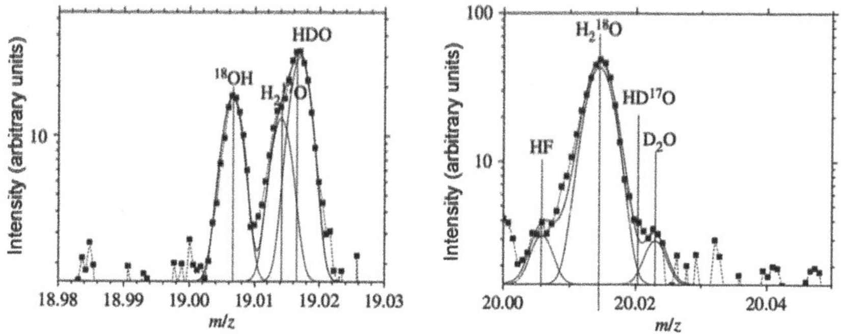

Figure 2.18. *An example of mass spectra: molecules from the water family observed at atomic masses 19 and 20 by Rosetta/Rosina in 67P/Churyumov-Gerasimenko (source: based on (Altwegg* et al. *2017))*

Glycine $NH_2CH_2CO_2H$ is the simplest amino acid. It is not among the molecules detected in the interstellar medium, but it is found (with many other amino acids) in carbonaceous chondrite meteorites. Traces of glycine have been found on the structure of the Stardust collector after its passage through the coma of 81P/Wild 2. This detection was confirmed in 67P by the Rosina mass spectrometer on Rosetta, with an abundance compared to water between 0 and 0.15%. The analysis of the distribution of glycine in the coma of comet 67P suggests that it does not come from the sublimation of ice from the nucleus, but rather from grains that escaped. Rosina also detected another ingredient that is essential for life, phosphorus, possibly in its PO form.

We must insist on another aspect of the chemistry of cometary atmospheres: the existence of "extended sources" or "distributed sources" of molecules. By observing their distribution in the coma, we notice that some molecules do not come (or do not all come) directly from the sublimation of ice from the nucleus, or from the decomposition of parent molecules. This is the case for formaldehyde H_2CO, carbon monoxide CO, and the HNC isomer of hydrogen cyanide. These chemical species can come from the sublimation of icy grains or from thermal or photolytic decomposition of grains rich in organic matter. Polyoxymethylene H-$[CH_2O]_n$-OH, a formaldehyde polymer, has been mentioned as a possible source of the latter.

The nature of the nonvolatile compounds on the surface of cometary nuclei is not well known. The Philae lander from the Rosetta mission failed to collect a sample of this matter for analysis. Its Cosac mass spectrometer did manage to survey the immediate environment of the nucleus's surface, observing molecules similar to those found in the coma by Rosina. For now, we must base our work on remote diagnostics using infrared spectroscopy. We thus find insoluble organic matter, kerogen, which is similar to terrestrial coal. This matter is also similar to the insoluble organic matter that makes up some meteorites (see Chapter 3).

2.1.6.5. *Daughter molecules, radicals, and ions*

Historically, only the radicals (daughter molecules) – C_2, CN, etc. – had been observed. We had to guess what their parent molecules were.

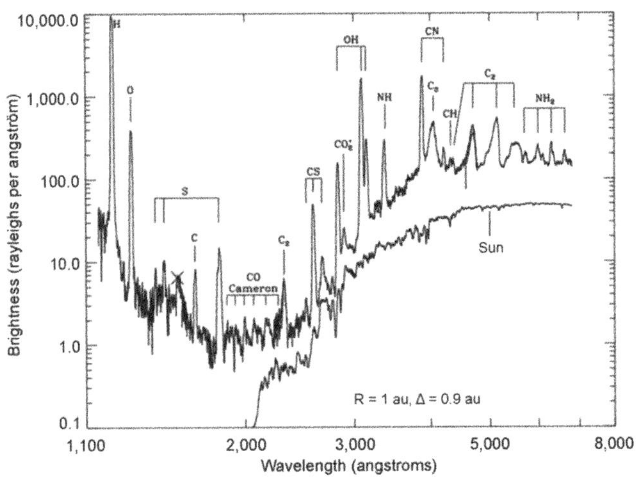

Figure 2.19. *The composite spectrum from visible to UV for the comet 103P/Hartley 2 observed by the Hubble Space Telescope (source: based on (Weaver and Fedman 1992))*

A number of spectroscopic observations in the ultraviolet and visible ranges (see Figure 2.19) have enabled a more or less complete analysis of the products of dissociation observed in comets (see Table 2.5). The excitation mechanism is resonant fluorescence, and the source of pumping is the solar flux. In some cases (in particular for the radical CN), we observe a very marked change in the fluorescence spectrum as a function of the comet's heliocentric velocity. Because the solar spectrum is very irregular, the rate of fluorescence varies drastically according to this velocity because of the Doppler effect. This is the "Swings effect", named for Belgian astronomer Pol Swings (1906–1983), who was the first to study this phenomenon. The same effect is found in the observation of radio lines of OH at the 18 cm wavelength. Observations of these lines via heterodyne spectroscopy (with a spectral resolving power of 10^6 at 18 cm wavelength), makes it possible to measure, via the Doppler effect, the velocity field inside the coma. They also led to the systematic monitoring of cometary activity rates since the abundance of OH makes it possible to work backward to H_2O. A fundamental parameter of the emission of lines and bands of fluorescence is the rate of fluorescence – the number of photons emitted per second by a molecule under the influence of solar radiation: known as the "*g* factor" in specialist's jargon. For a resonance line:

$$g_{lu} = \frac{1}{8\pi \mathrm{h}\sigma_{lu}^3} \frac{\omega_u}{\omega_l} A_{ul} J(\sigma_{ul}) \qquad [2.6]$$

where A_{ul} is the Einstein coefficient for spontaneous emission, $J(\sigma_{ul})$ is the density of the radiation at the frequency (wavenumber) σ_{ul} of the line, and ω_u and ω_l are the statistical weights of the upper and lower levels. If the radiation field is that of the black body at temperature T_{bb} and is received over a solid angle Ω_{bb}:

$$g_{lu} = \frac{\Omega_{bb}}{4\pi} \frac{\omega_u}{\omega_l} A_{ul} [\exp \frac{\mathrm{hc}\sigma_{ul}}{\mathrm{k}T_{bb}} - 1]^{-1} \qquad [2.7]$$

for the Sun at 1 au, $T_{bb} \sim 5{,}700$ K, $\Omega_{bb}/4\pi = 5.42 \times 10^{-6}$. g is therefore typically between 10^{-3} and 10^{-2} s^{-1} for the electronic bands of cometary radicals in the visible spectrum, and on the order of 10^{-4} s^{-1} for the fundamental vibrational bands of molecules in the infrared range. The Lyman α line for hydrogen at 121.5 nm has a g of 1.4 10^{-3} s^{-1}. For the doublet of the yellow line of sodium at 589 nm, the resonance fluorescence is particularly intense, with $g = 15$ s^{-1}. Fluorescence thus is accompanied by an efficient thrust that accelerates the sodium atoms to produce the characteristic tail.

In the visible range, the most intense cometary emissions are due to C_2, CN, and CH. We also find C_3, NH_2, and some prohibited transitions of O (1D). The ions observed in the visible range are CO^+ and H_2O^+ as well as N_2^+ and CH^+. The ultraviolet range has been studied extensively thanks to the IUE satellite and then thanks to the HST (see Figure 2.19).

Among the neutral species, the most abundant elements are H and OH. The coma of cometary hydrogen was identified by its Lyman α line by the first satellite-based ultraviolet observations, in the comet C/1969 T1 (Tago-Sato-Kosaka) by *OAO-2*. Lyman α is observed at great distances from the nucleus, up to several million km. We can thus consider that these cometary hydrogen clouds are the largest objects in the Solar System.

Estimates of the H/OH ratio tend toward a value on the order of 2; this is what we would expect from the dissociation of H_2O. H is both a primary product of the dissociation of water, following the reaction:

$$H_2O + h\nu \rightarrow OH + H$$

and a secondary product, following the reaction:

$$OH + h\nu \rightarrow O + H$$

These reactions produce an excess of energy leading to the high-speed ejection of the hydrogen atom: 20 km s^{-1} for the first reaction, 8 km s^{-1} for the second. These high velocities explain the large scale length of cometary hydrogen. The radical OH has an ejection velocity of only 1 km s^{-1}, leading to a scale length Vt on the order of 10^5 km; and its rate of fluorescence varies, as theory predicts, as the inverse square of the heliocentric distance.

The oxygen atom exhibits the particularity of being observed in its forbidden transitions, in the visible range, at 630–636.4 nm (^1D-^3P) and at 557.7 nm (^1S-^1D). These lines seem necessarily due to the formation of O in an excited state (^1D or ^1S) by dissociation of a parent molecule, and the atom subsequently falls to its fundamental state of ^3P. The first candidate parent molecule is of course H_2O; but the calculations predict that only 10% of H_2O should dissociate according to the reaction:

$$H_2O + h\nu \rightarrow H_2 + O(^1D)$$

while some measurements seem to indicate a ratio of $O(^1D)/H_2O$ on the order of 35%. The existence of a second parent molecule thus seems necessary; a possible candidate could be CO_2, detected in Halley's comet and in several subsequent comets.

Another radical is also of particular interest: CN, easily observed by its bands at 388 and 787 nm with a scale length of 3×10^5 km. Its presumed parent molecules are HCN, with a scale length of 2×10^4 km, detected by radio on a large number of comets, along with other observed cyanides (see Figure 2.17). Moreover, the distribution of jets of CN observed on some comets could imply that this radical is at least partially produced by outgassing of the organic mantle of cometary grains.

The detection of Swan bands in the comet C/1864 N1 (Tempel), identified later to the radical C_2, was the first result of spectroscopic observation of a comet. The origin of the radicals C_2, CH and C_3, long unexplained, is attributed today to hydrocarbons, particularly C_2H_2 and C_2H_6, detected in comet Hale-Bopp. Even more complex carbon chains have been detected in 67P by Rosina, as well as aromatic polycyclic molecules (benzene, toluene, and possibly naphthalene), whose presence had been suspected but were difficult to prove unambiguously, in the infrared spectra by their features near 3.3, 6.2, and 7.6 μm.

2.1.6.6. *Isotopic ratios*

The isotopic varieties of a single molecule are subject to the same chemical reactions, but with different velocities due to differences in molecular mass. This results in fractionation effects. Isotopic ratios thus provide important information about the conditions of the formation of the ices that became part of the comets. In the case of interstellar molecules, isotopic fractionations are attributed to ion-molecule reactions at low temperatures. This effect is particularly important for deuterium because of the large difference in mass presented by the H-D pair; we observe in the interstellar medium deuterium enrichments up to a factor of 10 compared to the local interstellar value.

The value of the D/H ratio in the primitive solar nebular is 2×10^{-5}, similar to that of the interstellar medium. We find it in the gas giants. It is much higher in terrestrial oceans: 1.56×10^{-4}. This enrichment reveals the history of the formation of Earth (see sections 4.3.3.7 and 5.2.2.2, Volume 1).

One of the hypotheses advanced for the origin of terrestrial water is the bombardment of our planet by asteroids and comets. To test this, what better than to compare the terrestrial D/H ratio to that of water in asteroids and comets? However, the determination of the D/H ratio of cometary water is not easy. Only two comets have had their D/H ratio measured *in situ* with mass spectrometers (1P/Halley by Giotto and 67P by Rosetta). For remote observation, the most sensitive measurements have been made from the sub-millimeter lines of water. But their interpretation requires a good model of the excitation of water including a correction for the effects of opacity.

Currently, this ratio has only been determined precisely for a few comets. The observed values have successively blown hot and cold. The first obtained values, for Halley's comet and other Oort cloud comets, were twice the terrestrial value. Then, the Jupiter-family comet 103P/Hartley 2 was observed by the satellite Herschel with a terrestrial value. Then, Rosetta measured and remeasured on 67P a value equal to three times the terrestrial value. Next, a terrestrial value was again found by the Sofia stratospheric airplane observatory for a hyperactive comet of the Jupiter family, 46P/Wirtanen (Lis *et al.* 2019).

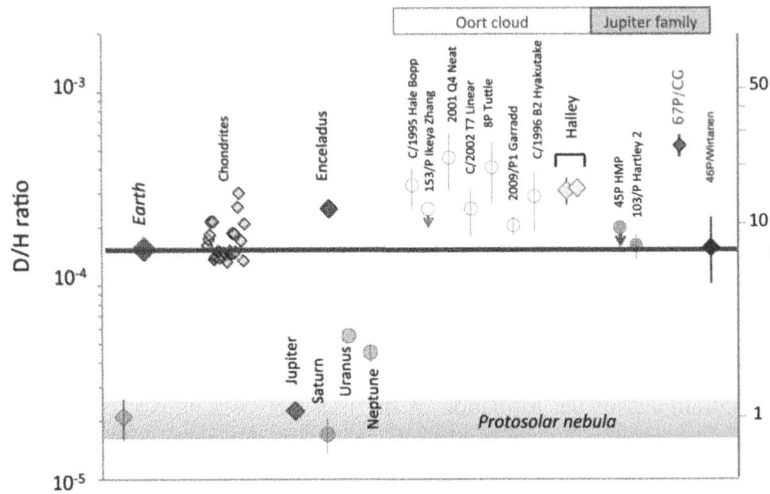

Figure 2.20. *The D/H ratio observed in cometary water. Many of the measurements come from radio observations. This ratio ranges from one (comets 103P/Hartley 2 and 46P/Wirtanen observed by radio with Herschel and Sofia) to three times (comet 67P/Churyumov-Gerasimenko observed with the Rosina mass spectrometer on Rosetta) the value in terrestrial oceans (source: adapted from (Altwegg* et al. *2015)). For a color version of this figure, see www.iste.co.uk/encrenaz/solar2.zip*

We see that the problem remains open, and that its solution will require the measurement of the D/H ratio for more representative samples, both for comets and asteroids (see (Altwegg *et al.* 2015) and Figure 2.20).

2.1.7. *Dust and the tail*

The Germans Heinrich Olbers (1758–1840) and Friedrich Bessel (1784–1846) were the first to suggest, at the beginning of the 19th century, that cometary tails could be particles emitted by the comets and pushed back by solar radiation.

We know now that the large, curved tails, previously called "type II", which develop near the perihelion, are indeed made up of dust. On some particularly large comets, a coma of dust was detected at heliocentric distances up to seven astronomical units, too high for the sublimation of water to be the cause.

As shown by observations of the comet Hale-Bopp, it is the sublimation of CO and CO_2 that is responsible for beginning this activity at large heliocentric distances.

2.1.7.1. *Kinematics of dust tails*

The analysis of the isophotes of a dust tail was addressed in detail in 1968 by Michael Finson and Ronald Probstein, who continued the work of the Russian Fiodor Bredikhine (1831–1894). Cometary dust is ejected from the nucleus at a given velocity V and is subjected to an acceleration due to two contrary factors: solar attraction and radiation pressure.

The acceleration F_{rad} due to radiation pressure is proportional to the section d^2 of dust grains of size d, while solar attraction F_{grav} is proportional to their mass $\approx \rho\, d^3$, where ρ is the density. The ratio of these two accelerations, which we typically write $1 - \mu$, is:

$$F_{rad}/F_{grav} = 1 - \mu = C^{ste} / (\rho d) \qquad [2.8]$$

We can thus calculate the density distribution of the particles in the dust tail; when the initial velocity V is null, the calculation is then particularly easy. In this case, the isophotes in a dust tail represent a two-dimensional consequence of two clearly distinct parameters:

– the distribution of particle sizes varies along each *synchrone* (set of dust emitted by the nucleus at a given moment);

– the production rate of particles of a given size varies along the *syndynes* (set of dust having a given size), as a function of the rate of emission.

Figure 2.21 shows the distribution of dust along synchrones and syndynes, illustrating the formation of the dust tail. In practice, the ejection velocity of the dust is not null, which leads to an enlargement of the tail; observations of this enlargement have made it possible to estimate the ejection velocity of the dust at 0.3 km s^{-1} on average, which is about three times less than the ejection velocity of gases. Next, the dust is accelerated by the gas until attaining its final velocity (on the order of a km s^{-1} for the smallest pieces), which is reached in the first hundred kilometers.

Beyond 10^4 km, the neutral particles are only influenced by radiation pressure. As the diffusion efficiency does not depend on the radius, the smaller the particles the more effective the radiation pressure; in dust tails that have been observed, the ratio F_{rad}/F_{grav} typically varies between 0.1 and 1.

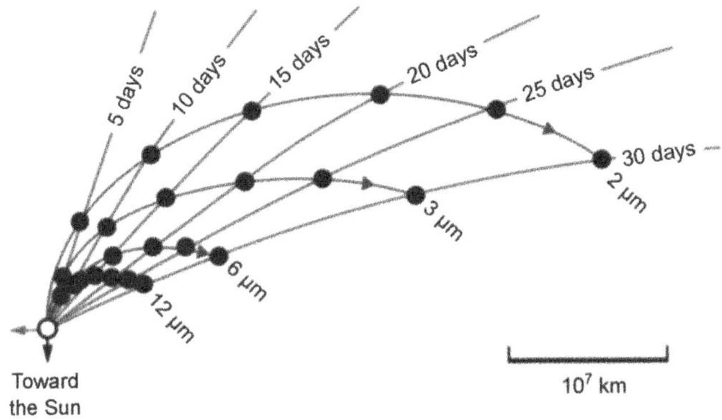

Figure 2.21. *The dynamics of a dust tail. The synchrones are in red and the syndynes are in blue (source: (Lequeux and Encrenaz 2015), adapted from (Delsemme 1982)). For a color version of this figure, see www.iste.co.uk/encrenaz/solar2.zip*

When a rotating comet nucleus exposes an active region at each rotation, the jet of escaping dust traces expanding sections of spirals. Temporal image analysis thus makes it possible to determine the rotation period of the nucleus (see section 2.1.5.1).

In addition to the particles observed in the dust tail, the comet nucleus ejects larger grains (d > 30 µm). We observe them as antitails, whose existence results from a geometric effect: these particles are ejected at almost null velocity, distributed in the comet's orbital plane; when the Earth crosses this orbital plane, this set of particles, seen edgewise, presents as a very narrow tail pointing toward the Sun. The large grains diffuse along the length of the comet's orbit to form a *cometary dust trail*, observed in infrared by the satellites Iras, ISO (22P/Kopff, 2P/Encke (see Figure 2.22), and Spitzer. These trails are the source of meteor showers, the sometimes spectacular rain of shooting stars that occurs when the Earth crosses their orbit. Thus, as early as 1866, Giovanni Schiaparelli (1835–1910) connected the Perseid meteor shower to the comet 109P/Swift-Tuttle (see Chapter 3).

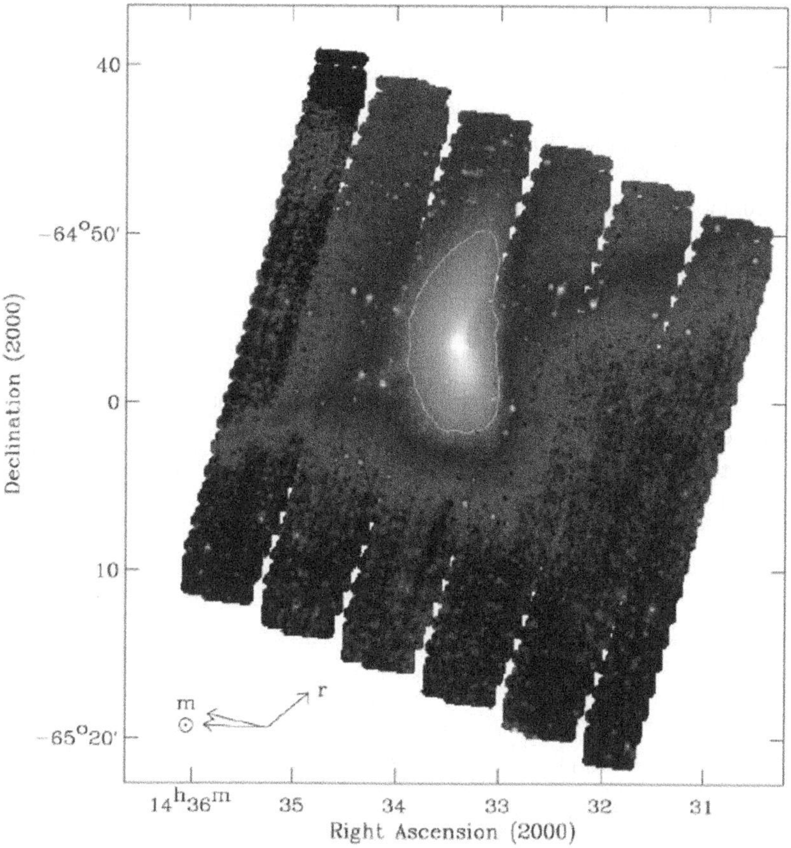

Figure 2.22. *A picture of the comet 2P/Encke obtained at a wavelength of 11.5 μm with the Isocam camera on the ISO satellite; we can see the thin cometary trail superimposed on the wide fan-shaped smudge of the tail (source: based on (Reach* et al. *2000)). For a color version of this figure, see www.iste.co.uk/encrenaz/solar2.zip*

2.1.7.2. Composition of the grains

Initial information came from the infrared spectroscopy of comets: in many cases, the characteristic emissions of silicates at 10 μm and 18 μm appear in the cometary thermal spectrum. In a few cases, the signature of H_2O ice was observed at 3 μm. The ISO satellite made it possible to measure the spectrum of the comet Hale-Bopp and of another periodic comet, 103P/Hartley 2, with a high degree of precision.

It thus determined the nature of the silicates that make up the grains in these comets (see Figure 2.23); they are forsterite, a crystallized silicate of olivine type, rich in magnesium: Mg_2SiO_4, as well as pyroxene crystals $(Mg,Fe)SiO_3$ and amorphous silicates. In interstellar grains, silicates are amorphous. The presence of silicate crystals in cometary dust bears witness to heating during passage through the inner Solar System, where ice could no longer exist. Cometary matter thus comes from compounds with different origins, intermixed because of their turbulent diffusion.

In addition, the observation of the visible spectrum of the comet C/1965 S1 (Ikeya-Seki), which approached the Sun in 1968, revealed emissions due to a number of metallic atoms produced through the volatilization of dust near the Sun (Ti, V, Cr, Mn, Fe, Co, Ni, Cu); their relative abundances agree with solar abundances. Emissions from metals, as well as from lithium, were also detected on the comet Shoemaker-Levy 9 during its collision with Jupiter (see Box 5.1, Volume 1).

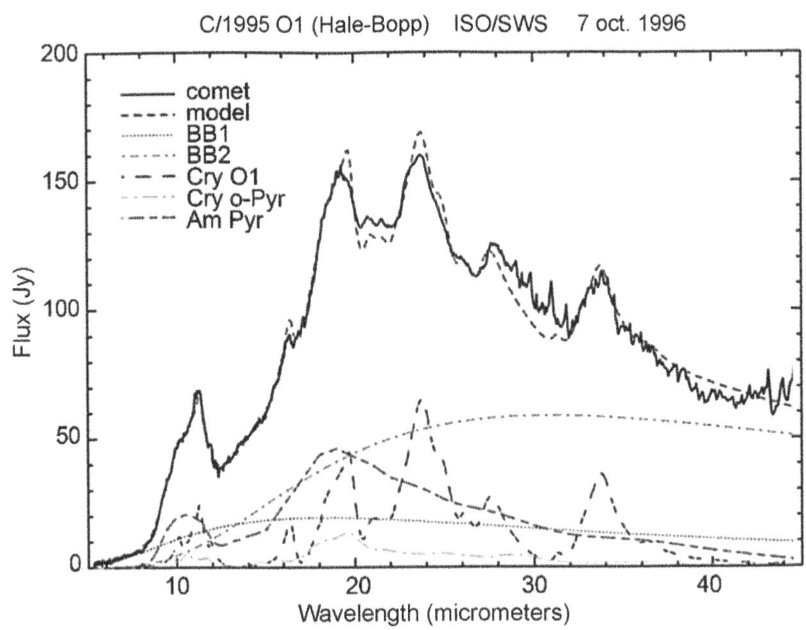

Figure 2.23. *The spectrum in the mid-infrared of the comet C/1995 O1 (Hale-Bopp) observed by ISO and a model of it by mixture of silicates (forsterite and pyroxene) and a black body emission (source: adapted from (Crovisier* et al. *1997). For a color version of this figure, see www.iste.co.uk/encrenaz/solar2.zip*

This spectroscopic information was supplemented by the results of direct observation:

– the analysis of dust collected in the terrestrial stratosphere, whose origin is supposed to be cometary and whose composition is similar to that of C1 chondrites (see Chapter 3);

– measurements by mass spectrometer and infrared spectrometry, carried on probes that flew by Halley's comet, Giotto and Vega, which indicate the presence of both silicate and carbon grains;

– the collection by the Stardust probe of grains collected from the coma of the comet 81P/Wild 2, brought back to Earth and studied in the laboratory (see Figure 2.24);

– the *in situ* analysis of grains from comet 67P by three instruments on the Rosetta probe – Cosima, Giada, and Midas – which made it possible to measure the composition, mass, volume, and flux of these particles (see Figure 2.25).

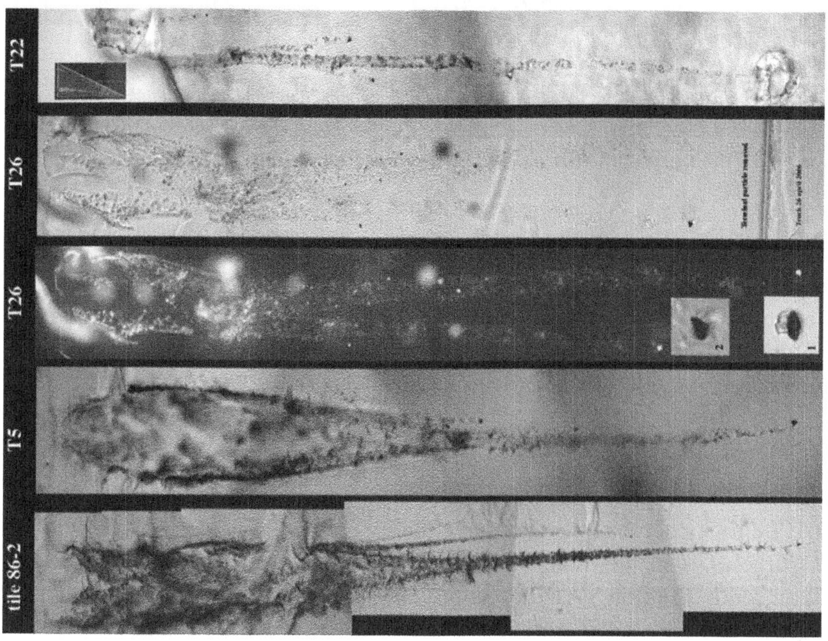

Figure 2.24. *Traces of dust grains from the comet 81P/Wild collected in aerogel by the Stardust probe (source: NASA). For a color version of this figure, see www.iste.co.uk/encrenaz/solar2.zip*

2.1.7.3. *Size of the grains*

The polarimetry and photometry of comets, as well as of the zodiacal cloud, as a function of their phase angle, in the visible and the infrared, provides us with constraints on the size and nature of the particles, using Mie's theory. While remote observations rarely make it possible to separate the two parameters, they allow us to constrain distribution models.

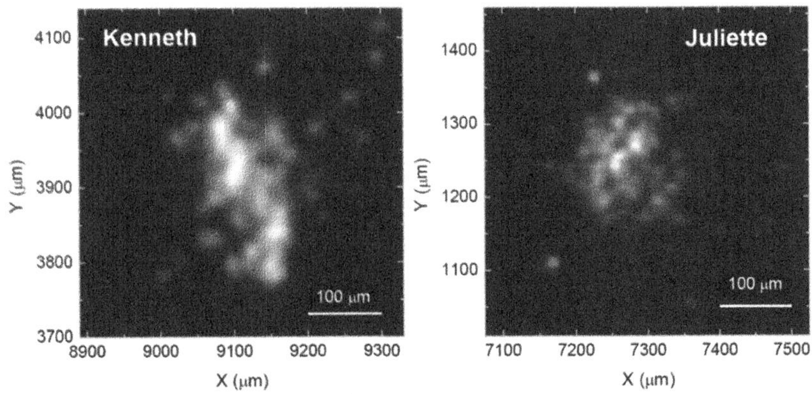

Figure 2.25. *Two grains of cometary dust collected by Rosetta/Cosima from 67P/Churyumov-Gerasimenko (source: based on (Fray* et al. *2016))*

Furthermore, direct measurements made *in situ* give access to the flux and the size of the particles encountered as a function of distance from the nucleus. We noted in particular the existence of a higher proportion of grains with very small masses (< 10^{13} g), at large distances from the nucleus in the direction of the Sun, which shows that they are essentially insensitive to radiation pressure.

2.1.7.4. *Dust ejection rates*

Estimates of production rates based on cometary observations in the visible range are tricky because they depend strongly on the diffusion index and the size of the particles. On the other hand, we can estimate the mass of the grains independently from their size by simply measuring their infrared flux; this is valid for grains that are smaller than the wavelength. The production rates measured vary, on average, from 10^4 g s^{-1} (for short-period comets, whose activity is slowed) to 10^6 g s^{-1} (for long-period comets, just beginning their activity). As for the ratio of dust/gas production rates, it has been estimated in a few cases based on analysis of the dust tail using the Finson-Probstein method; the first ratios measured were on the order of 0.7. However, the uncertainties are large, and it is not impossible that this ratio varies from one comet to another.

Practically, as an indicator for dust production, we use the parameter $Af\rho$, where A is the albedo of the grains, f is their filling factor, and ρ the linear radius of the field of view in cm. It is evaluated using the reflected solar radiation on the grains in the ultraviolet, visible, or near infrared range. For a simple model of grain distribution (constant velocity, stationary production) this parameter is independent from the field of view, and an $Af\rho$ of 1 cm corresponds to a dust production of the order of 1 kg s^{-1}.

2.1.8. *The chemical diversity of the comets: a relationship to their origins?*

It is possible to make a taxonomic classification of comets based on their orbits, as we have seen (see section 2.1.2), but also on their chemical composition (Mumma and Charnley 2011; Cochran *et al.* 2015; Dello Russo *et al.* 2016; Biver and Bockelée-Morvan 2017).

When we compare the compositions observed for a variety of comets, we have to be careful. The relative sublimation of the different components of cometary ice depends on the heliocentric distance (see Figure 2.10 and Table 2.5). In the coma, we thus see the CO/H_2O and CO_2/H_2O abundance ratios increase with distance from the Sun! It is therefore normal to find such differences between comets that are far away from or near the Sun because the composition of the gas in the coma does not directly reflect the composition of the nuclear ice. The comparison thus must be made for compositions obtained at similar heliocentric distances (usually one astronomical unit).

Systematic studies have been conducted based on spectrophotometry. They use standard filters (a spin-off of the International Halley Watch) centered on the bands of the main radicals in the visible range: CH, C_2, C_3, and the lines of O, sometimes also OH, NH, NH_2, and CH. O and OH serve as tracers for H_2O. Current databases contain around 250 comets. Taxonomic studies based on principal component analysis or on *cluster analysis* – as is done for asteroids – have been undertaken.

These studies show that while a majority of comets, the "typical comets", present similar compositions, 25 to 30% are different. In particular, many of them have C_2/CN and C_3/CN abundance ratios that are especially low: these are the "carbon-chain depleted comets". These may be linked to the Jupiter family, where they are more common. Or it may be the result of an evolutionary effect. None of this has been firmly confirmed.

It would of course be desirable to conduct a similar study based directly on the abundances of parent molecules. This is now possible thanks to progress in radio and infrared spectroscopy, which gives us access to H_2O, HCN, CH_3OH, CO, CH_4, and H_2S (to only mention the most widely studied molecules). But the sample we currently have is still limited, containing only a few dozen comets (Biver and Bockelée-Morvan 2016; Dello Russo *et al.* 2016).

We should note the existence of some atypical comets, like 43P/Wolf-Harrington, which lacks bands of C_2 and C_3; C/1988 Y1 (Yanaka), which exhibits particularly intense bands of NH_2; and C/2016 R2 (Panstarrs), which is exceptionally rich in CO and in N_2. The interstellar comet 2I/Borisov is also found to be very rich in CO, which is a possibly clue to its different origin.

In conclusion, a relationship between the chemical variety and the dynamic class remains to be found, which would allow us to determine the relationship between composition and location of formation.

2.1.9. *The interaction of comets with solar wind*

The most spectacular manifestation of the interaction of comets with solar wind is the development of ion gas tails, previously called "type I" tails, thin and straight, clearly distinct from the dust tail (see Figure 2.1). To explain this tail, Ludwig Biermann (1907–1986), followed by Hannes Alfvén (1908–1995), postulated the existence of a solar wind in 1957.

Ion tails develop at distances closer to the Sun than dust tails: typically, at less than 2 au. They are composed of molecular ions (H_2O^+, CO^+, CO_2^+, OH^+, etc.) which are accelerated by interaction with the solar wind and the interplanetary magnetic field to speeds of several hundred km s^{-1}.

The physical processes that govern interactions with the solar wind are described in Chapter 3 of Volume 1 (see in particular section 3.2.2.1 and Figures 3.17 to 3.19 in Volume 1). Comet nuclei, which have low mass, belong to the class of objects that lack a magnetic field, as confirmed by measurements taken *in situ* by space missions.

Impacts occur between high-speed particles of plasma in the solar wind, especially protons and ionized helium atoms, and low-speed ions in the cometary atmosphere, essentially CO^+ and H_2O^+, resulting from the photo-ionization of neutral molecules originating from ice in the nucleus. But for weakly active comets (whose water production rate is lower than 5 × 10^{26} molecules s^{-1}), the solar wind is not pushed back or slowed by the coma, and it hits the surface of the nucleus with full force.

The contact area between the solar wind and the coma is called the "bow shock". Its position depends on the comet's activity. It is typically located 5,000 km beyond the nucleus for a water production rate of 10^{28} molecules s^{-1}.

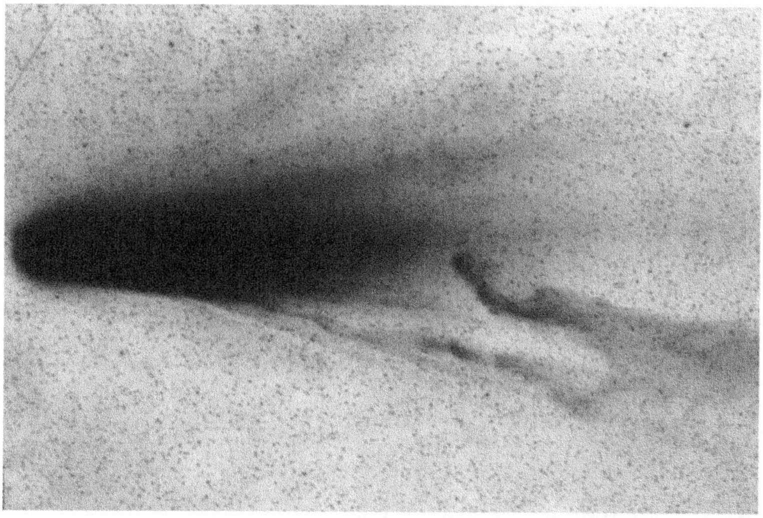

Figure 2.26. *The disconnection of the plasma tail from Halley's comet on March 10, 1986 (source: ESO)*

A characteristic phenomenon affecting ion tails that has been observed on several occasions is the apparent disconnection of this tail behind the coma, followed by the reformation of a new tail from ionized lines that converge toward it (see Figure 2.26).

We attribute this phenomenon to the passage of the comet nucleus through a region where the magnetic field changes direction. We thus observe a new plasma tail, while the piece of the old tail, of reverse polarity, is detached.

2.2. The "historical" asteroids

The German astronomer Johann Daniel Tietz (known as Titius, 1729–1796) published in 1766 an article indicating that the radii of planetary orbits increase geometrically, with an empty area between Mars and Jupiter. This observation is known as the empirical Titius-Bode law (see Box 1.2, Volume 1), because Johann Elert Bode (1747–1826), director of the Berlin observatory, popularized it in his books in 1772, thus inciting observers to seek the "missing planet".

On January 1, 1801, the Sicilian Giuseppe Piazzi (1746–1826) discovered a planetoid with a diameter of less than one thousand kilometers, orbiting at about 2.8 au, on an orbit corresponding to that of the "missing planet". He called the new planet Ceres (see Figure 2.27).

Figure 2.27. *Portrait of Giuseppe Piazzi and the cover of his book announcing the discovery of Ceres (source: Wikimedia Commons)*

The discovery in the following six years of three objects similar to Ceres – Pallas, Juno, and Vesta – with relatively similar orbits, showed that there was not a single planet, but a group of small objects, for which Bode proposed the name asteroids, for their star-like appearance (*asteri* in Greek).

The discoveries then followed in exponential fashion, at the pace of progress in observational techniques: today, almost a million asteroids have been inventoried. For more than 550,000 of them, a precise orbit has been determined, which makes it possible to locate them for observation (see section 1.3.4.2, Volume 1).

At the time of their discovery, the asteroids are provisionally named by the year of discovery followed by two capital letters indicating, respectively, the fortnight (A: January 1–15; B: January 16–31, etc.) and the order of discovery in that period (A first, ... Z twenty-fifth; the letter I is not used to avoid confusion with the number 1). If there are more than 25 discoveries in a fortnight, the letter A is used again and a number (1, 2, 3, etc.) is added as a subscript, which indicates that there are respectively 25, 50, or 75 additional discoveries. For example, 1992 QB_1 was the 27th asteroid discovered in the second fortnight of August 1992. It followed the 25th, 1992 QZ, the 26th, QA_1, and preceded the 28th, 1992 QC_1.

An identification number has been assigned to each of the 550,000 objects that have a clearly determined orbit, corresponding to when it was added to the list of known asteroids, followed by a name (chosen by the discoverer) and/or by a provisional description.

Box 2.6. *How to name an asteroid*

The asteroids are composed of rocks, metals, and ices, and their size varies from approximately 10 meters to several hundred kilometers. They are the remains of primitive material that were unable to form a planet because of their destructive interactions with Jupiter. They are present throughout the inner Solar System, up to the orbit of Jupiter, but they travel on orbits that are relatively circular in the *main belt*, a region located between the orbit of Mars and that of Jupiter, between 1.7 and 3.5 au from the Sun. They form a heterogeneous set of small bodies with a wide spectrum of physical and dynamic properties.

To date, 17 of these objects have been approached by space probes. (433) Eros, (4) Vesta, (25143) Itokawa, (162173) Ryugu, and (101955) Bennu have been studied for several months during orbital missions with, for the last three, a return of samples to Earth. The 11 other asteroids – (253) Matilda, (243) Ida and its satellite (243.1) Dactyl, (951) Gaspra, (5535) Annefrank, (4179) Toutatis, (21) Lutetia, (2685) Masursky, (2867) Steins, (132524) APL, and the trans-Neptunian (456958) Arrokoth – have only received flybys.

2.2.1. *The asteroids in the main belt*

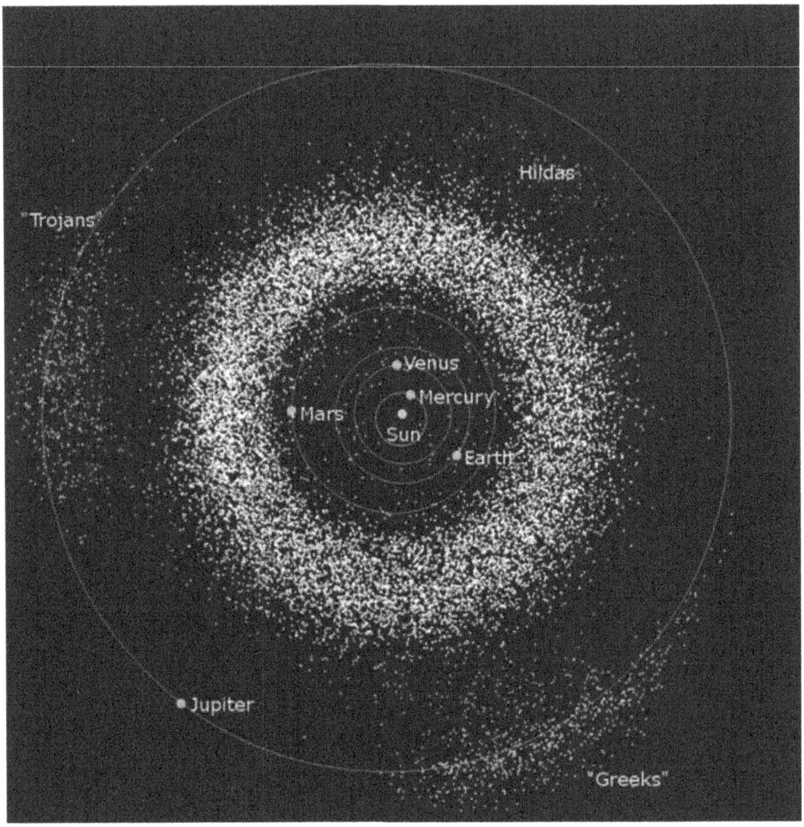

Figure 2.28. *The main belt and the Trojan asteroids (source: Wikimedia Commons). For a color version of this figure, see www.iste.co.uk/encrenaz/solar2.zip*

Most of the billions of asteroids that form the main belt have sizes on the order of dozens of meters; a million asteroids are kilometer-sized, 200 asteroids have diameters larger than 100 km, and only 30 asteroids have a diameter of more than 200 km, with the largest being Hygiea (429 km), Pallas (538 km), and Vesta (576 km). The total mass of the main belt is estimated to be 3.0-3.6 x 10^{21} kg.

The semi-major axes of the ellipses traveled by the asteroids around the Sun are included in a toroidal space between the orbits of Mars and Jupiter: their spatial distribution shows a maximum for an eccentricity on the order of 0.07 and an inclination below 4°. The central portion of the main belt contains more than 90% of these small bodies.

The distribution of the asteroids in the main belt is not uniform; some parts of the semi-major axis distribution are almost empty (these are the *Kirkwood gaps*) separated by concentrations of objects. The main gaps correspond to the asteroidal orbits in resonance with Jupiter or Saturn. An orbit is in resonance if its orbital period is in a small, whole number ratio with that of another body: for example, an asteroid is in 2:1 resonance with Jupiter if it completes two orbits around the Sun while Jupiter completes one (see section 1.2.3, Volume 1). The two bodies regularly find themselves in the same relative position along their orbits and the effects of gravitational interaction increase each time. An asteroid that has an orbit close to resonances is subject to strong perturbations from Jupiter (and, to a lesser extent, Saturn) and its orbit is pushed inward (or outward). This change of orbit implies that the asteroid cannot remain in the vicinity of the gap, which empties completely in a short time.

In 1918, the Japanese astronomer Kiyotsugu Hirayama (1874–1943), in studying the distribution of asteroids in terms of *proper orbital elements* (which are the semi-major axes, eccentricity, and inclination, where the effect of planetary perturbations has been eliminated), found that a certain number of asteroids had very similar orbital elements, forming groups that he called families of asteroids. He gave each family the name of its largest member: Themis, Eos, Koronis, and Flora. He also advanced the hypothesis that these families were the result of a collision between two asteroids. The energy released by the impact is almost all used to fragment the asteroids, and yet the trajectories of the debris will not be much changed in relation to that of the destroyed asteroid, even though gravitational perturbations tend to disperse them. The fact that the members of each family exhibit a spectrum with specific signatures indicates a common origin.

Today, several dozen dynamic families of asteroids have been cataloged as being groups of asteroids in terms of orbital elements (see Figure 2.29), but not all are the result of a violent collision between the parent body and another asteroid. The Phocæa asteroid group, located between 2.3 and 2.4 au, is separated from the main belt: this family was formed by the depopulation of neighboring regions due to strong, ancient resonances.

The 15,000 Hungaria asteroids (from the name of the largest object among them, (434) Hungaria) represent the inner boundary of the main belt: their orbits have eccentricities below 0.18, their semi-major axes are between 1.78 and 2.00 au, and their inclinations span between 16° and 34°. Their orbital periods around the Sun are about 2.5 years. All are in (9:2) resonance with Jupiter and (2:3) resonance with Mars, and so they are subjected to periodic effects of gravitational perturbations caused by these two planets: every 5 years when they are aligned with the Sun and Mars, and every 22.5 years when they are aligned with the Sun and Jupiter. This dynamic configuration has transported them from the innermost regions of the main belt and confined them in the border region they now occupy.

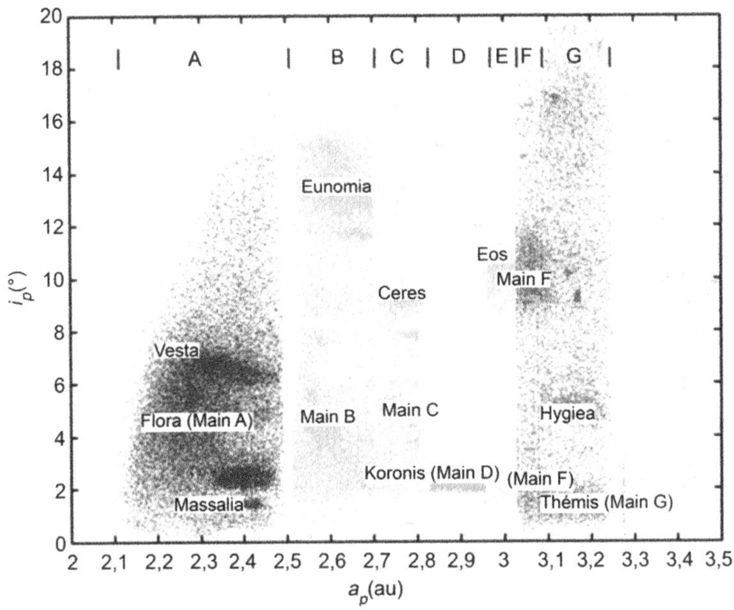

Figure 2.29. *The main families of asteroids (source: Wikimedia Commons). For a color version of this figure, see www.iste.co.uk/encrenaz/solar2.zip*

The Hilda group of asteroids (see Figure 2.30), which are in 3:2 resonance with Jupiter, constitutes the outer boundary of the main belt. Their semi-major axes are between 3.7 and 4.2 au, their eccentricities are lower than 0.3, and their inclinations are not more than 20°. At any given moment, the Hildas' position in space constitutes an equilateral triangle, called the "Hildas triangle", with sides about 1 au thick, slightly convex toward the exterior (see Figure 2.28). The angles of this triangle coincide with the positions of aphelion of their orbits (the opposite of Jupiter in relation to the Sun, at 60° before and 60° after the position of Jupiter). In the course of three successive orbits, each Hilda asteroid passes successively through each of these three points. Jupiter always follows one of the "sides" of the triangle; each Hilda, when it is close to perihelion, thus remains aligned with the Sun and Jupiter: this implies that each Hilda remains at a distance of more than 1.5 to

2.0 au from Jupiter, reducing perturbations to a minimum and thus avoiding being ejected. The triangle is in rigid rotation with Jupiter, and the Hilda group is stable.

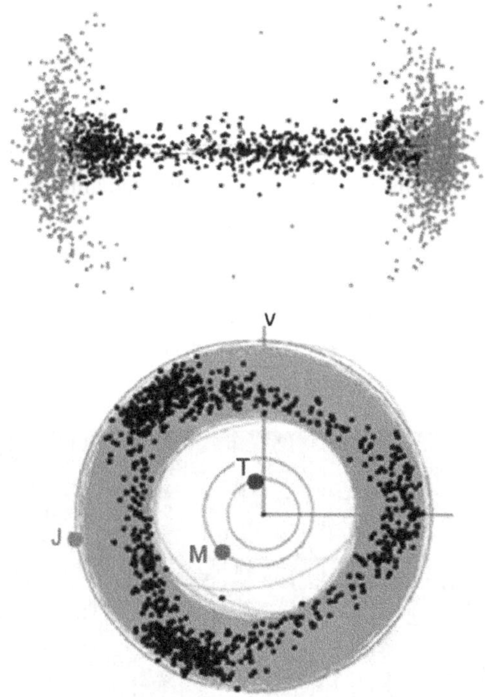

Figure 2.30. *(a) Hildas (black dots) and Trojans (gray dots) seen from the plane of the ecliptic at 190° longitude on January 1, 2005; (b) the Hildas triangle on the background of their orbits (in gray), between the orbits of Jupiter (J) and Mars (M); T indicates the Earth (source: based on V.N. L'vov, R.I. Smekhachova, S.S. Smirnov, S.D. Tsekmeyster (Glavnaya (Pulovskaya) astronomicheskaya observatoriya RAN)). For a color version of this figure, see www.iste.co.uk/encrenaz/solar2.zip*

Beyond, at 4:3 resonance (which corresponds to a distance from the Sun of 4.26 au), orbits the asteroid Thule, apparently alone.

2.2.2. *The asteroids that cross the orbit of the terrestrial planets*

In the inner Solar System there is a population of small bodies whose orbits cross those of the terrestrial planets (referred to as *planet crossers*). When they cross Earth's orbit, we call them *near-Earth objects* (NEO). The impacts of near-Earth objects with Earth have played an important role in the geological evolution of our planet and on the conditions for the emergence of life.

The action of the resonances with Saturn and Jupiter causes an increase in the eccentricity of some asteroids, which can reach the orbits of Earth, Mars, and Venus, and sometimes even plunge into the Sun: the time necessary to become an asteroid that crosses the orbit of an inner planet is around a half-million years. After several million years spent in the vicinity of the inner planets, most of them plunge into the Sun or are expelled from the Solar System, and the remaining asteroids adopt stable elliptical orbits. The perturbations of Neptune on the Kuiper belt can also inject trans-Neptunian objects into the inner Solar System.

The distance at perihelion of all of these asteroids is less than or equal to 1.3 au, depending on their orbital features; they are divided into four dynamical families, named for one of their members (see Figure 2.31).

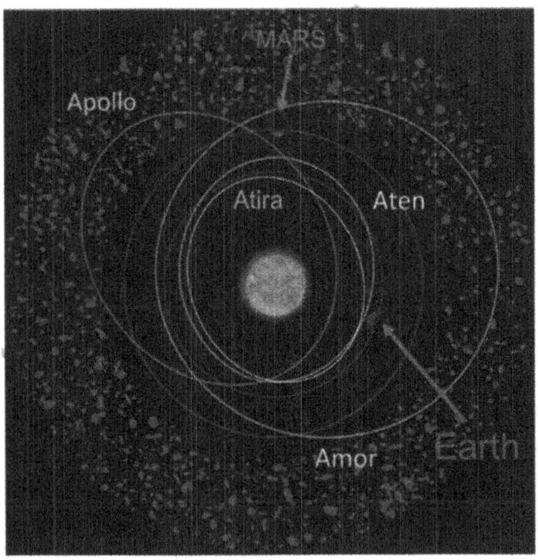

Figure 2.31. *Orbits of the asteroids after which we name the groups of "planet crossers" – objects that cross the orbits of the terrestrial planets (source: based on a photo from ESA). For a color version of this figure, see www.iste.co.uk/encrenaz/solar2.zip*

The Apollos (named after the asteroid (1862) Apollo) have a semi-major axis greater than that of Earth ($a > 1.0$ au) and a perihelion of $q \leq 1.017$ au. They represent 60% of the population of NEOs. This group includes just under 18,000 known members.

The Atens (named after the asteroid (2062) Aten) have a semi-major axis smaller than that of Earth ($a < 1.0$ au) and an aphelion of $Q > 0.983$ au. They regularly cross Earth's orbit, but being most often inside the terrestrial orbit, they are difficult to observe: to date, we know of around 1,500 Atens.

The Amors (named after the asteroid (1221) Amor) constitute more than 30% of NEOs (almost 9,000 objects). They have a semi-major axis larger than that of Earth ($a > 1.0$ au) and a perihelion 1.02 au $< q \leq$ 1.3 au. Their orbits are outside Earth's – this group of objects crosses Mars's orbit.

The Atiras (named after the asteroid (163593) Atira) are also called the Apohele asteroids or inner earth objects (IEO). These objects have $a < 1.0$ au and $q < 0.983$ au, and they thus orbit inside the terrestrial orbit. To date, only about 20 of the Atiras have been identified because they are very hard to observe. (163593) Atira is a synchronous binary asteroid.

2.2.3. *The Trojan asteroids*

The Trojan asteroids are small bodies that share their orbit with a planet or a satellite (we say that they are in 1:1 orbital resonance with the planet or the satellite). They are grouped around one of the two L4 and L5 Lagrange points (see section 4.3.1) corresponding to the points where the gravitational attraction of the Sun and the planet on the third body are in equilibrium, which ensures the stability of its orbit. From the earliest discoveries, these objects have been named for the heroes Homer sings about in the *Iliad*.

Jupiter's Trojans are the two highly populated groups of asteroids of this type that have long been known (in green in Figure 2.28). During the last few years, we have discovered Trojans of Neptune and Mars as well as of Uranus and Earth. Nineteen of the 22 Neptunian Trojans and 8 of the 9 Martian Trojans are in L5. The Earth's Trojan is in L4, as is Uranus's. This is the result of selection effect, due to the small number of discoveries.

The Trojans of Jupiter have orbits with an average eccentricity of 0.074 and an average semi-major axis of 5.20 au, with inclinations that are relatively high on the plane of the ecliptic. Today, we know of 4,603 in L4 and 2,476 in L5. The population of these objects with a diameter greater than 1 km is estimated to be 320,000.

NASA's Lucy space mission includes the first probe to explore Jupiter's Trojan asteroids. It is scheduled to launch in October 2021 and will visit six of these asteroids between 2027 and 2033, including five objects that are found in L4 and one in L5. Beginning in August 2027, Lucy will explore (3548) Eurybates (L4, type C) with a near flyby, followed one month later by (15094) Polymele (L4, type P). In April and November 2028, respectively, (1131) Leucus (L4, type D) and (21900) Orus (L4, type D) will be explored. A flyby of the binary system (617) Patroclus/(617.1)-Menoetius (L5, type P) will conclude the mission in March 2033.

2.2.4. *The properties of asteroids*

The asteroids share common distinguishing properties: *in situ* observations by space probes or ground-based telescopes make it possible to determine the physical and chemical characteristics of their population, trace their history, and understand the nature of their members.

The physical properties that can be measured directly are mass, volume, rotation period, temperature, and derived values such as density, porosity, gravity (which are the only information about the internal structure of the objects), as well as thermal inertia, roughness of the surface, and the granulometry of the regolith.

Photometric and spectral data from the asteroids obtained from telescope-based observations remain the most powerful means we have to analyze the chemical and mineralogical nature of the asteroidal population.

2.2.4.1. *Physical properties*

The luminosity of an asteroid varies according to its distance from the Sun and from the Earth. To these variations are added those caused during its rotation on its axis because of its more or less irregular shape: it transmits the sunlight reflected by the portion of its surface that is visible to the observer. These variations in luminosity are the light curves. The rotation period of asteroids are known precisely based on the periodicity of their light curve (see Figure 2.32).

We can determine the mass of an asteroid in different ways:

– if we can measure the effects of perturbations caused by a close encounter between two asteroids along the orbit of the smaller one (error $\geq$ 100%);

– if there is at least one satellite, by applying Kepler's third law (error $\approx$ 20% to 30%);

– by measuring perturbations in the trajectory of space probes in orbit or by carrying out a near flyby (error ≈ 1%–5%).

Several hundred measurements of the asteroids' mass have been obtained using the first two methods, and only five with the third method (see Table 2.6).

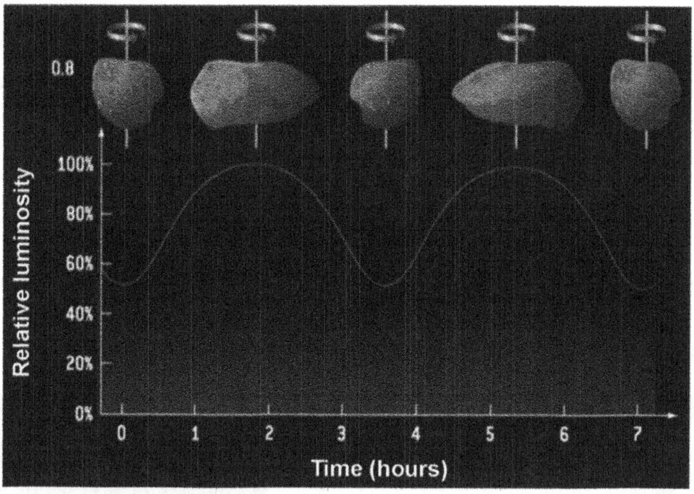

Figure 2.32. *The light curve of an asteroid makes it possible to determine its rotation period and contains information about its shape and its orientation in space (source: Wikimedia Commons)*

Asteroid	Type	Mass (kg)	Dimensions (km)	Density (kg/m^3)	Rotation period (h)
(4) Vesta	V	2.59076 ± 0.00001 10^{20}	572.6 × 557.2 × 446.4 ± 0.2	3,700 ± 300	5.34212766 ± 0.00000001
(433) Eros	S	6.6869 ± 0.0015 10^{15}	34.4 × 11.2 × 11.2 ± 0,2	2,670 ± 30	5.270 ± 0.005
(25 143) Itokawa	S	3.1464 ± 0.0094 10^{10}	0.535 × 0.294 × 0.209 ± 0.001	1,950 ± 130	12.1324 ± 0.0001
(162 173) Ryugu	C	4.5 ± 0.06 10^{11}	0.865 ± 0.015	1,190 ± 30	7.827 ± 0.007
(101 955) Bennu	B	7.329 ± 0.009 10^{10}	0.564 × 0.536 × 0.498 ± 0.001	1,190 ± 30	4.296057 ± 0.000002

Table 2.6. *Main physical properties of 5 asteroids explored by space probes in orbit*

The images of asteroids taken by cameras on-board the probes of space missions are the only ones to show their exact shapes (see Figure 2.33). We can obtain less precise indications about the shape of asteroids by using ground-based observations. For asteroids whose shape has been determined, we can calculate the volume using a three-dimensional model: the precision of the estimate of the volume of an asteroid thus depends on the precision with which we can reconstruct its shape.

Thermal models (Harris and Lagerros 2002) of bodies without atmospheres, constrained by the measurement of the flux in the mid-infrared, make it possible, by assuming a spherical object, to estimate its albedo and its size; we thus obtain a very general indication of its volume. With measurements carried out by NASA's Wise satellite, we have information estimating the diameter (and therefore the volume) of several thousand asteroids.

Radar observations make it possible to obtain information about asteroids that are near-Earth objects: shape, simple or binary, surface characteristics, etc. We have hundreds of radar observations of near-Earth objects (Ostro *et al.* 2002).

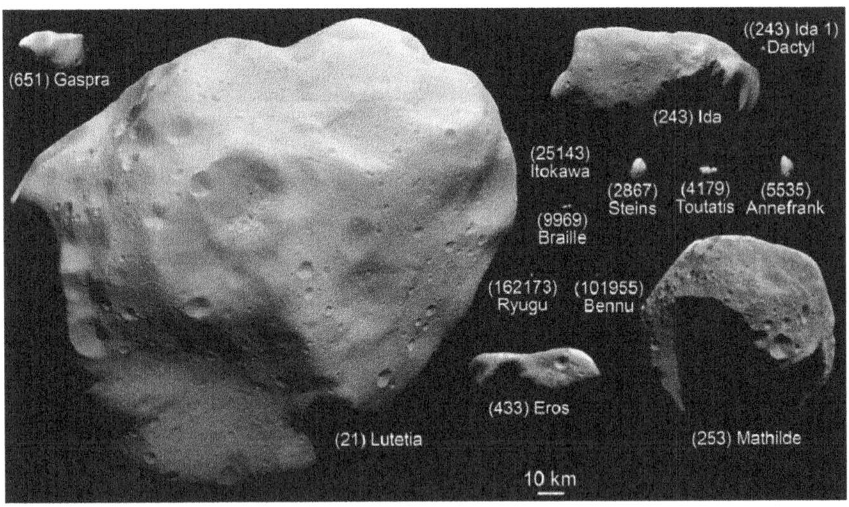

Figure 2.33. *The shape of several objects encountered by space probes*

COMMENTS ON FIGURE 2.33.– *(253) Mathilde is a spheroid flattened at the poles; (243) Ida, (951) Gaspra, (5535) Annefrank, (9969) Braille and (4179) Toutatis are elongated; (21) Lutetia, ((243) Ida 1) Dactyl, and (2685) Masursky have irregular shapes; (433) Eros is shaped like a peanut; (25143) Itokawa has the shape of two joined bodies as a result of a nondestructive collision, and (2867) Steins, (162173) Ryugu, and (101955) Bennu look like spinning tops (source: NASA and JAXA).*

We have seen that light curves contain information about the shapes of asteroids. Digital techniques to invert light curves have been developed (Kasalainen 2002) and have been widely used to obtain a relatively precise reconstruction of the shape of an asteroid, as long as it does not contain too many concavities. The shape of several hundreds of objects has been defined in this way.

Asteroids larger than 200 km have a relatively regular spheroidal shape and are probably the remains of planetesimals that formed planetary cores. Objects whose size is on the order of 1–10 km are generally *rubble piles*, created by a catastrophic impact and reassembled by gravity: they are shaped like spheroids, spinning tops, or can be more or less elongated. Intermediate-sized bodies, between 10 and 200 km, are the large, monolithic remains of the destruction of an asteroid, and they have irregular shapes.

The density (ratio between an asteroid's mass and volume) can vary between 900 kg/m^3 for the most porous asteroids and 3,500 kg/m^3, which is typical for massive, metal-rich asteroids.

The temperature of a celestial body is obtained by measuring the radiation flux in the mid-infrared, and thermal models of bodies without atmospheres also provide indications of their thermal inertia, surface roughness, and emissivity.

2.2.4.2. *The geology of asteroids*

The surface of the asteroids conserves traces of all of the interactions with their environment that have occurred over time.

The consequences of the collision of two asteroids of comparable size depends on the energy involved in the impact: if the energy is high, the two bodies will fragment and may disperse, forming a family, or may re-aggregate, giving rise to a *rubble pile*; if the energy is low, the two bodies may fuse and form a single asteroid characterized by a bi-lobed shape, as happened with the small asteroid (25143) Itokawa.

Non-catastrophic impacts of smaller bodies (projectiles) on a larger asteroid (target) are at the origin of craters, fractures, and ridges that alter the shape of the target and induce seismic phenomena responsible for the motion of surface material. Some asteroids have been subjected to structural modifications on a global scale, as illustrated by the presence of fractures and ridges that span their hemispheres.

The shape and size of craters can allow us to estimate the body's state of cohesion; their number also gives an indication of its age (if we hypothesize about the flux of the impacts) and their spatial distribution tells us how long the different regions of the surface remained exposed to impacts compared to each other, thus supplying a relative chronology of the surface's history. There are enormous craters that cover more than a third of the surface (such as Rheasilvia, D = 475 km, superimposed on Veneneia, D = 400 km, which altered Vesta's south pole, Figure 2.34) and craters a few centimeters in diameter, the result of the impact of micro grains of dust, which erode the surface materials.

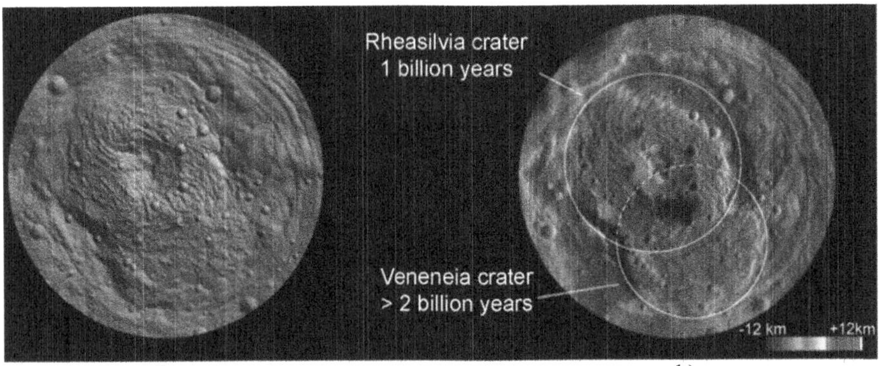

a) b)

Figure 2.34. *(a) Map of the relief of the south pole of (4) Vesta. (b) The walls of the Rheasilvia and Veneneia craters are indicated in white on the topographic map of the south pole (source: NASA). For a color version of this figure, see www.iste.co.uk/encrenaz/solar2.zip*

On the *rubble piles* (see Figure 2.35), we find fewer craters than we have observed on other objects, which indicates a more recent formation, and these craters are in general less deep: indeed, due to the high porosity of these bodies, more energy is released by the projectile to compact the target's surface materials than to excavate the crater.

The surfaces of asteroids are covered with blocks of rock, with sizes ranging from a few centimeters to dozens of meters. These are materials ripped away by impacts and ejected from craters, or the last fragments resulting from a collision that have accumulated to form a rubble pile.

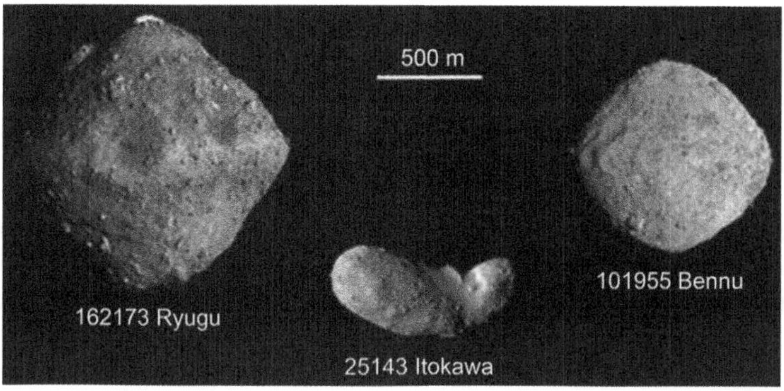

Figure 2.35. *The three rubble-pile asteroids explored by the JAXA space missions Hayabusa 2 (Ryugu) and Hayabusa 1 (Itokawa) and the NASA mission OSIRIS-REx (Bennu; source: JAXA and NASA)*

2.2.4.3. *Chemical and mineralogical composition of the asteroids*

Today, for conducting thorough laboratory analyses of the composition of an asteroid, we only have a single sample: a few micrograms of the surface of the asteroid (25143) Itokawa collected during the Japanese Hayabusa 1 mission. In December 2020, the Japanese Hayabusa 2 mission brought back samples from asteroid (162173) Ryugu that have not yet been analyzed, and the American OSIRIS-REx mission is expected to bring back fragments of asteroid (101955) Bennu in September 2023.

Most meteorites are fragments of asteroids that are near-Earth objects, and with the help of similarities between the absorption spectra of different meteorites and different kinds of asteroids, we have a good starting point for characterizing the latter's composition.

The absorption spectrum provides most of the information about the chemical and mineralogical composition of the materials on the asteroid's surface. If there is no absorption, the spectrum is flat (we say that it is neutral); if there is a positive slope with wavelength, we say that the spectrum is red, and if the slope is negative, we say that it is blue. The position of the absorption bands, if they are present, indicates which minerals are absorbent, and their intensity is proportional to the abundance of these minerals.

A taxonomic classification, obtained from the absorption spectra of asteroids in the visible and near infrared range (Demeo *et al.* 2009), separates the asteroid population into several compositional types. Those with spectra exhibiting the same global behavior but with differences in their details have been grouped in the same "complex": each complex is designated by a letter (see below), and the types that it includes are distinguished by an index added to this letter. Most of the asteroids belong to the three complexes C, S, and X. Asteroids with spectra characterized by particular signatures have been grouped into specific types.

The C complex (for "carbonaceous") include almost 75% of all known asteroids. Generally, they orbit the central portion of the main belt. Their color is neutral with respect to that of the Sun, and their albedo is low (0.03 on average). They all have very flattened spectra, with a very low intensity in the ultraviolet range. In a large number of cases (70%), they have absorption bands called "aqueous alterations", which show that some minerals in the surface materials have been in contact with liquid water. Carbonaceous chondrites (see Chapter 3) seem to be the most likely meteoric analog of these asteroids, which are thus their parent bodies. These meteorites contain carbon (up to 5%), water (20%), hydrated silicates, magnetite, and sulfurs (see Chapter 3). Only 5% of meteorites that have been cataloged are carbonaceous chondrites, while objects in the C complex represent three quarters of the asteroid population: this indicates that these very fragile meteorites generally do not survive their passage through the terrestrial atmosphere. Indeed, the C asteroids have very low densities, indicating a very porous and thus very fragile internal structure. Objects of this type are rich in carbonates, which are opaque and photo-absorbent minerals, and during their evolution, there has been liquid water on their surfaces.

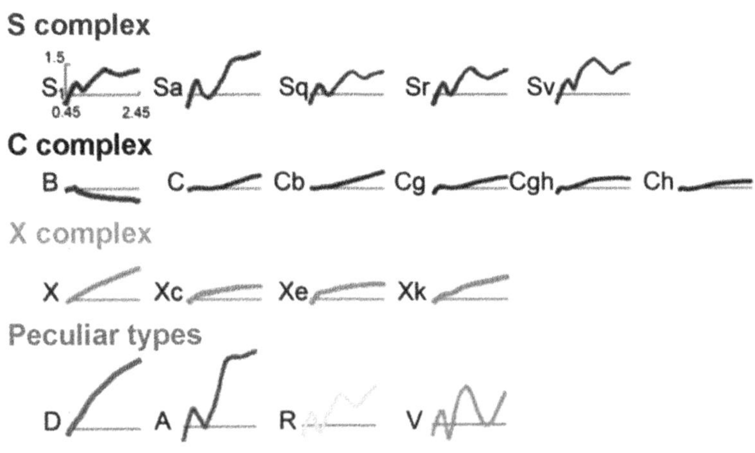

Figure 2.36. *Representative absorption spectra (between 0.45 μm and 2.45 μm) of the main types of asteroids (source: based on (Demeo* et al. *2009)). For a color version of this figure, see www.iste.co.uk/encrenaz/solar2.zip*

Of the known asteroids, 17% belong to the S complex (for "siliceous"); they are highly reflective (albedo between 0.10 and 0.22). Their spectra present the spectral features of siliceous rocks: absorption bands in the visible range (centered around 0.7 µm) indicating the presence of iron oxides, and especially bands in the near infrared (centered around 1 µm and 2 µm), typical of olivines and pyroxenes, which contain iron and magnesium. Their reddish color is due to the presence of relatively high levels of iron and metallic nickel. They generally occupy the central portion of the main belt and are the parent bodies of the meteorites most frequently found on Earth, ordinary chondrites.

In the X complex are grouped asteroids that are also classified as type M, P, and E. Their spectra are relatively flat, slightly red, and sometimes exhibit bands at 490 and 600 nm attributed to the presence of iron sulfide FeS. The albedo is the value that differentiates these three types: 0.02–0.08 for P, 0.1–0.2 for M, and 0.25–0.6 for E. The P-type objects are most common in the outer portion of the main belt and may be older than the C asteroids. Their spectra generally show no signature, and we do not know their meteoric analog. Their characteristic redness in probably due to a process of space weathering.

The spectra of the M (for "metallic") asteroids recall that of iron-nickel alloy; they are analogous to enstatite chondrites, meteorites composed of iron-nickel grains encased in a matrix of enstatite, which is a siliceous mineral rich in magnesium ($MgSiO_3$).

The E type (for "enstatite") includes a small group of objects such as the Hungaria family, characterized by the highest albedo of all of the asteroids. Their surface composition is dominated by enstatite and siliceous minerals with little iron, analogous to achondrites. This type is often found in the inner portion of the asteroid belt, although some, such as (64) Angelina, are located fairly far from this region.

The origin of these objects is not clear: indeed, types M and E meteoritic analogs suggest that these asteroids are very small fragments of mantle from a differentiated asteroid, but the fact that the spectrum of most of these objects (>30 km) exhibits the 3 µm water absorption band indicates rather an origin typical of primitive objects. The solution to this problem should be provided by the NASA mission Psyche, which will explore Psyche, the largest M type asteroid, beginning in 2026.

All of the Jovian Trojans are classified as type D asteroids (for dark), and in total, 10% of all asteroids belong to this class. These D objects are very dark (0.04 average albedo), and their spectra have no special signature, nor any analog with the meteorites, but are the reddest of all the asteroids in the main belt. Their color and albedo may indicate the presence of dark, organic materials (containing carbon, hydrogen, and oxygen) formed by *space weathering*. The nature and composition of the materials that formed the Jovian Trojans will become known thanks to NASA's Lucy mission.

The V type concerns the asteroid (4) Vesta, the vestoids, fragments expelled from Vesta when the giant basins that have shaped the south pole were formed, and a few rare objects in the outer main belt, such as (1459) Magnya. All of the V asteroids have a basaltic composition, and their spectra show deep absorption bands at 1 μm and 2 μm due to pyroxenes, minerals that on Earth are common components of igneous and metamorphic rocks. Their meteoritic analog is a family of meteorites, the HED (Howardites, Eucrites, Diogenites), which is believed to have come from vestoids that have landed on Earth (see Figure 2.37).

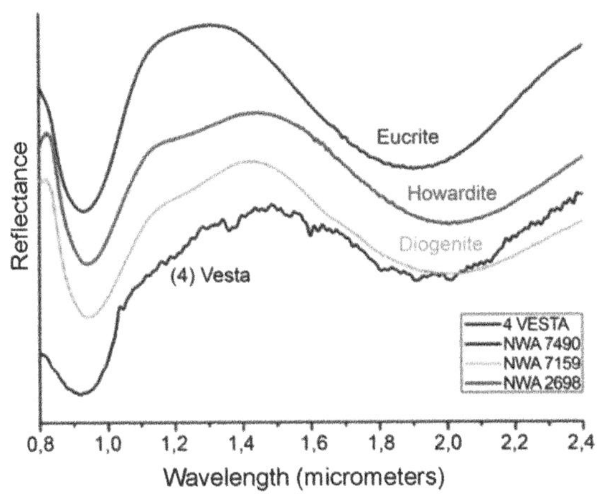

Figure 2.37. *Spectra of Vesta and the HED meteorites (source: based on (Ferrari* et al. *2017)). For a color version of this figure, see www.iste.co.uk/encrenaz/solar2.zip*

The composition of type A asteroids, an uncommon type in the asteroid belt, is characterized by a spectrum that shows a reflectance spike in the red at 0.7 μm, and an absorption band around 1.05 μm. The spectra show a characteristic shape and an increasing slope (with increasing wavelength), which is explained by the high abundance (> 80%) of olivine on the surface of these small bodies. It is reasonable to think that the rock of an asteroid so rich in olivine represents the mantle of a differentiated asteroid.

Type R is represented by asteroid (349) Dombrowska, which has a spectrum that indicates a composition dominated by olivines (a group of iron and magnesium silicates typical of basaltic rocks), with very little or no metals. These resemble the silicates that remain after metals are extracted from the upper mantle of an incompletely differentiated body. This type is intermediate between the V and A types.

The distribution of these types in the asteroid belt does not reflect the variations of composition of the original nebula from which the Solar System formed, but is linked to the temperatures to which the different bodies were subjected in the different locations where they formed. The darkest materials, rich in carbon and water, condensed originally in the coldest regions that were farthest from the proto-Sun, while the brighter, less volatile rocky materials were able to aggregate and subsist in the hottest regions near the Sun. Migratory events then contributed to displacing these small bodies, causing a chaotic redistribution of compositional types.

2.3. The "new" asteroids

Trans-Neptunian objects (TNO), Centaurs, and interstellar asteroids belong to the group of "new" asteroids: they have all been discovered in the last 45 years, thanks to progress made in observational approaches and instrumentation.

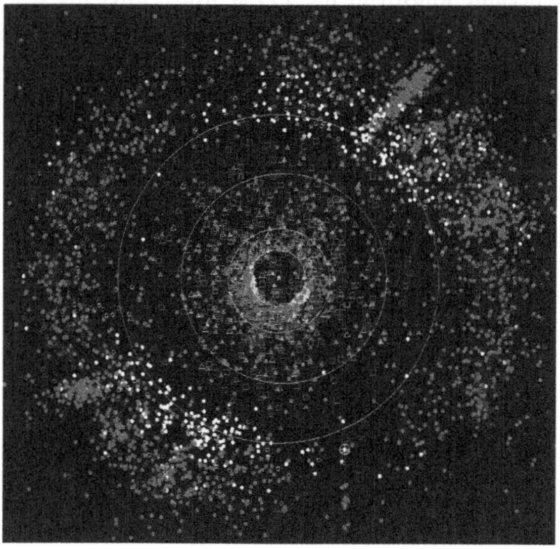

Figure 2.38. *Position of the trans-Neptunian asteroids on January 5, 2019. For a color version of this figure, see www.iste.co.uk/encrenaz/solar2.zip*

COMMENTS ON FIGURE 2.38.– *The orbits of Jupiter, Saturn, Uranus, and Neptune are shown. The red dots are classic trans-Neptunian objects, the white circles are plutinos, including Pluto, (the larger white symbol in the lower right). The purple dots are objects in the scattered disk, the blue triangles are objects with high eccentricity, the orange triangles are Centaurs, and the blue squares are periodic comets (source: based on the Minor Planet Center).*

TNOs are all very rich in volatile products, but because of their distance from the Sun, they do not exhibit any activity and are considered as asteroids. On the other hand, some of the Centaurs that orbit Jupiter and Neptune exhibit activity when they approach the Sun, at perihelion, and because of this they are considered both as asteroids and comets. Interstellar objects come from outside the Solar System, which they travel through at high speed.

2.3.1. *The Centaurs*

The Centaurs are small bodies that travel along orbits whose perihelion is located between the orbits of Jupiter and Neptune; their semi-major axes are between 5 and 30 au. We have discovered several hundred Centaurs, but we estimate the number of Centaurs with a diameter greater than 100 km to be 44,000. The existence of this population only became known in the 1990s when, after the discovery of (2060) Chiron in 1977, the number of these objects discovered increased significantly.

With its diameter of 250 km, (10199) Chariklo is the largest known Centaur. Discovered in 1997, it is the first asteroid found to have a ring system: its two rings are 396 km and 405 km away from Chariklo and are about 7 km and 3.5 km wide, respectively. The discovery was made during a stellar occultation by the Centaur, which occurred on June 3, 2013 (see section 1.5.7.1).

Some Centaurs, including (2060) Chiron, have developed sporadic activity, exhibiting a cometary nature, but in general, these objects exhibit no activity and are considered as asteroids: they are named based on the general rules for naming asteroids. The objects that exhibit activity have, on the other hand, received names like the comets: on October 18, 1977, when Charles Kowal discovered the first Centaur, it was designated 1977UB; then, when its activity was observed, the object was named comet 95P/Kowal. Finally, since this body seems to be both a comet and an asteroid, Kowal chose the name of the mythological Centaur Chiron for this 2,060th asteroid. The entire population thereby gained the name of the "Centaur group", and all of its members are named after the mythological being, half man and half horse.

The Centaurs are subject to intense perturbations caused by the giant planets, which make their orbits unstable. The orbital instability produces variations in eccentricity that range from almost circular to extremely elongated orbits. A Centaur remains a member of the population for an average of between 1 and 10 million years; then it may be expelled into interstellar space, or Jupiter may capture it, turning it into a comet in its family; it may also collide with a planet or plunge into the solar photosphere and end up vaporized.

The objects in the Kuiper belt are the main source of the Centaurs, which have similar physical surface features. They are in dynamic transition from the outer Solar System to the inner: they represent a sample of the Kuiper belt and are the parent bodies of the Jupiter-family comets. Their spectra exhibit a bimodal distribution, including objects with flat spectra and objects with red spectra. Space weathering may be the cause of the redness of the surfaces, while cometary activity, by recovering their surfaces with fresh material, may refresh them, giving them a neutral spectral character.

The visible and near infrared spectra of several Centaurs have been modeled with mixtures of different compounds whose spectra have been obtained in the laboratory (water ice, amorphous carbon) and tholins, organic nitrogen compounds with sepia red-brown coloration and a poorly understood structure, which we find on the planetary surfaces of icy bodies in the outer Solar System such as Titan (see section 1.3.1) or Triton.

The presence of water ice has been confirmed on several Centaurs (such as (2060) Chiron, (10199) Chariklo, and (5145) Pholus). The surface of Chariklo may contain water ice and a mixture of tholins of Triton and Titan types with amorphous carbon. The ice on Pholus may have been covered by a mixture of Titan-type tholins, carbon black, olivine, and methanol ice. It has been suggested that the surface of (52872) Okyrhoe was a mixture of kerogens, olivines, and a small percentage of water ice. (8405) Asbolus has been suggested as having a mixture of 15% Triton tholins, 8% Titan tholins, 37% amorphous carbon, and 40% ice. Chiron seems to be the most complex. The observed spectra vary as a function of the observation period. The signature of water ice was detected during a period of low activity and disappeared during intense activity.

2.3.2. *Trans-Neptunian objects*

K.E. Edgeworth (1880–1972) in 1949 and then G.P. Kuiper (1905–1973) in 1951 proposed the existence of numerous objects beyond Neptune. Progress in instrumentation and the refinement of observational techniques and research strategies allowed David Jewitt and Jane Luu to discover, in 1992, another object besides Pluto orbiting beyond Neptune: the object 1992QB1 (see (Jewitt and Luu 1993), and Figure 2.39). Since then, more than 3,000 additional celestial bodies have been discovered beyond Neptune's orbit and, for this reason, they are called "trans-Neptunian objects" (see section 1.3.4.3, Volume 1).

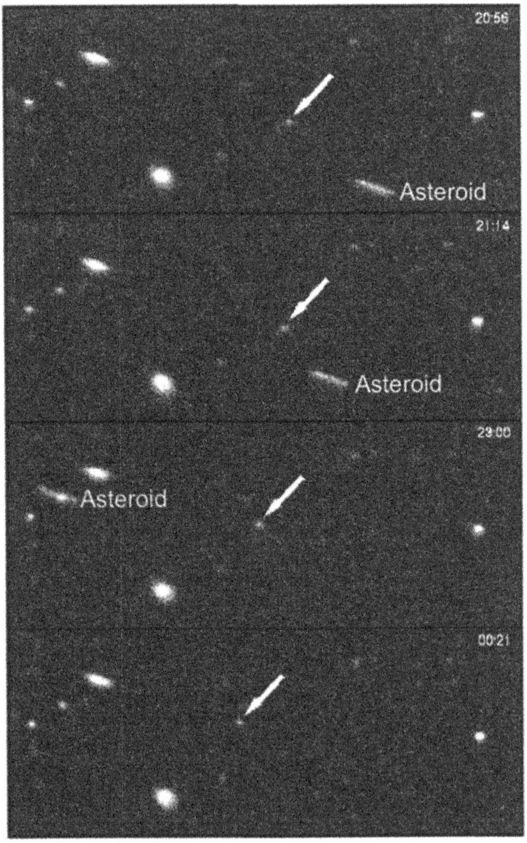

Figure 2.39. *Sequence of photos that allowed D. Jewitt and J. Luu to discover 1992 QB1: the object's small motions (indicated with an arrow) between one photo and the next indicate that they move at a velocity typical of objects that orbit beyond Neptune (source: based on (Jewitt and Luu 1993)). For a color version of this figure, see www.iste.co.uk/encrenaz/solar2.zip*

These objects are located in the Kuiper belt, a ring of small bodies with orbits between 30 and 55 au from the Sun. In the scattered disk, we find the trans-Neptunians, which have highly eccentric orbits ($e > 0.2$), semi-major axes greater than 48 au, and perihelion beyond Neptune's orbit. Trans-Neptunian objects are primarily made up of ice (water, methane, nitrogen) that does not sublimate because the temperature is on the order of 50 K (-223°C). Because of the absence of activity, they are considered as asteroids, and they receive a name based on the rules established for asteroids.

The Kuiper belt has the shape of a torus centered on the ecliptic, stretching between 30 and 55 au from the Sun. Inside, at 39.5 and 48 au, are found objects that are respectively in 2:3 and 1:2 orbital resonance with Neptune. It is estimated that there are more than 100,000 KBOs (*Kuiper belt objects*) with a diameter greater than 50 km, and that the global mass of these objects is between a tenth and a hundredth of the Earth's mass. The orbit of most of the KBOs have an inclination less than 10°, but several have larger inclinations. The presence of one or more satellites is fairly frequent around known KBOs.

The KBOs are divided into two groups: resonant objects and classic objects, or *Cubewanos*. Resonant objects have orbits linked to Neptune's: the orbits of the *plutinos* are in a 2:3 ratio with Neptune's orbit, while the *twotinos* have a 1:2 ratio, although other resonances (4:3, 5:3, etc.) also contain members of the Kuiper belt.

The 3:2 resonance in relation to Neptune is the site of several KBOs: the name of *plutinos* was given to these bodies because Pluto is the first body to have been found in this resonance. Plutinos are protected from Neptune's perturbations by the resonance because, during their passage to perihelion, which is also the point closest to Neptune's orbit, the planet is always on the other side of the Sun.

The *Cubewanos*, or classic KBOs, get their name from the first trans-Neptunian object, 1992 QB1 (pronounced *cu-be-wane*). Their perihelion is far from Neptune's orbit, they have low eccentricity, and their semi-major axes are between 42 and 48 au. We divide them into two sub-groups based on their orbital inclination: the dynamically "cold" KBOs have an inclination $i < 4.5°$, and the "hot" KBOs have an $i > 4.5°$. The cold objects are planetesimals that formed in the area where they are currently located, preserved from Neptune's perturbations by their distance from the planet; the hot objects formed in the area where the giant planets are currently located, and they were ejected by gravitational interactions beyond Neptune during the migration of the giant planets.

On January 1, 2019, NASA's New Horizon probe flew over a trans-Neptunian object for the first time: (486958) Arrokoth, which had not even been discovered when the mission was launched. It is a classic KBO, characterized by a bi-lobe shape (see Figure 2.40). The larger lobe is around 21 × 20 × 9 km in size, and the smaller lobe, which is less flat, is 15 × 14 × 10 km in size. Arrokoth measures 36 km along its longest axis and is around 10 km wide, and the centers of the lobes are separated by a distance of 17.2 km. The two lobes are not spherical: the larger one seems to be the result of the aggregation of smaller objects, while the smaller lobe seems monolithic. No signs of violent collision have been found at the junction between the two lobes: it seems that the two halves of the object, which formed separately, orbited around each other for so long that their longest axes aligned and then, very slowly, touched, at a velocity of around 3 m/s, to end up united.

Arrokoth's absorption spectrum exhibits a strong spectral slope extending from the red to the infrared wavelengths at 1.2–2.5 μm. This spectrum revealed the presence on the surface of methanol, hydrogen cyanide, and water ice, as well as organic compounds; an absorption band at 1.8 μm has not been identified. Arrokoth's spectrum resembles that of (2002) VE95 and the Centaur (5145) Pholus, which both exhibit equally strong red spectral slopes as well as signs of methanol on their surfaces.

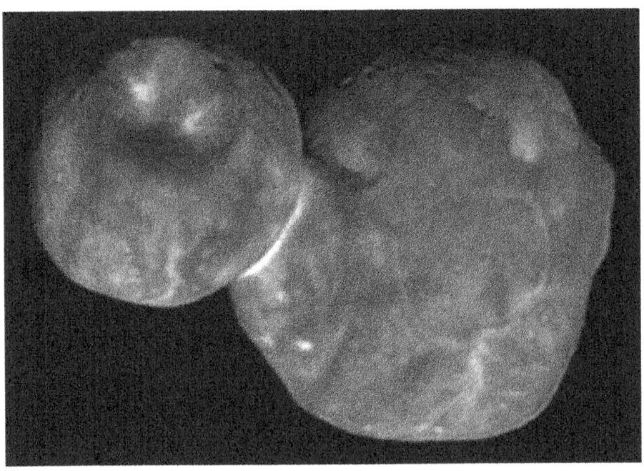

Figure 2.40. *Image of the classic KBO (486958) Arrokoth (source: NASA/JHU/APL/SWRI)*

The scattered (or faint) disk contains objects that were probably pushed onto irregular orbits because of the gravitational influence of the gas giants, particularly during Neptune's migration. These objects, called SDOs (*scattered disk objects*) are characterized by orbits whose perihelion is in the Kuiper belt and whose aphelion is at more than 150 au from the Sun, with extremely high eccentricities and inclinations.

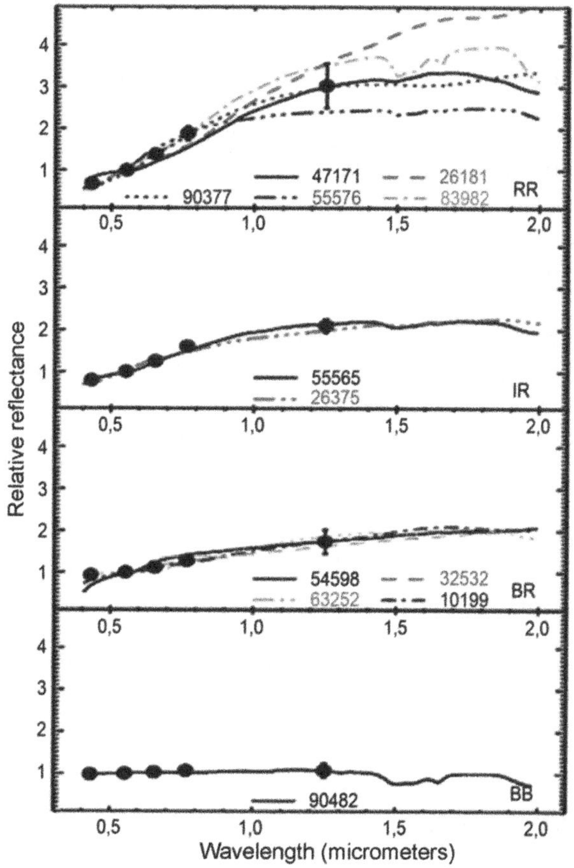

Figure 2.41. *Spectra of several trans-Neptunian objects (indicated by their number, categorized in four groups defined based on V and IR spectrophotometry). For a color version of this figure, see www.iste.co.uk/encrenaz/solar2.zip*

Several trans-Neptunian objects, called "detached objects", orbit at a distance from the Sun that varies between a hundred astronomical units at perihelion to more than a thousand astronomical units at aphelion, with orbital periods of several thousand years. These objects receive a millionth of the solar energy received by Earth, and their temperature is extremely low. The first object of this type to be observed was (90377) Sedna.

Because of their distance, it is very difficult to observe these trans-Neptunian objects and, in particular, obtaining good spectra requires using the world's largest telescopes. The visible spectra of the TNOs and Centaurs have no significant signatures, but on the basis of their slope in the range of 0.4 μm–1.25 μm, they can be divided into four groups: BB, BR, IR, and RR (Fulchignoni *et al.* 2008).

The BB (blue) group contains objects with flat spectra, thus neutral in relation to the Sun, while BR, IR, and RR include objects that are increasingly red (see Figure 2.41). (5145) Pholus, 1992 QB_1, and the very distant Sedna, all RR, are among the reddest objects in the Solar System. All of the dynamically cold objects in the Kuiper belt are red; the hot ones are blue. The diffuse objects are almost all BB.

These differences in color may indicate different initial compositions for the TNOs or different degrees of evolution on their surfaces. The progressive reddening of the surfaces may be the consequence of space weathering, while the neutral bodies may have been the result of collisions, or show a comet-type activity (observed in Centaur (2060) Chiron), which could have rejuvenated their surfaces, covering over the effects of space weathering.

The spectra in the near infrared between 0.9 and 3.5 μm (see Figure 2.41) contain more information in the absorption bands of several compounds present on the surfaces of TNOs: we have detected the presence of several types of ice, methane, ammonia, carbon oxide, and crystalline and amorphous water, but the number of objects observed is modest and the observations only concern the largest objects.

2.3.3. *Interstellar objects*

'Oumuamua ("messenger" in the Hawaiian language), discovered on October 19, 2017, is an asteroid traveling from interstellar space along a hyperbolic orbit, at a maximum velocity of 22.6 km/s. The first calculations indicated that this object came from the constellation Lyra; it was classified by the International Astronomical Union as an "interstellar object" and named *1I/'Oumuamua* ("1I" for "first interstellar object").

The luminosity of 1I/'Oumuamua varies enormously during its rotation, which has a period of 7.3 hours, indicating that this asteroid is very long. We have calculated that it is about 800 m long and 100 m wide. It is flattened and dark red in color. It is probably made up of refractory materials in which cosmic rays have eliminated any trace of water.

On August 30, 2019, another interstellar object was observed, a comet named 2I/Borisov in its discoverer's name, coming from the constellation Cassiopeia and traveling across the Solar System at a speed of 30.7 km/s at infinity. We estimate that such an interstellar object enters the Solar System once per year, but these celestial bodies are very small and difficult to spot.

2.3.4. *The origin and evolution of the asteroids*

The origin of the asteroids is directly linked to the formation of the planetary system (see sections 2.6 and 2.7, Volume 1 and section 4.9 in this volume).

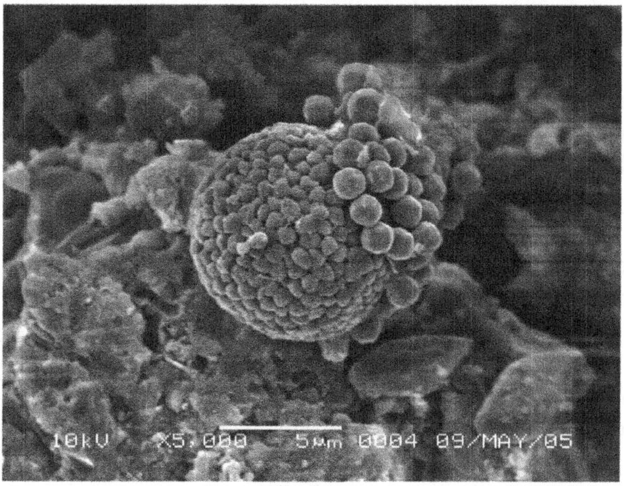

Figure 2.42. *Aggregate of small grains in the Orgueil meteorite (source: (Mukhopadhyay* et al. *2009))*

The proto-solar nebula, which is composed in mass of 99% hydrogen and helium and 1% dust grains on the order of 0.1 µm, began to collapse in on itself following perturbations in its galactic region five billion years ago. The gas accumulating in the barycenter of the nebula began to form the Sun, and dust grains settled on the evolving nebula's equatorial plane, forming a disk around the central star where the gas/grain ratio was relatively high. During sedimentation, the grains may have been able to grow in size by adhering to other grains, following a crossing of their trajectories, forming very thin and unstable aggregates with sizes from 0.1 µm to a few centimeters (see Figure 2.42). The sedimentation time for these aggregates ranges between a thousand years for aggregates of 1 mm size to 1 million years for those on the order of a micrometer.

Particles were slowed down by the gas, which increased the likelihood of encounters between them: the particles adhered due to surface and electrostatic forces, and were thus able to form bodies of several meters in size. As the number of these bodies increased, the disk decreased in thickness and its surface density increased. The solid bodies were in Keplerian orbit around the still-forming Sun, and their relative velocities decreased, leading to increased constructive collisions. The disk became gravitationally unstable and, as a result of these instabilities, formed progressively larger objects thanks to mutual interactions between the bodies and the gas. These processes gave rise to the formation of bodies a few kilometers in diameter, held together by their own gravity: planetesimals (see section 2.5, Volume 1).

The growth of the largest planetesimals accelerated as their cross-section increased, thanks to gravitational focalization, and this led to progressively faster formation of large asteroids and planetary embryos. Most of the planetesimals were encompassed by the embryos to form terrestrial planets and the cores of the giant gas planets. In the dynamic chaos that followed, planetesimals from outer regions were injected into the region < 4 au, and objects originally in this region were ejected beyond Neptune.

At that time, there was 1,000 times more solid matter in the region of the asteroids than there is today. An explanation of this difference is that Jupiter indirectly provoked the emptying of this area by interacting with the largest bodies in the main belt, increasing their eccentricities; the effects of their perturbations eliminated most of the primitive asteroids. This effect prevented the formation of a planet between the orbit of Mars and Jupiter.

Some models suggest the four giant planets formed closer to the Sun than their current positions; they migrated in the disk (see section 1.2.6, Volume 1, section 4.9 and Figure 4.9 in this volume) by first approaching and then moving away from the Sun. Jupiter may have migrated inward to 1.5 au from the Sun, followed by Saturn, which approached almost as closely, and it is during this phase that the large majority of bodies in the asteroid region at the time may have been ejected. During their reverse trip, the interactions between the two giant planets may have thus set off their migration outward, as well as that of Uranus and Neptune. During these round trips, some of the planetesimals from the outer regions were introduced into the main belt, which was thus populated by a mixture of bodies from different parts of the Solar System. Moreover, a population of objects migrated outward and constitute the dynamically "hot" component of the KBOs.

The survivors continued to evolve after internal modifications (differentiation, outgassing), processes of mutual interaction (collisions), or changes in their spatial environment (resonances, space weathering, etc.). Larger bodies, heated by the energy released from the radioactive elements with short half-lives (such as aluminum 26) that they contained, were able to differentiate into a metallic core, a mantle, and a siliceous crust.

The impacts of smaller bodies were the principal agent of the erosion of the larger asteroids' surfaces: they fragmented and pulverized the surface, forming a thick layer of regolith. Because of the low gravity, a non-negligible portion of the dust thus formed was ejected and ended up in the Sun or in interstellar space. These impacts were also responsible for changes in the morphology of the asteroids' surfaces, which were shaped by craters of all sizes, ejected rocks, fractures, and landslides.

Collisions between objects of similar size resulted in the two bodies' fragmentation: we estimate that, with the exception of asteroids larger than 100 km, all asteroids in the main belt are fragments of larger bodies. This fragmentation continues today.

2.4. The dwarf planets

A dwarf planet is a planetary body gravitating around a star whose mass is sufficient to give it an almost spherical shape through hydrostatic equilibrium, but that was not able to "clear" its orbit of all other objects.

Resolutions adopted by the International Astronomical Union during its General Assembly held in Prague on August 20, 2006 concerning the definitions of "planet", "dwarf planet", and small bodies.

Recent observations have changed our vision of planetary systems, and it is important that the nomenclature of objects reflects our current understanding. This applies in particular to the definition of a "planet". The word "planet" initially designated the "vagabonds" of the sky; in other words, the points of light that moved in relation to the stars. Recent discoveries led us to a new definition corresponding to our new understanding.

Resolution 5: definition of a planet in the Solar System

As a result, the International Astronomical Union (IAU) decided to redistribute the planets and other bodies in the Solar System into three categories as follows:

– a "planet" is a celestial body that (a) is in orbit around the Sun, (b) has sufficient mass for its gravity to prevail over the forces of cohesion of the solid body and to maintain it in hydrostatic equilibrium, in an almost spherical shape, (c) has eliminated all bodies likely to travel along a near orbit;

– a "dwarf planet" is a celestial body that (a) is in orbit around the Sun, (b) has sufficient mass for its gravity to prevail over the forces of cohesion of the solid body and to maintain it in hydrostatic equilibrium, in an almost spherical shape, (c) has not eliminated all bodies likely to travel along a near orbit, (d) is not a satellite;

– all other objects in orbit around the Sun are called "small Solar System bodies".

Resolution 6: Pluto

In accordance with the above definition, Pluto is a dwarf planet. It was identified as the prototype for a new category of trans-Neptunian objects.

Box 2.7. *Official definition of a dwarf planet*

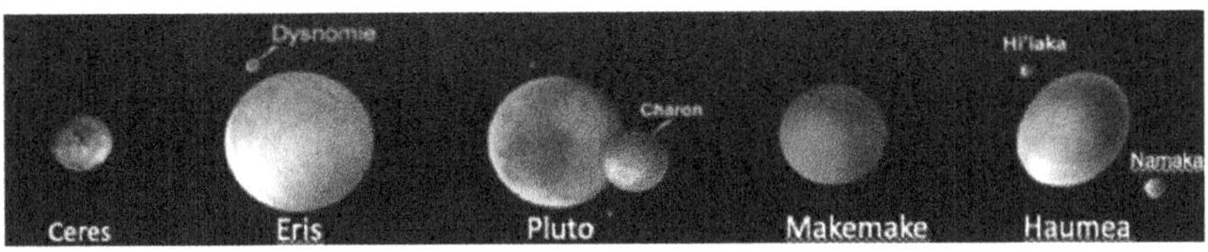

Figure 2.43. *The dwarf planets (source: based on NASA images). For a color version of this figure, see www.iste.co.uk/encrenaz/solar2.zip*

The International Astronomical Union currently recognizes five dwarf planets: (1) Ceres (asteroid), (134340) Pluto (plutino), (136108) Haumea (cubewano), (136472) Makemake (cubewano) and (136199) Eris (scattered object). They are identified by name, but also keep the number attributed to them during the confirmation of their discovery (or, in the case of Pluto, in the IAU resolution).

The dwarf planets beyond Neptune are called "Plutoids".

2.4.1. *Ceres*

The dwarf planet Ceres (see Figure 2.44) is the largest object in the asteroid belt (25% of the total mass) and is the only dwarf planet located in the inner Solar System.

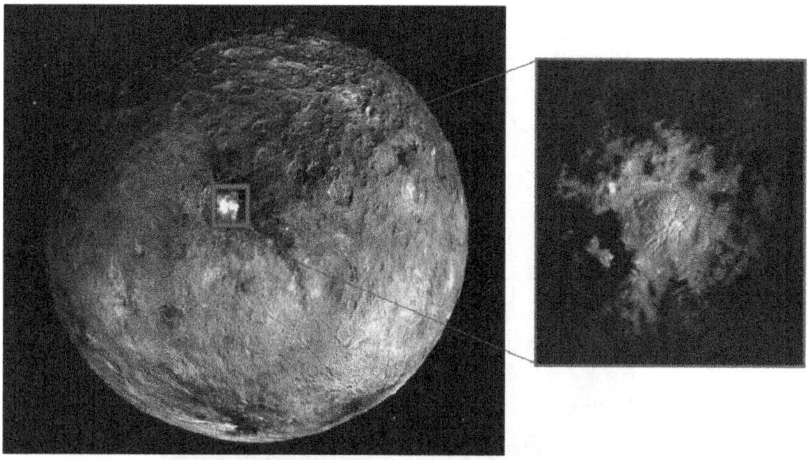

Figure 2.44. *The dwarf planet Ceres: in the center of the photo (false colors), the crater Occator (D = 92 km) is characterized by the presence of vast deposits of bright material, the Cerealia Faculae (source: NASA). For a color version of this figure, see www.iste.co.uk/encrenaz/solar2.zip*

Ceres formed at the same time as the rest of the Solar System, around 4.5 billion years ago, when gravity forced gas and dust from the protoplanetary disk to become a planetary embryo. Ceres thus began to grow, but its size was limited by the strong gravitational perturbations of Jupiter, which depopulated the swarm of planetesimals that had been fueling its growth.

NASA's Dawn mission launched in 2007 to explore the two largest objects in the main asteroid belt, the protoplanet (4) Vesta and the dwarf planet (1) Ceres. Dawn arrived at (1) Ceres in July 2011 and continued to observe for 3 years and 9 months, using its ionic engines to travel along orbits optimized for the dwarf planet.

The data collected by Dawn made it possible to obtain a detailed description of surface morphology, mineralogy, gravity, and topography.

Ceres probably has a solid core, a mantle made of water ice, and a rocky and dusty crust with large deposits of various salts, such as magnesium sulfates (see Figure 2.45). Ceres may contain at least 25% water (much more than all the water on Earth). In its internal structure, which is not yet well known, Ceres resembles terrestrial planets more than the neighboring asteroids, but it is less dense.

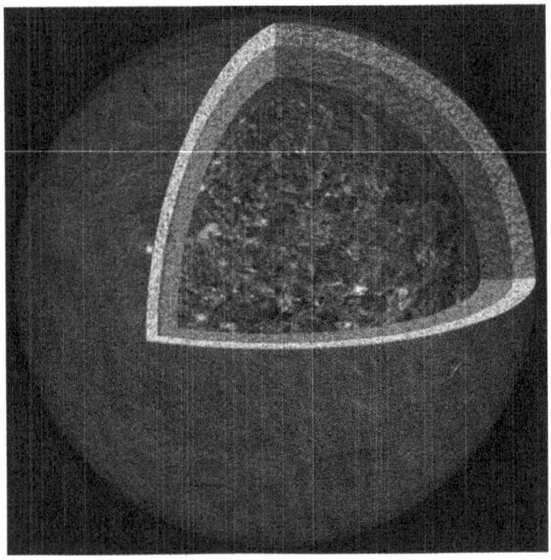

Figure 2.45. *Internal structural model of Ceres. Ceres probably has a solid core and a mantle made of water ice. In fact, Ceres may contain up to 25% water and thus would have more water than Earth. Ceres's crust is rocky, dusty, and punctuated with salt deposits (source: NASA). For a color version of this figure, see www.iste.co.uk/encrenaz/solar2.zip*

The data collected by the instruments on-board the Dawn probe revealed a generally homogeneous surface, with numerous craters, punctuated by bright spots, such as the complex of bright areas (*Cerealia Faculae*) inside the crater Occator (see Figure 2.44). The Cerealia Faculae are made of a sort of brine containing magnesium sulfate hydrate ($MgSO_4$, $6H_2O$) associated with ammonia-rich clay. The near infrared spectra of these bright areas also indicates a large quantity of sodium carbonate (Na_2CO_3) and smaller amounts of ammonium chlorides (NH_4Cl) and ammonium bicarbonate (NH_4HCO_3). These materials seem to come from a recent crystallization of brine that reached the surface whose water was sublimated.

The overall map of the surface reveals a complex landscape with young and old terrains. Repeated observations suggest that Ceres exhibits some activity, with physical and/or chemical transformations of the surface materials. The presence of water ice, already shown by measurements carried out by the Herschel space observatory, was confirmed inside some of the craters at high latitudes, with a distribution of the ice on the ground typical of runoff.

Ceres is covered with innumerable small young craters, but none of them is larger than 280 km. This is surprising, given that the dwarf planet should have been hit by many relatively large asteroids during the course of its 4.5 billion years of existence. The absence of large craters could be due to the presence under the surface of layers of ice or other low-density materials like salts, which would have attenuated the effect of the impact. It is also possible that past hydrothermal activity, possibly volcanic, has erased some large craters. The distribution and geomorphology of Ceres's craters seems to indicate the presence of a solid crust and an upper mantle composed of ice and possibly salt hydrates. Furthermore, the existence of ancient craters up to 280 km in diameter implies that this mixture of ice and rock (or salt hydrates) could have enough robustness to conserve them for billions of years. The upper mantle may periodically crack and produce characteristic flows, bright deposits, and extrusive episodes, like the ascending plume made of a mixture of salt water and rock that must have risen to build up Mount Ahuna, 4 km high and 17 km wide; the mountain is geologically young, as implied by the recent flow on its slopes, the absence of debris at its base, and the rarity of craters at its summit (see Figure 2.46). Although this process occurred at Ceres's very cold temperatures, probably below 0°C, it is similar to those that form volcanoes on Earth.

This new data suggests that the dwarf planet Ceres is a geologically active world, which provides support for the hypothesis that Ceres could possess a frozen ocean. A new approach is required to model the evolution of Ceres's interior in order to explain its physio-chemical conditions and their manifestations on the surface in the form of salts and carbonates.

Ceres has a very faint atmosphere that contains water vapor, possibly produced by cryovolcanic phenomena or sublimation of ice near the surface.

Figure 2.46. *The mountain Ahuna Mons, on Ceres, seen in simulated perspective; the elevation is exaggerated by a factor of 2 (source: NASA). For a color version of this figure, see www.iste.co.uk/encrenaz/solar2.zip*

2.4.2. *Pluto and its satellites*

NASA's New Horizon mission completed a flyby of Pluto (see Figure 2.47) on July 14, 2015. It has five known satellites, the largest of which is Charon, whose size is half that of Pluto. For this reason, Pluto and Charon are often designated as a double system.

Figure 2.47. *The dwarf planet Pluto (source: NASA/JHU/APL/SWRI). For a color version of this figure, see www.iste.co.uk/encrenaz/solar2.zip*

Pluto's diameter is two-thirds that of the Moon; it probably has a rocky core covered by a mantle composed of water ice and other substances such as methane and nitrogen. Because of its low density, Pluto's mass is only one-sixth of the Moon.

Pluto possesses a thin atmosphere, discovered in 1985 during a stellar occultation, primarily made up of molecular nitrogen; methane and carbon monoxide were also detected. Pluto's low gravity means that this atmosphere extends much farther than Earth's does. This atmosphere changes periodically: when Pluto is closer to the Sun, the surface ice partially sublimates, temporarily forming a tenuous atmosphere, but when Pluto is in orbit farther from the Sun, thus becoming much colder, the constituents of its atmosphere fall to the surface in the form of snow. Paradoxically, a measurement of pressure carried out in 2002 during another stellar occultation indicated a value of 0.3 Pa, twice as high as in 1985, when Pluto was farther from the Sun. This enrichment was due to the sublimation of nitrogen that occurred when the south pole emerged from shadow in 1987 for the first time in 120 years (Sicardy *et al.* 2003).

The surface of Pluto exhibits mountains, valleys, plains, and craters (see Figure 2.48). The mountains can reach heights of 2 to 4 kilometers, for example, Hillary Montes and Tenzing Montes, and are made up of water ice, sometimes covered by a layer of frozen gas (methane, nitrogen, etc.). Other characteristic structures of the Plutonian landscape are depressions and valleys that can reach 600 km in length.

Craters with diameters as much as 260 kilometers punctuate Pluto's landscape. They show signs of erosion and refilling: this suggests that tectonic forces are the source of slow resurfacing.

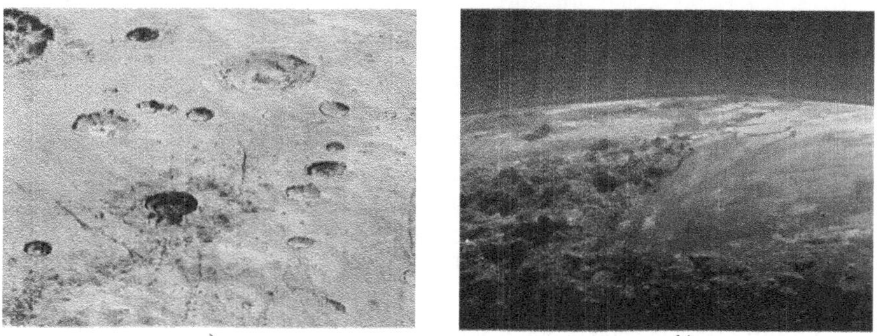

a) b)

Figure 2.48. *Craters and mountain chains on Pluto. (a) Most of the craters in this photo are found inside the Burney basin (250 km in diameter), whose outer wall forms a line of hills. (b) Sputnik Planum is flanked on the west (left) by steep mountains up to 3,500 m high, the Tenzing Montes in the foreground, and the Hillary Montes on the horizon (source: NASA/JHU/APL/SWRI)*

The largest plains observed on Pluto seem to be covered with nitrogen ice; they do not exhibit craters but rather irregularly shaped structures, segments that are about 20 km wide, bordered by shallow depressions that resemble fissures in frozen mud that are found on Earth. Two hypotheses have been proposed to explain the formation of these features: the irregular shapes could be the result of the contraction of surface materials, like when mud dries, or they could be the result of convection caused by the little amount of heat in Pluto's interior in materials with different densities, such as a superficial layer of carbon monoxide, methane, and nitrogen ice.

Some of these depressions are filled with dark material, while others are surrounded by hills that seem to rise above the surrounding terrain, such as in the region unofficially named *Sputnik Planum* after Earth's first artificial satellite (see Figures 2.48 and 2.49). Elsewhere, the surface seems punctuated by a large number of small holes formed by the sublimation of nitrogen.

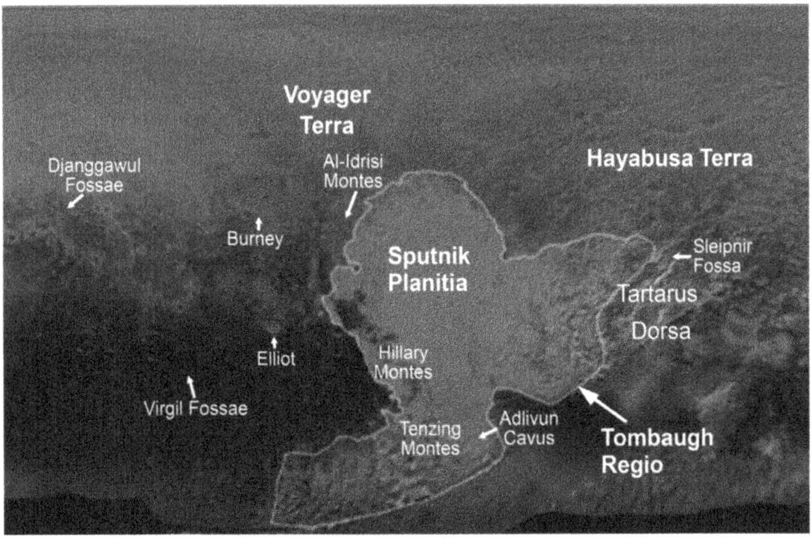

Figure 2.49. *Location of some structures on the surface of Pluto (source: NASA/JHU/APL/SWRI). For a color version of this figure, see www.iste.co.uk/encrenaz/solar2.zip*

An uncommon geological structure that resembles a giant spider stretches across Pluto's glacial landscape. This structure is made up of six extensional fractures (indicated by white arrows in Figure 2.49) that converge toward a point located near the center. The longest fractures are aligned approximately north-south, and the longest of all, unofficially named Sleipnir Fossa, is 580 km long. The east-west fracture is shorter, less than 100 km long. To the north and west, the fractures extend through rolling plains at high latitudes, and to the south, they cut through the terrain unofficially called the Tartare Dorsa. The "spider's" curious configuration may be caused by constraints in the crust below the point where the fractures converge, possibly a lifting of underground material.

Pluto has five known satellites: Charon, Nix, Hydra, Kerberos, and Styx (see Figure 2.50). This system of satellites may have been formed at the beginning of the

Solar System by a collision between Pluto and another body of similar size: the materials ejected following the collision accumulated, giving birth to Pluto's family of satellites.

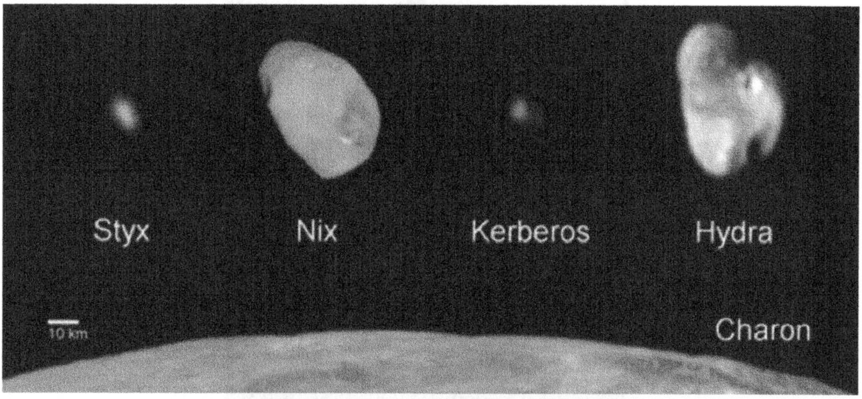

Figure 2.50. *The five satellites of Pluto (source: NASA/JHU/APL/SWRI)*

Charon, by far the largest of the satellites, orbits around Pluto at a distance of only 19,640 km in 153 hours and always shows the same side to Pluto, which also shows its same side to Charon, because of spit-orbit resonance (see section 4.7). Seen from Pluto, Charon never rises or sets, always floating in the same place in Pluto's sky, and vice versa.

Charon's surface (see Figure 2.51) is densely cratered, and it is estimated to be 4 billion years old. We can see flows of ammonia and water ice that dominate the planes in the southern hemisphere, and at the north pole there is a red cap that seems to be made up of methane and nitrogen. Charon's surface is crisscrossed by rifts, ridges, escarpments, and canyons. These geological structures seem to indicate that a liquid ocean was present under the surface in the past and that it then froze: its expansion under the surface led to fracturing of the upper layers and the formation of the structures that characterize the satellite's morphology.

Pluto's four other satellites are much smaller than Charon. They have irregular shapes and, contrary to many other satellites in the Solar System, are not tide-locked to Pluto, and they therefore do not always show it the same side.

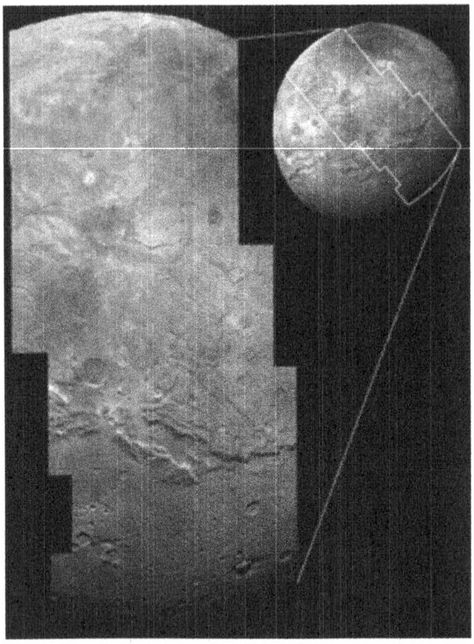

Figure 2.51. *Charon and its surface (source: NASA/JHU/APL/SWRI). For a color version of this figure, see www.iste.co.uk/encrenaz/solar2.zip*

	Semi-major axis (au)	Mass (kg)	Dimensions (km)	Density (kg(/m^3)	Satellites
Ceres	2.76534	9.3835 $\pm$ 0.0001 10^{20}	964.4.6 × 964.2 × 891.8 ± 0.2	2,162 ± 8	
Pluto	39.582	1.303 $\pm$ 0.003 10^{22}	2,376.6 ± 1.6	1,854 ± 6	Charon, Styx, Kerberos, Hydra, Nyx
Eris	67.864	1.66 $\pm$ 0.02 10^{22}	2,326 ± 12	2,520 ± 70	Dysnomia
Haumea	43.182	4.006 $\pm$ 0.04 10^{21}	2,100 × 1,680 × 1,074 ± 8	1,885 ± 80	Namaka, Hi'iaka + a ring
Makemake	45.43	7.329 $\pm$ 0.009 10^{10}	1,502 ± 45 × 1,430 ±9	1,190 ± 30	MK 2 (provisional)

Table 2.7. *Main properties of dwarf planets*

2.4.3. *Eris, Haumea, and Makemake*

The year 2005 saw the unprecedented discovery of three of the largest trans-Neptunians: (136108) Haumea, (136472) Makemake, and (136199) Eris. When the main properties of these bodies became better known, the International Astronomical Union placed them in the category of dwarf planets (see Table 2.7).

2.4.3.1. *Eris*

Discovered on October 21, 2003, Eris is three times farther away from the Sun than Pluto and at first seemed to be larger than Pluto. This set off a debate in the scientific community that led the International Astronomical Union to define Pluto, Eris, and other similar objects as "dwarf planets". Originally classified as a trans-Neptunian object and designated 2003 UB313, Eris is named after the Greek goddess of discord and conflict. The name is suitable because Eris was at the center of the debate.

Eris orbits at 68 au from the Sun and takes 557 Earth years to complete its revolution. The orbit's eccentricity is very high (0.44177) and the plane of its orbit is highly inclined along the ecliptic ($i = 44.187°$). It rotates on its axis in 25.9 hours, which makes its day similar to ours. With a radius of around 1,163 km, Eris is one of the largest known dwarf planets in our Solar System. In comparison, it is about the same size as Pluto, a bit smaller than the Moon.

We know very little about Eris's internal structure. Eris probably has a surface similar to Pluto's, but its surface temperature varies between around 56 K (-217°C) and 30 K (-243°C). It is often so far from the Sun that its atmosphere freezes, falling to the surface in the form of snow. When it approaches the Sun along its distant orbit, the atmosphere reappears.

Eris has a very small satellite named Dysnomia, after the daughter of the goddess Eris. Dysnomia has an almost circular orbit with a duration of 15,744 days, and its presence has allowed astronomers to calculate Eris's mass, which is 0.00278 of the Earth's mass.

2.4.3.2. *Haumea*

Haumea, named after the Hawaiian goddess of fertility, has a complex system made up of two satellites and a ring, and was discovered in 2017 when Haumea occulted a star (see section 1.5.7.1).

Namaka is the inner satellite, and Hi'iaka the outer; the two are named after Haumea's mythological daughters. Hi'iaka is the matron goddess of the island of

Hawaii and of hula dancers. Namaka is a water spirit in Hawaiian mythology. It is possible that a massive impact billions of years ago set Haumea rotating and created these satellites.

The ring has an opacity of 0.5, is 70 km wide, and has a radius of around 2,287 km. It is co-planar with Haumea's equator and with the orbit of the satellite Hi'iaka. The ring's radius puts it near the 3:1 resonance with Haumea's rotation; Haumea rotates three times on its axis while a particle in the ring rotates once.

Initially designated 2003 EL61, Haumea is about the same size as Pluto, with a radius of around 620 km. The object orbits at a distance of 43 au from the Sun, and its orbital period is 285 years. While Haumea revolves around the Sun, it rotates around its axis every four hours, which makes it one the largest objects whose rotation is also the fastest in our Solar System. As a result, Haumea's shape is a triaxial ellipsoid of $2{,}322 \times 1{,}704 \times 1{,}138$ km.

Thanks to the presence of its satellites, Haumea's mass was determined to be $4{,}006 \times 10^{21}$ kg, for an average density of 2,600 kg/m^3, which indicates that Haumea consists of rock covered with ice.

Haumea's surface composition is fairly unusual. Its spectrum in the near infrared shows that its surface is composed of almost pure water ice, with a mixture of two phases, crystalline and amorphous. The presence of water ice is not exceptional, but the fact that the presence of other species is limited to less than 8% of the surface composition is unique. This composition is only shared with a small group of TNOs with orbital parameters that are similar to Haumea, which points to the conclusion that these objects constitute a collisional family. The absence of volatile ice detected on Haumea may be due to the collision that formed the family and to the heating to which Haumea was thus subjected.

2.4.3.3. *Makemake*

Makemake was named after the Rapanui god of fertility. Slightly smaller than Pluto, Makemake is the second brightest object in the Kuiper belt, as seen from Earth. It takes 305 years for it to orbit the Sun, and the orbit has an eccentricity of 0.15: its perihelion is at 38.71 au and its aphelion is at 52.57 au. Makemake completes a rotation every 22.5 hours, which makes the length of its day similar to that of the Earth and Mars.

Makemake is almost spherical, with a flattening of only 0.00837 (equatorial diameter of 1,434 km and polar diameter of 1,422 km).

We cannot see details on the surface of Makemake because of its distance. It is red-brown in color, similar to Pluto. Frozen methane and ethane on its surface have also been detected in absorption spectra in the near infrared, and the presence of solid nitrogen has been deduced from the displacement of the methane band. Balls of frozen methane a centimeter in diameter may lie on Makemake's frozen surface, where the average temperature is 32 K. The presence of these components on Makemake may enable the existence of an atmosphere during a portion of its orbit. Observations made during an occultation in 2001 did not reveal an atmosphere. The absence of a detectable atmosphere today, combined with the high albedo of this object (0.81), provides solid proof of active resurfacing on this dwarf planet as a result of the seasonal condensation of the atmosphere.

In 2016, the Hubble Space Telescope located a small satellite in orbit at around 20,000 km from Makemake, named S/2015 (136 472) 1, and nicknamed MK 2. It is more than 1,300 times less bright than Makemake, and its radius is estimated at around 80 km. Thanks to its discovery, Makemake's mass was estimated to be 3-4 × 10^{21} kg, putting the density value between 2,000 and 6,000 kg/m^3, analogous to that of other plutoids.

2.4.3.4. *The candidate dwarf planets*

	Dwarf planets	Candidate dwarf planets	
	Accepted by the IAU	D > 900 km	450 < D < 900 km
Object denomination	Ceres Pluto Eris Haumea Makemake	2007 OR$_{10}$ Quaoar Orcus 2002 MS$_4$ Sedna Salacia	Varda, 2002 AW$_{197}$, 2003 AZ$_{84}$, 2002 UX$_{25}$, 2004 GV$_9$, 2005 RN$_{43}$, 2003 UZ$_{413}$, Varuna, Ixion, Chaos, 2007UK$_{126}$, 2002 TC$_{302}$, 2002 XW$_{93}$, 2002 XV$_{93}$, 2003 VS$_2$, 2004 TY$_{364}$, Dziewanna, 2005 QU$_{182}$, 2005 UQ$_{513}$, 2005 TB$_{190}$, 2004 NT$_{33}$, 2004 PF$_{115}$, 2004 XA$_{192}$, 2003 FY$_{128}$, 2003 QW$_{90}$, 2002 GJ$_{32}$, 2002 KX$_{14}$, 2002 VR$_{128}$, 2001 KA$_{77}$ Huya, 10 Hygea

Table 2.8. *List of candidate dwarf planets*

The number of dwarf planets is expected to increase with improvements in understanding of the physical properties of the objects already discovered and those that will be.

The International Astronomical Union must decide if a certain number of objects, including some asteroids and the largest TNOs, should be added to the definition of dwarf planet. Let's recall the criteria:

– the candidates share their orbital area with other objects;

– they must be in isostatic equilibrium;

– they are not satellites.

Observations make it possible to immediately verify criteria (a) and (c), but it is difficult to determine if an object is in hydrostatic equilibrium using ground-based observations. This property is linked to the shape and size of the body. The combination of internal pressure (caused by the body's gravity) and sufficient plasticity (to allow for gravitational relaxation) will also allow bodies larger than a critical size to overcome their internal resistance and to adopt the characteristic shape of a body in hydrostatic equilibrium, typically a triaxial ellipsoid.

By assuming density values appropriate for icy bodies ($\rho \sim 1.3$ g cm^{-3}, typical of Uranian and Saturnian satellites) and compatible with the small sample of densities known for TNOs, we obtain the critical diameter of 450 km. This density criteria supplies enough information to construct a list of candidate dwarf planets. A possible list, developed using this criterion and the diameter and density values obtained from observations in the infrared from the Herschel space observatory, is provided in Table 2.8.

2.5. References

A'Hearn, M.F. and Johnson, L.N. (2015). Deep Impact and related missions. In *Handbook of Cosmic Hazards and Planetary Defense*, Pelton, J.N., Allahdadi, F. (eds). Springer, Berlin.

Altwegg, K., Balsiger, H., Bar-Nun, A. *et al.* (2015). 67P/Churyumov-Gerasimenko, a Jupiter family comet with a high D/H ratio. *Science*, 347, 1261952.

Altwegg, K., Balsiger, H., Berthelier, J.J. *et al.* (2017). D_2O and HDS in the coma of 67P/Churyumov-Gerasimenko. *Philosophical Transactions R. Soc.*, A375, 20160253.

Altwegg, K., Balsiger, H., Fuselier, S.A. (2019). Cometary chemistry and the origin of icy solar system bodies: The view after *Rosetta*. *Annual Reviews of Astronomy and Astrophysics*, 57, 113–155.

Biver, N. and Bockelée-Morvan, D. (2016). Chemical diversity in the comet population. *Astronomy in Focus*, 1, 228–232.

Biver, N. and Bockelée-Morvan, D. (2019). Complex organic molecules in comets from remote-sensing observations at millimeter wavelengths. *ACS Earth Space Chem.*, 3, 1550–1555.

Biver, N., Bockelée-Morvan, D., Colom, P. *et al.* (2002). The long-term monitoring of comet C/1995 O1 (Hale-Bopp) at radio wavelength. *Earth, Moon and Planets*, 90, 5–14.

Biver, N., Bockelée-Morvan, D., Debout, V. *et al.* (2014). Complex organic molecules in comets C/2012 F6 (Lemmon) and C/2013 R1 (Lovejoy). *Astronomy and Astrophysics*, 566, L5.

Biver, N., Bockelée-Morvan, D., Hofstadter, M. *et al.* (2019). Long-term monitoring of the outgassing and composition of comet 67/Churyumov-Gerasimenko with the Rosetta/MIRO instrument. *Astronomy and Astrophysics*, 630, A19.

Bockelée-Morvan, D. and Biver, N. (2017). The composition of cometary ices. *Philosophical Transactions R. Soc.*, A375, 20160252.

Bockelée-Morvan, D., Crovisier, J., Mumma, M.J. *et al.* (2004). The composition of cometary volatiles. In *Comets II*, Festou, M., Keller, H.U., Weaver, H.A. (eds). University of Arizona Press, Tucson.

Bockelée-Morvan, D., Calmonte, U., Charnley, S. *et al.* (2015). Cometary isotopic measurements. *Space Science Reviews*, 197, 47–83.

Borg, J. and Levasseur-Regourd, A.-C. (2018). *L'exploration cométaire de l'Antiquité à Rosetta*. Nouveau Monde Éditions, Paris.

Brownlee, D., Tsou, P., Aléon, J. *et al.* (2006). Comet 81P/Wild 2 under a microscope. *Science*, 314, 1711–1716.

Cochran, A.L., Levasseur-Regourd, A.-C., Cordiner, M. *et al.* (2015). The composition of comets. *Space Science Reviews*, 197, 9–46.

Combi, M.R., Kabin, K., Dezeeuw, D.L. *et al.* (1999). Dust-gas interrelations in comets: observations and theory. *Earth, Moon and Planets*, 79, 275–306.

Combi, M.R., Harris, W.M., Smyth, W.H. (2004). Gas dynamics and kinetics in the cometary coma: Theory and observations. In *Comets II*, Festou, M., Keller, H.U., Weaver, H.A. (eds). University of Arizona Press, Tucson.

Crovisier, J., Biver, N., Bockelée-Morvan, D. *et al.* (1995). Carbon monoxide outgassing from Comet P/Schwassmann-Wachmann 1. *Icarus*, 115, 213–216.

Crovisier, J., Leech, K., Bockelée-Morvan, D. *et al.* (1997). The spectrum of comet Hale-Bopp (C/1995 O1) observed with the infrared space observatory at 2.9 astronomical units from the sun. *Science*, 275, 1904–1907.

Dello Russo, N., Kawakita, H., Vervack, R.J. *et al.* (2016). Emerging trends and a comet taxonomy based on the volatile chemistry measured in thirty comets with high-resolution infrared spectroscopy between 1997 and 2013. *Icarus*, 218, 301–332.

Delsemme, A.H. (1982). Chemical composition of cometary nuclei. In *Comets*, Wilkening, L.L. (ed.). University of Arizona Press, Tucson.

Demeo, F.E., Binzel, R.P., Silvan, S.M. *et al.* (2009). An extension of the Bus asteroid taxonomy into the near infrared. *Icarus*, 202, 160–180.

Despois, D., Crovisier, J., Bockelée-Morvan, D. *et al.* (1986). Observations of hydrogen cyanide in Comet Halley. *Astronomy and Astrophysics*, 160, L11–12.

Ferrari, M., Dirri, F., Palumba, E. *et al.* (2017). FT-IR and μ-IR characterization of HED meteorites in relation to infrared spectra of Vesta-like asteroids. *EPSC2017*, 859.

Festou, M., Keller, H.U., Weaver, H.A. (2004). *Comets II*. University of Arizona Press, Tucson.

Fray, N., Bardyn, A., Cottin, H. *et al.* (2016). High-molecular-weight organic matter in the particles of comet 67P/Churyumov-Gerasimenko. *Nature*, 538, 72–74.

Fulchignoni, M., Belskaya, I., Barucci, M.A. *et al.* (2008). Transneptunian objects taxonomy. In *The Solar System Beyond Neptune*, Barucci, M.A., Boehnardt, H., Cruikshank, D.P. *et al.* (eds). University of Arizona Press, Tucson.

Gérard, E., Bockelée-Morvan, D., Bourgois, G. *et al.* (1987). 18-cm wavelength radio monitoring of the OH radical in comet P/Halley 1982i. *Astronomy & Astrophysics*, 187, 455–461.

Harmon, J.K., Nolan, M.C., Ostro, S.J. *et al.* (2004). Radar studies of comet nuclei and grain comae. In *Comets II*, Festou, M., Keller, H.U., Weaver, H.A. (eds). University of Arizona Press, Tucson.

Harris, A.W. and Lagerros, J.S.V. (2002). Asteroids in the far infrared. In *Asteroid III*, Bottke, W., Cellino, A., Paolicchi, P. *et al.* (eds). University of Arizona Press, Tucson.

Huebner, W.F. and Mukherjee, J. (2015). Photoionization and photodissociation rates in solar and blackbody radiation fields. *Planetary and Space Science*, 106, 11–45.

Jewitt, D.C. and Luu, J.X. (1993). Discovery of the Candidate Kuiper Belt Object 1992 QB1. *Nature*, 362, 730–732.

Jewitt, D., Hsieh, H., Agarwal, J. (2015). The active asteroids. In *Asteroids IV*, Michel, P., DeMeo, F., Bottke, W. (eds). University of Arizona Press, Tucson.

Jones, G.H., Knight, M.M., Battama, K. *et al.* (2018). The science of sungrazers, sunskirters and other near-sun comets. *Space Science Reviews*, 214, 20.

Kaasalainen, M., Mottola, S., Fulchignoni, M. (2002). Asteroid models from disk-integrated data. In *Asteroid III*, Bottke, W., Cellino, A., Paolicchi, P. *et al.* (eds). University of Arizona Press, Tucson.

Kronk, G.W., Meyer, M., Seargent, D. (1999–2017). *Cometography, A Catalogue of Comets*, Volumes 1–6. Cambridge University Press, Cambridge.

Lamy, P.L., Toth, I., Fernandez, Y.R. *et al.* (2004). The sizes, shapes, albedos and colors of cometary nuclei. In *Comets II*, Festou, M., Keller, H.U., Weaver, H.A. (eds). University of Arizona Press, Tucson.

Lis, D., Bockelée-Morvan, D., Güsten, R. *et al.* (2019). Terrestrial deuterium to hydrogen ratio in hyperactive comets. *Astronomy and Astrophysics*, 625, L5.

Meierhenrich, U. (2014). *Comets and their Origin: The Tools to Decipher a Comet.* John Wiley and Sons, New York.

Müller, T.G., Abraham, P., Crovisier, J. *et al.* (2005). Comets, asteroids and zodiacal light as seen by ISO. *Space Science Reviews*, 119, 141–155.

Mukhopadhyay, P.K., Mossman, D.J., Ehrman, J.M. (2009). A universal, unconventional petroleum system exists throughout our solar system: SPIE [Online]. Available at: https://spie.org/news/1699-a-universal-unconventional-petroleum-system-exists-throughout-our-solar-system?SSO=1.

Mumma, M.J. and Charnley, S.B. (2011). The chemical composition of comets – Emerging taxonomies and natal heritage. *Annual Review of Astronomy and Astrophysics*, 49, 471–524.

Mumma, M.J., Dello Russo, N., Disanti, M.A. *et al.* (2001). Organic composition of C/1999 S4 (LINEAR): A comet formed near Jupiter? *Science*, 292, 1334–1339.

Noll, K.S., Weaver, H.A., Feldman, P.D. (1996). *The Collision of Comet Shoemaker-Levy 9 and Jupiter*. Cambridge University Press, Cambridge.

Ostro, S.J., Hudson, R.S., Benner, A.M. *et al.* (2002). Asteroid radar astronomy. In *Asteroid III*, Bottke, W., Cellino, A., Paolicchi, P. *et al.* (eds). University of Arizona Press, Tucson.

Oumuamua (2019). The natural history of 'Oumuamua. *Nature Astronomy*, 3, 594–602.

Reach, W.T., Sykes, M., Lien, D. *et al.* (2000). The formation of Encke meteoroids and dust trail. *Icarus*, 148, 80–94.

Rickman, H. (2018). *Origin and Evolution of Comets. Ten Years After the Nice Model and One Year after Rosetta.* World Scientific, New Jersey.

Samarasinha, N.H., Mueller, B.E.A., Belton, M.J.S. *et al.* (2004). Rotation of cometary nuclei. In *Comets II*, Festou, M., Keller, H.U., Weaver, H.A. (eds). University of Arizona Press, Tucson.

Schulz, R., Alexander, C., Boehnhardt, H. *et al.* (eds) (2009). *Rosetta, ESA's Mission to the Origin of the Solar System*. Springer, Berlin.

Sicardy, B., Widemann, T., Lellouch, E. *et al.* (2003). Large changes in Pluto's atmosphere as revealed by recent stellar occultations. *Nature*, 424, 168–170.

Snodgrass, C., Argawal, J., Combi, M. *et al.* (2017). The Main Belt comets and ice in the Solar System. *Astronomy & Astrophysics Reviews*, 25, 5.

Weaver, H.A. and Fedman, P.D. (1992). Probing the nature of comets with the Hubble Space Telescope. *Science with the Hubble Space Telescope, ESO. Conference and Workshop Proceedings*. European Southern Observatory, Garching, 44, 475–480.

Yeomans, D.K. (1991). *Comets: A Chronological History of Observation, Science, Myth and Folklore*. John Wiley and Sons, New York.

3

Meteorites and Cosmochemistry

Brigitte ZANDA

Muséum national d'Histoire naturelle and Observatoire de Paris, Paris, France

3.1. Rocks falling from the sky

> As the Sun rose, a bright "star" appeared from nowhere and grew rapidly. In 5 seconds, its size had increased by a factor of 10 and it continued to grow. A trail formed behind it, then it divided in two. A muffled crack rang out and it became brighter than the eye could tolerate. After about thirty seconds, the luminous portion of the phenomenon ended. Two minutes passed before the shock wave arrived. The windows exploded, sending shards of glass up to 4 meters away!

This extract from the words of Nikolay Pankov recorded by J.-L. Dauvergne recounts the fall of the Chelyabinsk[1] meteorite, which occurred on February 15, 2013. This spectacular event was the second[2] largest in human memory to occur near a town. What happened? A 10,000-ton block of extraterrestrial matter the size of a five-story building (Brown *et al.* 2013) collided with Earth's atmosphere at high

1 The meteorite is named for the town near which it fell, written using English orthography.

2 On June 7, 1908, the Tunguska event destroyed the windows in the town of Vanavara. This cataclysmic event, which devastated the forest within a 20 km radius, was probably ~10 times more intense than the one in Chelyabinsk, but Vanavara was located farther from the epicenter of the phenomenon (60 km).

The Solar System 2,
coordinated by Thérèse ENCRENAZ and James LEQUEUX.

velocity (70,000 km/h). The abruptly compressed and heated air ionized, transforming into a plasma that surrounded the object, hotter than the surface of the Sun, which was responsible for the luminous phenomenon. Suddenly subjected to monumental pressure, the falling body fragmented catastrophically: it released some of its kinetic energy, which generated a shock wave. The shock wave sped toward the ground at the speed of sound, about 100 times slower than the bolide, and thus arrived after a delay. In the case of Chelyabinsk, the phenomenon was repeated 10 times, and it was the explosions that occurred at altitudes of 31.9 and 30.6 km that were the most violent (Borovicka *et al.* 2003). Most of the body vaporized as it passed through the atmosphere: only a ton of matter made it to the ground, in the form of hundreds of thousands of rocks, the largest of which, found nine months later in Chebarkul lake, weighed 500 kg, while the smallest ones weighed less than 1 g. Each of these rocks (which we call meteorites – see Figure 3.1) was coated with a thin black layer: the fusion crust, a glassy material left behind after the surface rock was melted by the plasma which literally ripped atoms and then droplets from this surface before it started cooling.

Figure 3.1. *One of the stones found near Chelyabinsk; this stone is broken and the black fusion crust that covers its surface is visible on the right (source: Wikimedia Commons; S. Buhl). For a color version of this figure, see www.iste.co.uk/encrenaz/solar2.zip*

While we estimate the quantity of extraterrestrial materials arriving and adding to the Earth each year as a few thousand tons (Rojas *et al.* 2019), the majority falls over time in the form of dust, a significant fraction of which results from the recondensation of incidental bodies vaporized in the atmosphere, with the meteorites reaching the ground only representing a few dozen tons on average (see Figure 3.2).

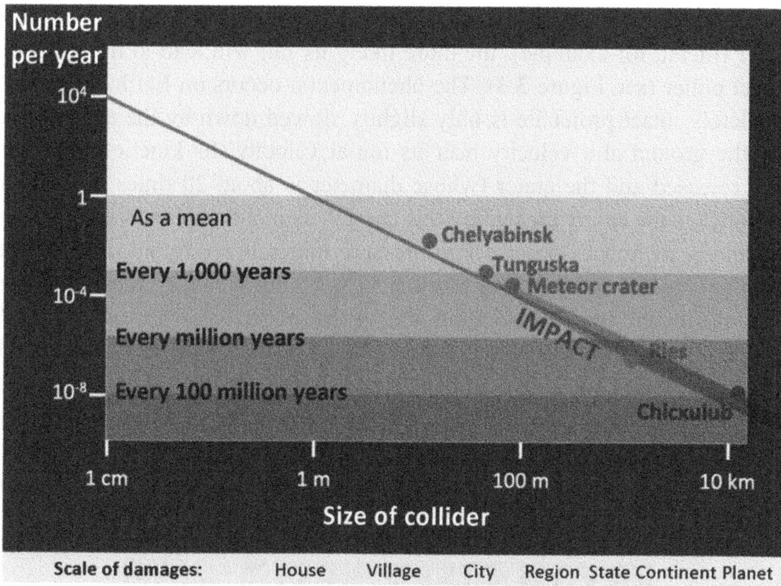

Figure 3.2. *Average number of objects colliding with Earth's atmosphere each year. The number decreases as their size increases. For a color version of this figure, see www.iste.co.uk/encrenaz/solar2.zip*

COMMENTS ON FIGURE 3.2.– *The logarithmic scale makes it possible to show on a single diagram objects ranging from centimeter-sized dust – which creates large shooting stars and never reaches the ground – to multi-kilometer-sized objects, responsible for the largest known impact craters on Earth and whose fall generates catastrophes on a planet-wide scale. Statistically, such an event occurs every 100 Myr. The last to occur formed the crater of Chicxulub (in the Gulf of Mexico), which is estimated to have been about 200 km in diameter (Connors* et al. *1996). It may have been caused by the collision with Earth of a carbonaceous asteroid (Trinquier* et al. *2006), almost completely evaporated during impact – only a few mineral grains survived: their analysis allowed the nature of the object to be determined. This event, which occurred 65.5 Myr ago (Schulte* et al. *2010), at the boundary between the Mesozoic and Tertiary eras, may have led to the extinction of many species, including the non-avian dinosaurs.*

The collision of large extraterrestrial bodies with Earth is rare (see Figure 3.2), which is fortunate because the damage they could cause increases proportionally with their size. We estimate that Earth encounters an object of the size of the one that generated the Chelyabinsk meteorite about every 100 years, and an object the size of Tunguska about every 1,000 years. The larger a body and the more resilient its matter (metal, for example), the more likely its fall will lead to the formation of an impact crater (see Figure 3.3). The phenomenon occurs on Earth when a mostly or completely intact projectile is only slightly slowed down by the atmosphere and reaches the ground at a velocity near its initial velocity. Its kinetic energy is then abruptly released and the crater (whose diameter is about 20 times larger than the falling body) is the cavity created by this energy, as in a bomb explosion. Moreover, in the vicinity of impact craters, high-pressure minerals are found, similar to those observed at the sites of nuclear explosions. The identification of these minerals has greatly contributed to understanding the origin of craters (Shoemaker 1959), a problem that was still being debated in the early 1960s, before the first human lunar flights. The surfaces of bodies without atmosphere such as the Moon are riddled with craters of all sizes (see Figure 3.3) because even the smallest particles hit the surface without having been slowed. That is why the lunar surface is covered with the very fine dusty material that captured the imprint of the astronauts' footprints: the regolith.

Figure 3.3. *Surface of the moon riddled with impact craters; the large crater that occupies most of the image and is overlapped by craters formed later is the Apollo crater; it is about 500 km in diameter (source: NASA (LRO))*

The impacts are none other than the extension of the accretion phenomenon at the origin of the planetary bodies, whose surfaces they shaped. They are a major geological process throughout the Solar System, in the very same way as volcanism. They can shatter small bodies or rip fragments away from the largest bodies, fragments that can in turn collide with other bodies. Because of this game of "cosmic billiards", we find fragments of asteroids, of the Moon, and of Mars on the surface of the Earth. Until 1969, these meteorites were the only samples we had to directly study the geology of planetary bodies other than Earth. Return of lunar samples (382 kg) has allowed for considerable advances in this domain:

– the traces of impacts of all sizes present in these rocks have made it possible to end the controversy over the origin of the craters[3] and to confirm the mechanism of accretion;

– the discovery that the lunar highlands are, for the most part, made up of a mineral (anorthite) formed by flotation above a magma made it possible to show that the Moon and probably the planets had largely melted after their formation,[4] which was one of the elements on which Hartmann and Davis (Hartmann and Davis 1975) relied to propose that the Moon was formed by a giant impact;

– the cutting-edge instruments intended for the analysis of lunar samples have also been used for meteorites, which has greatly contributed to the results presented in the rest of this chapter.

Since the Lunar samples were returned, particles of solar wind,[5] a few nanograms from a comet,[6] and a few micrograms from an asteroid[7] have been brought back. On December 5, 2020, the Hayabusa 2 mission (JAXA) brought back 4 grams of matter from asteroid Ryugu, and an American mission is soon to do the same for another primitive asteroid.[8] We may hope that the study of these samples, which are different from the ones we have had up to now, and of the micrometeorites identified in Antarctic collections that are very probably of cometary origin (Duprat *et al.* 2010), will lead to another decisive step in our understanding of the Solar System.

---

3 Galileo thought that the craters riddling the Moon's surface were of volcanic origin. The hypothesis of an external cause linked to impacts was proposed in 1665 by Robert Hooke, but he concluded that the hypothesis was improbable because it is difficult to imagine "whence those bodies should come".

4 We speak of a "magma ocean". This theory, originally proposed by Wood (Wood *et al.* 1970), is still applicable today: see (Elkins-Tanton 2012).

5 Genesis mission (NASA, 2004).

6 Stardust mission (NASA, 2006).

7 Hayabusa mission (JAXA, 2010).

8 The OSIRIS-REx mission (NASA) sampled asteroid Bennu. It is believed to have collected about 4.4 lb of material and is expected to return in 2023.

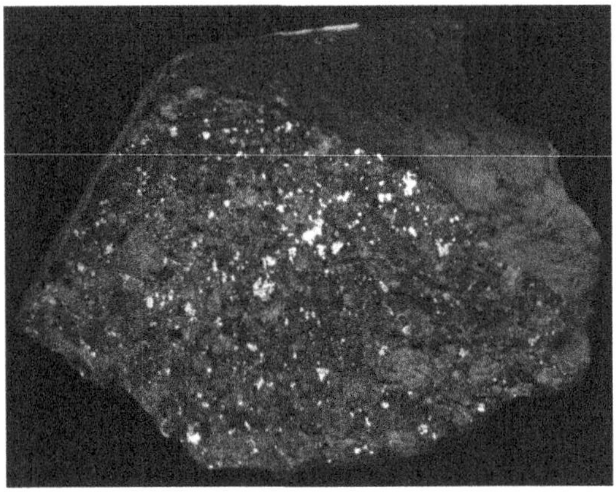

Figure 3.4. *Fragment of chondrite from Bremervörde (sample MNHN no. 108, 2.6 cm width). For a color version of this figure, see www.iste.co.uk/encrenaz/solar2.zip*

COMMENTS ON FIGURE 3.4. – *The fusion crust is clearly visible in the upper part on the right of the sample – it is marked in red on the left side: this is a number painted on the sample (a now discontinued curating practice). The broken front side of the sample shows some of the spherules of which it is constituted: the chondrules. It also shows bright points: these are grains of metal, some of which, slightly rusted, dot this gray-toned surface with brown (source: Zanda* et al. *2018, photo: L.-D. Bayle).*

3.2. Origin of meteorites

If the origin of meteorites and the way in which they come to Earth are now fairly well understood,[9] it was only at the beginning of the 19th century that their extraterrestrial nature was accepted.[10] Indeed, the small number of occurrences and

9 However, it is only in a limited number of cases that we can determine with certainty from which body a meteorite came: the Moon, Mars, and Vesta. For most other meteorites, spectroscopy and studies of trajectories allow us to determine the family of asteroids that a parent body belongs to, but it is difficult to identify the body itself with precision.

10 To convince the reader, we can cite this statement by American President Thomas Jefferson on December 23, 1803: "I find nothing surprising in the raining of stones in France, nor yet had they been mill-stones. There are in France more real philosophers than in any country on earth; but there are also a greater proportion of pseudo-philosophers there. The reason is that the exuberant imagination of a Frenchman gives him a greater facility of writing, & runs away with his judgment unless he has a good stock of it. It even creates facts for him which never happened, and he tells them with good faith" (Burke 1986).

their spectacular features when they were observed made them seem like wonders. Scholars of the age of enlightenment raised doubts about these observations – most often seen by lay people – and they considered these rocks that had supposedly fallen from the sky as only fossils, prehistoric tools, or concretions, and the idea of wandering rocks in space seemed impossible.[11] The German physicist Ernst Chladni (1756–1827), who is more well known for his work in acoustics, nevertheless published in April 1794 an innovative work (Chladni 1794) that highlighted the similarity in testimony about events that occurred in very different times and places. He also brought to the fore the absence of observations that would justify the rejection of an extraterrestrial origin for meteorites.[12]

Less than 10 years after the publication of Chladni's book, in 1803, the report at the Academy of Sciences by physicist Jean-Baptiste Biot (1774–1862) marked the end of the controversy. What had happened in the meantime? New rock falls had a significant effect, as did more sophisticated chemical analyses than those that had been possible previously. On June 16, 1794, a shower of stones whose mass ranged from a few mg to 3 kg fell to the southwest of the town of Sienna in front of many witnesses. This event proved the reality of rock falls, and Sir Joseph Banks (1743–1820), the president of the Royal Society, acquired a fragment of one of them. His attention was then attracted by its similarity to a rock that fell on December 13, 1795 at Wold Cottage (England) that its owner displayed at fairs, and by the announcement of new falls that occurred in Portugal in 1796 and in India in 1798. These events convinced him to initiate a serious study of the subject, a study for which he commissioned the chemist Edward Howard (1774–1816). Howard enlisted an immigrant French mineralogist, Jacques-Louis de Bournon (1751–1825), and procured other rocks that were supposed to have fallen from the sky as well as "native iron" (see Figure 3.5), also considered as such. In parallel, the French chemist Louis Nicolas Vauquelin (1763–1829) analyzed a rock that fell in 1798 at Salles (Beaujolais, France). The results of these analyses agreed. These rocks all had

11 Based on those of Aristotle and then Newton, representations of the cosmos in the 18th century agreed that between the stars, the Sun, its planets, and their satellites only the "fifth essence," ether, is found (Burke 1986).

12 "The idea that far away in space many additional, smaller celestial bodies exist appears to all so incredible that they reject the entirety of my theory. What is unbelievable in this assertion is only pretended; it is not based on the slightest reason but much more so on the fact that it is unusual and a little strange. Indeed, if one had to extract from commonly accepted physical theory the following assertion: 'there are no materials in space other than celestial bodies and the stars or another sort of "faint" elastic fluid,' this assertion would seem as presumptuous as the opposite assertion 'there are other materials in space'. Neither of these two assertions can be proved *a priori*; it is observation that must decide which of the two is correct" (Chladni 1794).

a common feature that distinguished them from terrestrial rocks: they contain significant quantities of small grains of iron in its metallic form scattered among the silicates (see Figure 3.4) or forming the majority of the rock in the case of "native iron" (see Figure 3.5). The iron was always associated with a few percent of nickel, an element discovered only about 50 years earlier. These results were convincing to scientists of the time and, following a stone shower at L'Aigle (Normandy, France) in 1803, the minister of the interior, Chaptal, sent Biot to the site, probably at Laplace's instigation (Burke 1986). Biot's very detailed inquiry vividly demonstrated the extraterrestrial origin of these rocks, but it is reasonable to think that his primary goal was to find clues that would allow their orbit to be reconstructed. Indeed, according to Burke, Biot especially wanted to test Laplace's hypothesis that these rocks originated in the Moon's volcanoes.

In fact, if meteorites were clearly acknowledged to have an extraterrestrial origin, their precise origin remained unknown, and an alternative theory posited that the rocks were generated in Earth's atmosphere (Burke 1986). These two theories were not abandoned until around 1860, when a new debate emerged, one that lasted for almost a century, pitting a planetary origin against an interstellar origin for meteorites (Marvin 1996). It was only in the second half of the 20th century that the origin of meteorites as fragments of planetary bodies was accepted by all on the basis of their elliptical orbits. The association of the large majority of them with asteroids, in the end, relies on the properties of these rocks (for example, their cooling rates indicating parent bodies of relatively small size), and finally, after 1970, on their spectral properties compared to those of asteroids (McCord *et al.* 1970; Chapman and Salisbury 1973).

Since then, the links between the different families of meteorites and classes of asteroids have gradually become better established, even if only a few meteorites can actually be associated with a precise body rather than with a family. There are only three cases: the HED meteorites (Howardites, Eucrites, Diogenites), which come from the asteroid 4 Vesta (McCord *et al.* 1970), and our collections also contain fragments of the Moon and Mars.

The first meteorite of lunar origin was discovered in 1982 in Antarctica and was identified thanks to a comparison of its mineralogical, chemical, and isotopic properties with those of samples brought back from the Moon by the Apollo 16 mission (Marvin 1996). The year 1982 also marked the acceptance of the Martian origin of another family of meteorites: the SNC meteorites (Shergottites, Nakhlites, and Chassignites) on the basis of a recent crystallization age for some of them (1.3 billion years, which indicates that they formed in a massive body that cooled slowly – Walker *et al.* 1979) and the similarity between the rare gases trapped in them and those measured by the Viking probe on the surface of Mars (McSween 1994; Bogard 1983).

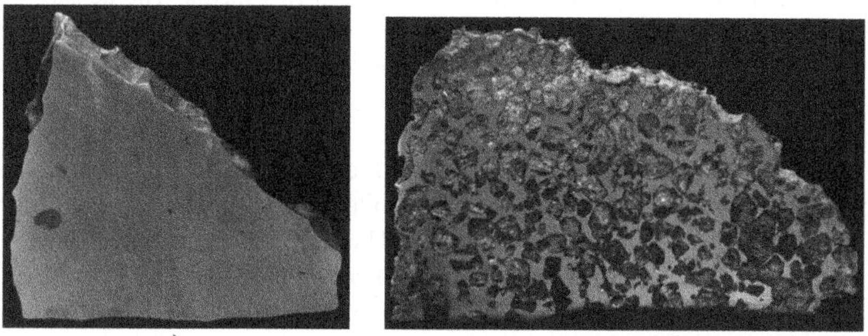

a) b)

Figure 3.5. *(a) Cut and polished slice of the Boguslavka iron meteorite (sample MNHN no. 3792, 20 cm width). Some inclusions of iron sulfide (beige) are visible on this slice of iron-nickel, particularly one on the left, which measures ~2 cm; (b) cut and polished slice of the Esquel pallasite (sample MNHN no. 3429, 22.8 cm width) (source: (Zanda and Jambon 2012), photos: L.-D. Bayle). For a color version of this figure, see www.iste.co.uk/encrenaz/solar2.zip*

3.3. Planetary differentiation and groups of meteorites

In the rocks found on the surface of the Earth, iron is found almost exclusively in association with oxygen in the form Fe^{2+} (ferrous ion) or Fe^{3+} (ferric ion), in oxides, or in silicate rocks and ores. The Iron Age (around 1,000 years BCE) corresponds to the time when humans began to master the extraction of iron from ore by heating it in the presence of carbon, whose affinity for oxygen is greater than that of iron. This reduction reaction, which releases CO_2, has long been carried out in bloomery furnaces. Today, it is produced on an industrial scale in blast furnaces. There are a few objects dating from the Bronze Age made of metallic iron: a dagger found in the tomb of Tutankhamen, the blades of Chinese weapons, etc. All were made using iron of meteoric origin.

Why do the large majority of meteorites contain iron and nickel in metallic form when terrestrial rocks do not? To understand this, we must remember that Earth is not a homogeneous sphere. It is made up of three main entities (see Figure 3.6d), which are themselves subdivided: a central core, 3,586 km in diameter; a thin, superficial crust no deeper than ~70 km under the continents and ~15 km under the oceans; and a mantle, which may be as much as 2,800 km thick and contains most of Earth's volume (~85%). This structure is well known, despite the inaccessibility of

Earth's[13] interior, thanks to seismic tomography: the study of the propagation of seismic waves, which are reflected or refracted at the boundaries between these areas, similar to light rays at the interface between two media. The density of the core indicates that around 80% of its mass consists of iron, supplemented by nickel, sulfur, and lighter elements such as oxygen, silicon, and carbon – a chemical composition that is similar to iron meteorites (see Figure 3.5).

On the other hand, Earth as a whole has a composition that is close to that of chondrites (see Figure 3.4), which is to say similar to that of the Sun – except for the most volatile elements (see section 3.4.1). This average composition consists of materials that accreted to form the Earth, but our planet, like the Moon, partially melted when it formed, under the combined effect of the heat given off by the disintegration of the highly radioactive nuclei[14] it contained and impacts related to accretion. The fusion of silicates that make up rocks generated silicate magma, while the metal grains generated metallic magma. These two immiscible liquids with different densities separated naturally due to gravity, in the same way that oil rises above vinegar when salad dressing unmixes. The denser metallic liquid flowed to the center of the Earth to generate its core, while the less dense silicate liquids rose to the surface of the planet to form a basaltic crust.[15]

This process of planetary differentiation (see Figure 3.6c) has not yet completely ended in the case of Earth: its external manifestations are volcanism and seismic activity, fueled by heat given off by radioactive nuclei with long decay periods such as ^{40}K (1.2 Ga), ^{238}U (4.5 Ga), and ^{232}Th (14 Ga).

For bodies smaller than Earth, on the other hand, whose surface-to-volume ratio means that they will give off their heat more efficiently, this process ended very early in the Solar System's history when it occurred – because it could also have been aborted at some point along the way. Indeed, for smaller bodies, and especially for those that formed late (at large heliocentric distances), the internal temperature elevation was insufficient to set off fusion: their accretion only occurred after the massive disintegration of more radioactive nuclei, and the heat produced by this process was given off in the gas of the protoplanetary disk. Such bodies are not differentiated (see Figure 2.6b): they are the source of the chondrites we find today.

13 The deepest drilling sites are barely more than a dozen kilometers deep.

14 Their period of decay is short on the geological time scale – typically a few hundred thousand years (^{26}Al) to a few million years (^{60}Fe).

15 The formation of the continental crust, less dense than the basaltic oceanic crust, is a phenomenon that occurred later.

These meteorites have preserved a chemical composition close to that of the materials that made up the protoplanetary disk and, hence, as we will see later in this chapter, they constitute a unique source of information about the physicochemical conditions that prevailed at the time and about the processes that were taking place. On the other hand, the meteorites from differentiated bodies (iron meteorites from the cores of bodies fragmented by collisions or basaltic meteorites from planetary crusts) can inform us about the internal evolution of these bodies, but most of the information relative to the conditions of the genesis of the original material was destroyed by fusion.

Meteorites are thus divided into two large main categories: chondrites (or "primitive" meteorites), which came from asteroids that did not melt, and the so-called "differentiated" meteorites. These two large categories are themselves subdivided as a function of mineralogy, chemistry, and the isotopic composition of rocks that come from distinct planetary bodies (see Figure 3.7). The study of different kinds of rocks is therefore a source of information about both the nature and the geology of the planetary bodies from which they originate and about the physicochemical conditions that governed their formation in the primitive Solar System. Unlike the samples brought back by space missions, the geological context from which these rocks were extracted is unknown, despite the increasingly precise work being done to determine their source regions. But they do sample a much larger number of bodies, despite the bias resulting from transmission through interplanetary space and passage through the atmosphere – a destructive event for the most fragile samples. Meteorites and samples brought back from space missions thus contribute in complementary ways to our understanding of the nature and genesis of the Solar System.

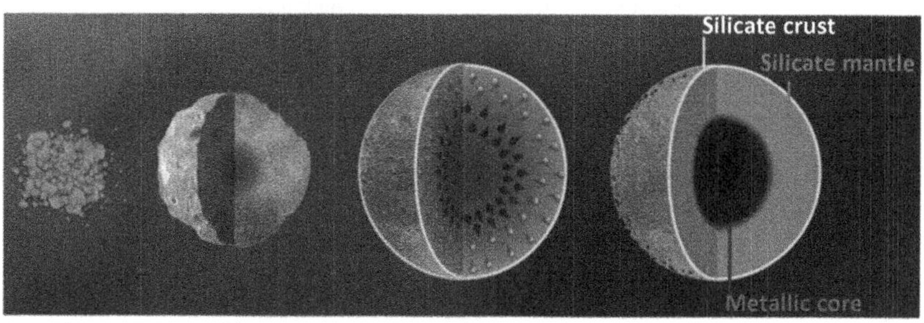

a) b) c) d)

Figure 3.6. *Diagram of planetary accretion and differentiation. For a color version of this figure, see www.iste.co.uk/encrenaz/solar2.zip*

COMMENTS ON FIGURE 3.6. – *(a) Accretion: genesis of the first bodies in the protoplanetary disk, first through coalescence of dust grains, then through gravitational capture. (b) Internal heating of the first bodies due to the heat produced by the disintegration of radioactive nuclei they accumulated. Primitive meteorites or chondrites (see Figure 3.4) are the result of bodies like this one, whose internal temperature elevation was insufficient to provoke fusion. (c) Differentiation: the internal temperature elevation is such that the rocks partially melt, forming immiscible silicate and metallic liquids that will separate by gravity. The densest metallic liquids sink toward the planet's center of gravity, while the least dense silicate liquids rise toward the surface to form the basaltic crust, leaving behind a residual silicate mantle. (d) Internal structure of the Earth and differentiated bodies (other terrestrial planets, the asteroid Vesta, and probably other asteroids). Iron meteorites (see Figure 3.5) come from the cores of such differentiated bodies, while stony differentiated meteorites come from a basaltic crust (primarily Vesta, the Moon, and Mars).*

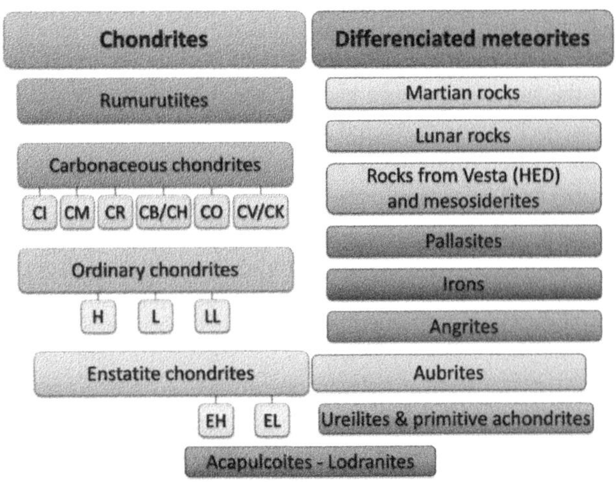

Figure 3.7. *Main groups of meteorites. These groups are subdivided into a number of subgroups not listed on this simplified figure. For a color version of this figure, see www.iste.co.uk/encrenaz/solar2.zip*

COMMENTS ON FIGURE 3.7. – *Groups and subgroups are defined on the basis of petrological (see Figures 3.10 and 3.11), chemical (see Figure 3.8), and isotopic (see Figure 3.13) criteria. It is generally agreed that samples in a single group have a common origin: a single parent body or a single family of parent bodies. On the basis of the variability of different criteria, and particularly the chemistry of iron meteorites, we estimate that at least a hundred different planetary bodies are represented in our meteorite collections: for the most part asteroids, but also the Moon and Mars.*

While the main distinction is the one that separates chondrites from differentiated meteorites, this is not as absolute as we may believe, because some bodies may have experienced partial differentiation. This may be the case for the parent bodies of primitive achondrites and especially acapulcoites-lodranites: these two groups of meteorites have mineralogical, chemical, and isotopic similarities that clearly indicate a common origin, and yet acapulcoites have a global chemical composition that is chondritic, while lodranites are comparatively lacking in Fe, S, Ca, Na, etc., which suggests the beginnings of differentiation, with extraction of a metallic and sulfurous liquid and a basaltic silicate liquid (Keil and McCoy 2018). In the same way, strong similarities exist between enstatite chondrites and enstatite achondrites (aubrites), without any determination of the real link between these two groups of objects.

3.4. Chondrites and the origin of the Solar System

Chondrites, which come from undifferentiated planetary bodies, can be identified because they contain grains of bright metal (as distinguished from grayish silicates – see Figure 3.4) and chondrules (from the Greek χονδροσ, *chondros*, which means "grain"), spherical, sub-millimeter particles that can constitute up to 95% of their matter and to which they owe their name. But these features are not present systematically, in particular because the geological processes that took place on the parent bodies (asteroids and possibly cometary nuclei) can have significantly modified the materials: circulation of fluids, which alters minerals and in particular oxidizes the grains of metal; heating – insufficient to cause fusion but sufficiently intense to cause the chemical diffusion of elements and the recrystallization of minerals,[16] leading to the disappearance of the original boundaries of the chondrules. Nevertheless, chondrites do possess one common, unique feature: their chemical composition is near that of the Sun, or even identical in some – with the exception of the most volatile elements (see Figure 3.8). This common chemistry does not however exclude a certain degree of variability, a variability that manifests at the mineralogical and isotopic levels, as will be described in the next section. A better understanding of these variations should allow us to better understand the genesis of chondrites and the first bodies in the Solar System.

16 Metamorphism, analogous to what affects terrestrial rocks, but, in the case of meteorites, occurs in the absence of water and at low pressures. We call this type of metamorphism "thermal".

3.4.1. *The chemical composition of chondrites*

A very rare group of meteorites, the CI chondrites, the largest of which is the Orgueil meteorite (14 kg), are rocks whose chemical composition is identical to that of the Sun, with the exception of the most volatile elements and lithium (see Table 1.5, Volume 1). Figure 3.8b compares the chemical composition of the Orgueil meteorite analyzed in the laboratory to the Sun's surface, measured by spectroscopy. In order to compare them, the two compositions are expressed as a ratio to 1 atom of Si, and the chart is drawn using logarithmic units (by a factor of 100). The data points line up along the diagonal: the chemical compositions in Orgueil and the Sun, as a ratio with Si, are almost identical to 8 orders of magnitude. A few points diverge notably, corresponding above all to the volatile elements, which do not enter easily into the minerals: H, N, and C are respectively 6,000, 30, and 10 times more abundant in the Sun than in the meteorite with respect to Si. We should note that O, which does enter into the minerals and is the most abundant atom in CI chondrites (44% in terms of number of atoms, 46% by mass), is only twice as abundant in the Sun as in Orgueil relative to Si. On the other hand, He, a noble gas with very low reactivity, is the second most abundant element in the Sun, and its abundance as a ratio of Si is 10 orders of magnitude lower in Orgueil. The Sun represents 99.86% of the Solar System's mass (most of the remaining matter consists of the gas giant planets). We thus consider that the Sun is chemically representative of the Solar System, and, as a result, that Orgueil and the other CI chondrites are chemically representative of the matter of interstellar origin that generated the Solar System.[17]

CI chondrites appear thus as the chemical "standard" of the Solar System, but they are the only ones with a chemical composition so similar to that of the Sun. Even if their chemical compositions are similar overall, all of the other chondrites are significantly different with regard to the volatile elements (see Figure 3.8c), but also the relative abundances of O, Fe, and Si, which (in addition to Mg, which follows Si exactly) are the elements that constitute almost 95% of the mass of the CI chondrites. This is true even for the other carbonaceous chondrites (see Figure 3.7), as is shown in Figure 3.8c, which represents the abundances of the major elements and that of a few minor elements in the four main groups of carbonaceous chondrites. To allow a comparison, these abundances have been normalized to aluminum and to the ratio of the same element with aluminum in the CI chondrite Orgueil. On the right, the caption indicates the colors corresponding to each

---

17 Lithium also deviates from the curve: this element is destroyed by reactions that occur in the Sun and result in the synthesis of He from H. The chondrites contain a record of the quantity of Li present in the beginning in the gas and dust cloud that generated the Sun and planets.

element, as well as the temperature at which 50% of this element would condense if a gas of solar composition cooled progressively in the physicochemical conditions of the protoplanetary disk.

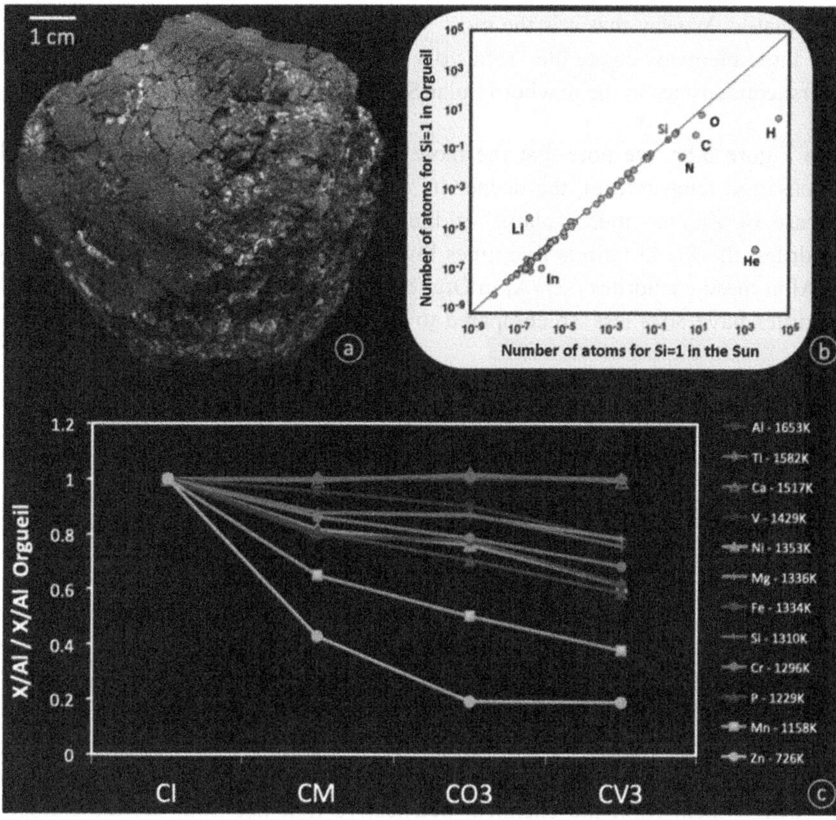

Figure 3.8. *The chemical composition of chondrites. For a color version of this figure, see www.iste.co.uk/encrenaz/solar2.zip*

COMMENTS ON FIGURE 3.8. – *(a) CI chondrites, the most studied of which is the one that fell at Orgueil (Tarn et Garonne, France) on May 14, 1864 (sample MNHN no. 362; source: (Zanda 2013), photo: L.-D. Bayle). (b) Chemical composition of Orgueil very close to that of the Sun, with the exception of the most volatile elements (source: data from (Lodders and Fegley 1998)). (c) The other chondrites, including carbonaceous chondrites, have compositions that deviate from CI chondrites (source: based on (Palme 2001)). For more detailed explanations of diagrams (b) and (c), refer to the text.*

When it reaches 1,653 K, half of the Al initially present in the gas is condensed into a mineral (oxide or silicate), while the temperature has to fall to 726 K for half of Zn to be in its condensed form: Zn is the most volatile element of those present in this diagram. On the other hand, Al (chosen as the element for normalization) is the least volatile. We say that it is the most "refractory," along with Ti and Ca. In fact, these three elements constitute "refractory inclusions", which are considered to be the first condensates in the newborn Solar System (see section 3.4.2).

In Figure 3.8c, we note that the more volatile an element (the lower its half-condensation temperature), the deeper its curve. What does this mean?[18] Consider the case of Zn, the most volatile of the represented elements. In CO and CV chondrites, the Zn/Al ratio is five times lower than in Orgueil, and in CI chondrites (Zn/Al in these chondrites / Zn/Al in Orgueil = 0.2). In other words, the CO and CV chondrites have 80% less Zn compared to Al, while CM chondrites lack only 60% of Zn.

This diagram allows us to see that:

– the more volatile an element, the less of it there is in CM, CO, and CV chondrites, which indicates that at least one of the processes that causes the chemical composition of chondrites to vary is linked to processes of evaporation and partial condensation (Grossman 1972);

– CM chondrites are less depleted than CO and CV chondrites. It just so happens that it is precisely the CM chondrites that contain the most matrix between the chondrules, grains of metal, and the refractory inclusions matrix, which is the main location for volatile elements (see section 3.4.2).

Carbonaceous CI chondrites are thus representative of the material that generated the Solar System from the chemical perspective (but not from the mineralogical perspective, particularly because they do not contain high-temperature compounds – see section 3.4.2). All of the other chondrites have chemical compositions that remain close to that of the Sun, although they deviate significantly as the result of processes linked to evaporation or condensation as well as to oxidation-reduction reactions.[19] The latter translate, notably, into a higher or lower abundance of metal (Fe^0), with an equal quantity of Fe in the rocks, because Fe can also be found in its oxidized form (Fe^{2+} or Fe^{3+}) in sulfides, silicates, or oxides.

18 It should be noted that all of the ratios are by definition equal to 1 in the CIs, which are the reference meteorites.

19 These fractionations are not described here, for lack of space. The three elements or groups of elements O, Fe, and Si + Mg have relative abundances that vary from one group of chondrites to another. For example, see Figure 1.1.7 in (Sears and Dodd 1988).

3.4.2. *The mineralogy of chondrites*

Grains of metal, chondrules, refractory inclusions, and a fine-grained interstitial matrix are the main components of chondrites (see Figures 3.4 and 3.9), rocks that may be considered as sorts of "cosmic sediments". Indeed, these diverse components, which have very different origins, found themselves united in a single rock because of random inter-particle collisions and under the effect of gravitational attraction – probably after transport, which suggests a sorting based on size and the variations of relative abundances between one group of chondrites and another (see Figures 3.10 and 3.11). The origin of these various chondritic components and their potential genetic links remains poorly understood (Krot *et al.* 2009). Apart from the matrix, they all formed or were transformed at high temperatures.

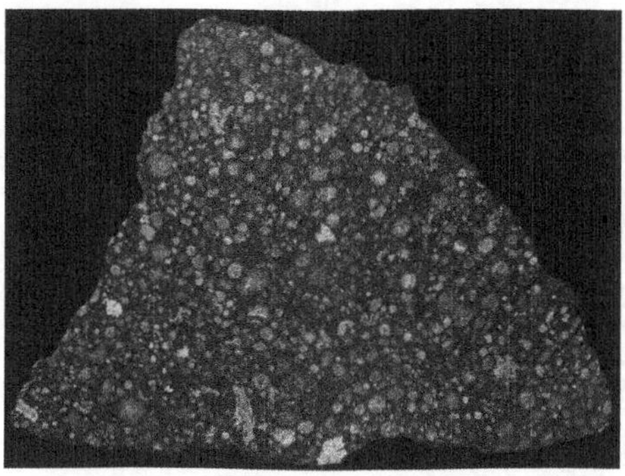

a)

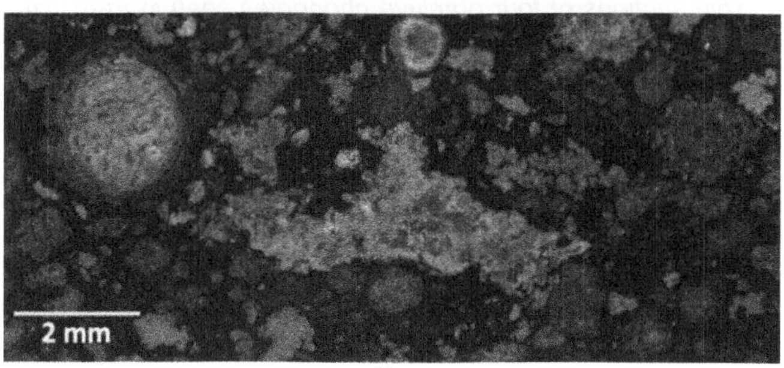

b)

Figure 3.9. *Overview and detailed view of two CV3 chondrites. For a color version of this figure, see www.iste.co.uk/encrenaz/solar2.zip*

COMMENTS ON FIGURE 3.9.– *(a) Slice of the Bali CV 3 chondrite (sample MNHN no. 3,754, 7 cm wide). This primitive chondrite contains many chondrules (whitish or grayish spherical particles) and refractory inclusions (irregular white particles) set in a dark fine-grained matrix* (source: (Zanda 2013)). *(b) Thin slice (30 μm thick) of the Allende CV3 chondrite seen in transmitted light in an optical microscope. Made up of very fine, packed grains (micrometer-sized), the matrix scatters light in all directions and appears black. The silicate crystals that make up chondrules and refractory inclusions are larger (10–100 μm) and are therefore transparent, bringing out these constituents. A refractory inclusion with an irregular shape occupies the center of the photo. We also see chondrules of more or less spherical shape (source: photo by B. Zanda).*

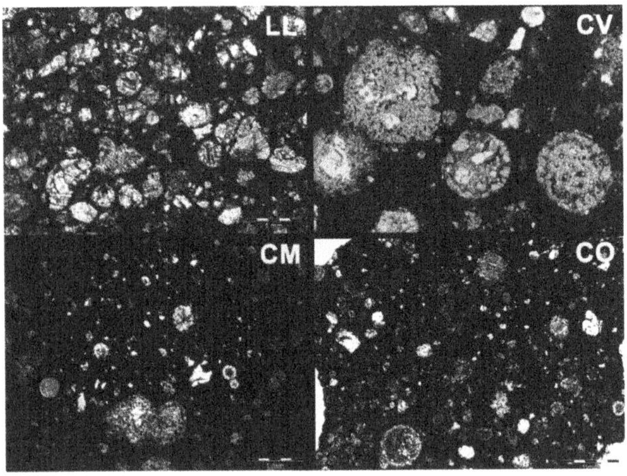

Figure 3.10. *Thin sections of four primitive chondrites seen in transmitted light in an optical microscope, at the same level of enlargement (1 mm scale bar). Source: photos by B. Zanda. For a color version of this figure, see www.iste.co.uk/encrenaz/solar2.zip*

COMMENT ON FIGURE 3.10.– *This montage highlights the differences in the size of chondrules (typically on the order of a millimeter in CVs, a half-millimeter in LLs, and a few hundred micrometers in CMs and COs), despite the presence of a few larger objects. We also note the different abundances for the matrix (which appears black, as do opaque metal and sulfur minerals – which are uncommon in the three carbonaceous chondrites, however). This matrix represents in volume 10 to 15% of the ordinary LL chondrite, 30 to 50% of carbonaceous CV and CO chondrites, and more than 60% of carbonaceous CM chondrites – see also Figure 3.12.*

We think that chondrules owe their spherical form to the forces of surface tension that affect a liquid in a microgravitational environment. To convince oneself of this, one can think of Captain Haddock's whisky, which forms droplets in the lunar rocket, or one can watch an astronaut play with a water droplet in the International Space Station.[20]

Chondrules are not only magma droplets solidified by rapid cooling, which created glass. They are essentially made up of silicate minerals (olivine and pyroxene) that crystallized at the expense of the liquid, which was preserved in the form of interstitial glass. Chondrules therefore formed from liquids that resulted from the fusion of pre-existing mineralogical aggregates, through one or more poorly understood processes that took time – the duration for crystal growth is at least on the order of an hour (Villeneuve *et al.* 2015). This process occurred repeatedly over the course of the few million years that followed the formation of the Sun (Krot *et al.* 2009; Connelly *et al.* 2012; Pape *et al.* 2019). The formation of chondrules thus could have occurred before the formation of the planets (Desch *et al.* 2012) and may even have been the triggering event: or, on the other hand, they could have formed in the vicinity of already formed planets (Pape *et al.* 2019). As for the grains of metal, they can be found inside chondrules or isolated, having sometimes been considered to be "metallic chondrules" (Bevan and Axon 1980). The large majority of these metal grains were formed and/or reworked during the processes that generated the chondrules (Campbell *et al.* 2005).

Refractory inclusions (see Figure 3.9) are made up in large part of oxides and silicates of refractory elements (in other words, of the least volatile elements in the conditions of the protoplanetary disk; see Figure 3.8 and section 3.4.1). They include two types of components: *Calcium-Aluminum-rich Inclusions* (CAI) and *Amoeboid Olivine Aggregates* (AOA), and are considered to be the first solids formed in the Solar System, by condensation or reworking of presolar materials (Larsen *et al.* 2020). This agrees with their age as obtained through radiochronological methods[21] (4,567.3 ± 0.16 10^6 years – (Connelly *et al.* 2012)). By convention, we consider this to be the age of the Solar System.

The matrix of the chondrites that were the least transformed on their parent body (see below) is made up of very fine-grained materials (< 1 μm), whose bulk composition is close to that of CI chondrites (Zolensky *et al.* 1993; Alexander 2005; Zanda *et al.* 2012), rich in volatile elements (H, C, and N – (Zanda *et al.* 2009)) that are primarily found in hydrated minerals (indicating the incorporation of ice) and organic compounds. The matrix also contains grains (see Figure 3.11) that were

20 See: https://www.youtube.com/watch?v=Ckh3TyNN6_Q.

21 We use the half-life of some radioactive nuclei as a "clock," in particular ^{238}U and ^{235}U, which respectively decay into ^{206}Pb and ^{207}Pb. The example of the dating of zircons from Mars is given in section 3.5.4.

formed in the vicinity of stars and trapped in the gas and dust cloud that was the precursor of the Solar System (Lewis *et al.* 1987; Bernatowicz *et al.* 1987). This suggests that the matrix is – in a very simplified manner – dust from the protoplanetary disk that came from cold areas beyond the ice line and accreted in variable proportions (see Figure 3.12) around high-temperature components (chondrules, grains of metal, and refractory inclusions) that lack volatiles. The data presented in Figure 3.8 and discussed in section 3.4.1 can thus be explained by the relative abundances of the matrix and the high-temperature components in carbonaceous chondrites – CI chondrites, which are closest to solar composition, being composed exclusively of matrix (secondarily hydrated on their parent body). The processes of evaporation/condensation and redox mentioned at the end of section 3.4.1 to explain chemical fractionations between the various groups of chondrites thus must result from the formation of chondrules and refractory inclusions (Zanda *et al.* 2018).

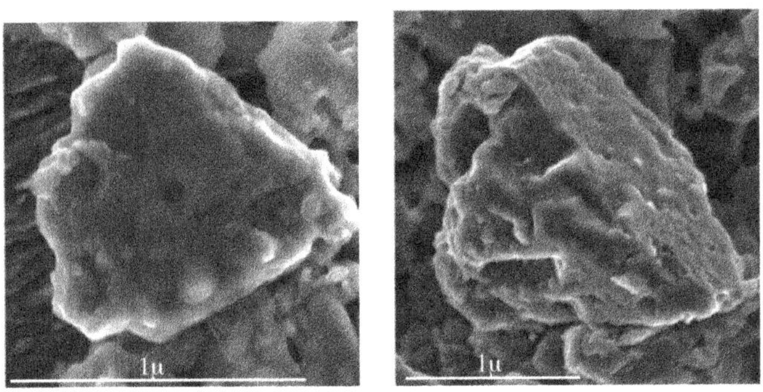

Figure 3.11. *The matrix of primitive carbonaceous chondrites contains presolar grains formed in the vicinity of stars that disappeared before the Sun formed. The isotopic study of these grains makes it possible to connect them to the category of stars from which they came and to refine models of stellar nucleosynthesis (Nittler and Ciesla 2016). These scanning electron microscope images show two presolar grains of SiC (silicon carbide) from the Orgueil meteorite (source: T. Bernatowicz)*

The study of these diverse chondritic components and their assembly informs us about the physicochemical conditions and the processes at work in the genesis of the first planetary bodies (asteroid-like planetesimals), one of the major problems being to understand how components with such highly variable chemical compositions (and also isotopic compositions – see section 3.4.3) could be found in the same rock. Furthermore, the processes that we are trying to study occurred almost 4.57 billion years ago, and the asteroids have had a brief but significant geological history, the marks of which have been superimposed over the primary features.

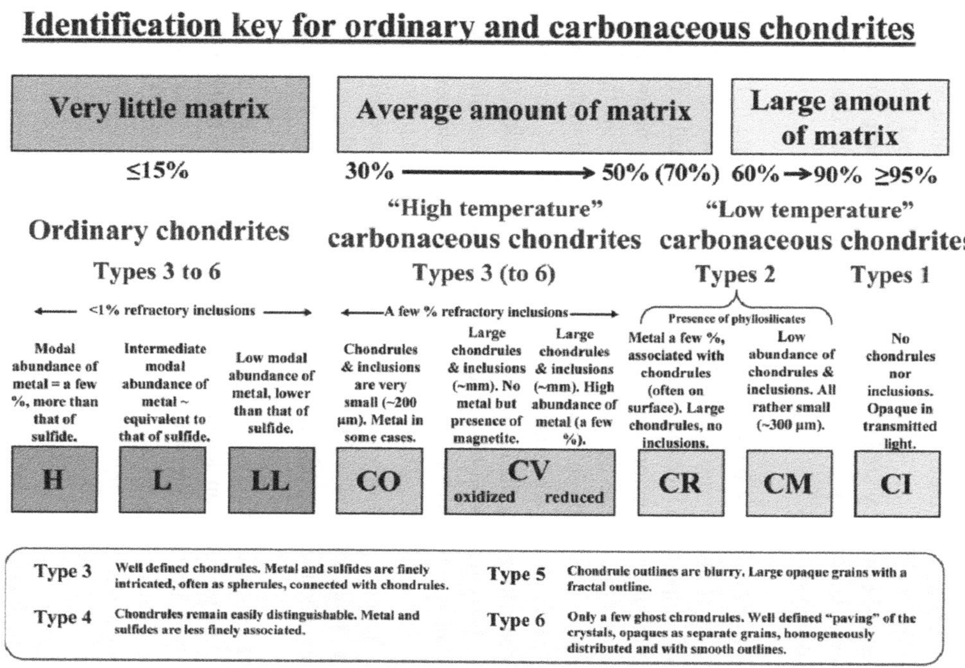

Figure 3.12. *The proportions and sizes of different chondritic components (matrix, chondrules, metal, refractory inclusions) are characteristic of each group of chondrites and make it possible to identify them (source: Zanda 2013)). For a color version of this figure, see www.iste.co.uk/encrenaz/solar2.zip*

The gravitational capture that determined the growth of planetesimals and planetary embryos released large amounts of energy, but the main source of heat in the first planetary bodies remains the disintegration of radioactive nuclei (see section 3.3 and Figure 3.6). The most stable among them (half-life > 10^9 years – typically ^{40}K, ^{238}U, ^{232}Th) still play a fundamental role in current volcanic and seismic activity on Earth. But their role in the initial heating of planetary bodies was very modest: initial heating resulted rather from the disintegration of the most unstable radioactive nuclei, particularly ^{26}Al, which is an isotope of a major element and has a decay period of 717,000 years. Formed by stellar nucleosynthesis, all of these nuclei were injected into the interstellar medium and were present in the cloud of dust and gas that generated the Solar System.

Chondrites came from planetary bodies that formed relatively late (~10^6 years) after the isolation of the cloud; having accumulated smaller quantities of ^{26}Al, they experienced a level of temperature increase that was not enough to lead to fusion but is likely to have profoundly transformed their mineralogy. The nature of this transformation depends both on the quantity of heat available and on the body's initial mineralogy, more specifically its water content which accreted in the form of grains of ice in the matrix. If this ice is not abundant, the temperature increase will quickly lead to its sublimation or to its fusion followed by its evaporation. The rocks from the same parent body will attain higher temperatures the more deeply they are buried in this body, which will induce reactions between minerals by diffusion in the chemical elements in the solid state. The minerals will thus acquire compositions in equilibrium from a chemical point of view (these compositions are specific to each group) and from the point of view of their texture: minerals in the matrix will recrystallize and grow, and the glass of the chondrules will also become crystalline.

These transformations will notably have the effect of erasing the boundaries of the chondrules, giving the rocks an appearance closer to that of a homogeneous metamorphic rock. We speak of "thermal metamorphism" because, in contrast to terrestrial rocks, this metamorphism occurs in the absence of water. The heated chondrites are assigned to a "petrographic type" that describes the intensity of this heating and, for historical reasons, varies from 3 (for the least heated chondrites)[22] to 6 (for the most heated chondrites) and even to 7 for the (rare) rocks that seem to have reached the beginnings of fusion. Rocks containing a high proportion of matrix (primarily CM and CI carbonaceous chondrites, but also CR and some of the so-called "anomalous" chondrites) have evolved differently on their parent body, probably as a result of their initial water content.

22 There are intermediate decimal subdivisions from 3.0 to 3.9.

Indeed, such carbonaceous chondrites have the distinction of having been less heated, but more altered on their parent body, as shown by the presence of hydrated minerals (phyllosilicates) that replace the minerals of the matrix and the glass of chondrules. A simple interpretation is that the heat produced by radioactive decay was largely consumed by the fusion of the ice grains, and that it is the water thus liberated that altered the primary minerals. This interpretation has the merit of explaining the paradox indicated by McSween (1999): although they are the meteorites whose chemical compositions is closest to that of the Solar System, the CI chondrites look, from a mineralogical point of view, like "dried mud". Indeed, as these rocks are exclusively composed of matrix,[23] they are both the ones that accreted the most water (in the form of ice) and the ones whose original materials were the most fragile (because of their micrometer size and possibly also because of their mineralogical nature).

The carbonaceous chondrites are also assigned to petrographic types. This type varies from 3 (for the rare objects that are both less heated and less altered) to 1 for the most altered objects (CIs), through type 2[24] for objects that are altered but that contain chondrules and refractory inclusions (CM and CR). The mineralogical features of the main groups of chondrites as well as the secondary transformations (metamorphism and alteration) that affected them on their parent bodies are summarized in Figure 3.12.

In addition to these transformations there are those that result from repetitive impacts on the surface of the parent bodies. We saw above (see section 3.1) that the surfaces of bodies without atmosphere, like the Moon, were riddled with craters, around which high-pressure minerals often formed. Collisions between bodies big enough to cause their fragmentation generate the future meteorites. In the textures of their minerals and in the form of impact veins,[25] these meteorites can preserve traces of the impact that ejected them and of the bombardment that preceded it. A number of them may even be formed from crushed materials that were then consolidated. We call such rocks "breccias". They can be composed of a unique material reassembled after fragmentation but, in a number of these breccias, we can find traces of one or more of the materials from the impact.

---

23 This is suggested by their high concentration of presolar grains. In fact, the abundances of these grains in the matrices of the least heated and least altered chondrites on their parent bodies are identical to the CI type (Alexander 2005).

24 These are also subdivided from 2.9 to 2.1 for an increasing degree of alteration.

25 These "shock veins" are fracture planes sealed together by minerals newly formed by local fusion of the rock: crystals, glass and micrometer-sized pellets of metal.

3.4.3. *The isotopic characteristics of bulk meteorites*

The groups of chondrites defined on the basis of mineralogical and chemical criteria (see Figures 3.7, 3.8, and 3.12) also have distinctive stable isotope characteristics (we speak of an "isotopic signature"). In this paragraph, we will focus in more detail on the case of oxygen, the first element for which evidence was found of variations in isotopic signatures between groups (Clayton 2003) and which also represents 44% of chondrites' atoms. The diagram shown in Figure 3.13 makes it possible to visualize the relative concentrations of ^{16}O, ^{17}O, and ^{18}O in samples. The different groups of chondrites sampled are each shown in a clearly defined region by a colored ellipse.[26] The straight Young and Russell line (Young and Russell 1998) of slope 1 constitutes a boundary in this diagram: with the exception of the R chondrites (rumurutites), all known bodies in the Solar System are to its right. We now know, thanks to the Genesis mission, that the Sun itself is on the Y-R line (near – 60‰), while planetary bodies are found instead in the vicinity of its intersection with Earth's fractionation line (in other words, close to enstatite chondrites[27] – EC in green in Figure 3.13).

Discovered in 1973 in the refractory inclusions of carbonaceous chondrites (Clayton *et al.* 1973), isotopic variations of oxygen independent of mass are the first "isotopic anomalies"[28] discovered in natural samples. Almost half a century later, their origin remains controversial. Already in 1973, Clayton *et al.* attributed the formation of isotopically distinct "reservoirs" to injection into the gas and dust cloud at the origin of the Solar System of ^{16}O freshly synthesized in a supernova – the explosion of which also caused the gravitational collapse of this cloud and thus led to the formation of the Sun. Of course, the addition of pure ^{16}O to a reservoir containing the three isotopes of oxygen will cause identical variation in the ratios of $^{17}O/^{16}O$ and $^{18}O/^{16}O$, distributing the points along a straight line with a slope of 1 in the diagram shown in Figure 3.13 (while fractionations linked to mass differences

26 CK chondrites are not shown: they have the same signature as CVs, to which they are related. In the same way, CH and CB chondrites have the same signature as CRs, both rich in metal. The ellipse corresponding to the L chondrites is visible under the H and LL chondrites.

27 Enstatite meteorites (enstatite chondrites and aubrites) have the same isotopic signature as the Earth, both for oxygen and for the other elements whose isotopic composition varies in extraterrestrial rocks. This suggests a common origin for the enstatite meteorites and the Earth.

28 This expression is no longer used commonly, because the origins of such "isotopic anomalies" are now fairly well understood in most cases. This is particularly the case with the excess of ^{26}Mg in refractory inclusions discovered only a few years after the isotopic variations of oxygen (Lee *et al.* 1976): it is known to result from the radioactive decay of ^{26}Al. Most of the other "isotopic anomalies" are due to the presence of presolar grains.

result in a slope closer to 1/2 – or more precisely: 0.52). Since then, Robert Clayton and his colleagues have shown that the isotopic fractionations independent from mass that they discovered in the chondrites' high-temperature phases were also found in bulk rocks. They showed that the groups of meteorites defined on mineralogical and chemical criteria also had isotopic specificities (Clayton 2003), with each of these groups occupying a specific "region" of the diagram, indicated by an ellipse in Figure 3.13.

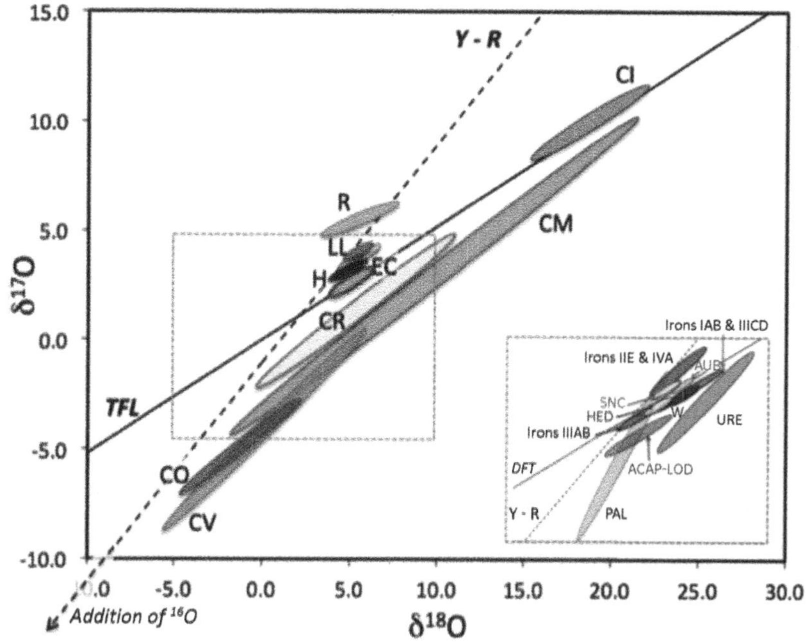

Figure 3.13. *Variations in relative concentration of ^{16}O, ^{17}O, and ^{18}O in groups of meteorites. For a color version of this figure, see www.iste.co.uk/encrenaz/solar2.zip*

COMMENTS ON FIGURE 3.13.– *These variations (in ‰), are normalized to ^{16}O and to the ratios $^{17}O/^{16}O$ and $^{18}O/^{16}O$ in a terrestrial standard called SMOW (*Standard Mean Ocean Water*): $\delta^{7}O = \{[(^{17}O/^{16}O)_{sample} / (^{17}O/^{16}O)_{SMOW}] - 1\} \times 10^3$; $\delta^{8}O = \{[(^{18}O/^{16}O)_{sample} / (^{18}O/^{16}O)_{SMOW}] - 1\} \times 10^3$. TFL (Terrestrial Fractionation Line) – line slope 0.52 along which occur all points corresponding to terrestrial rock samples.[29] Y-R = Young and Russell line, slope 1. We note that the area of the*

29 In the physicochemical processes that generate the rocks, the different isotopes behave in slightly different ways as a function of their mass (for example, ^{16}O diffuses a little more

diagram where the differentiated meteorites are found (box in the figure) is more constrained than in the case of chondrites: they show less variability. The groups are all listed in Figure 3.7 and/or mentioned in the text, except for W = Winonaites.

To the original hypothesis of Clayton *et al.*, of contamination of the protosolar cloud by pure ^{16}O synthesized by a supernova, we now tend to prefer the idea of self-shielding by UV[30] (Alexander *et al.* 2017) that caused the photodissociation of the CO molecules in the protoplanetary disk (Clayton 2003), or even in the protosolar molecular cloud.

These hypotheses are far from being proven, and the question of the origin of isotopic variations of oxygen between groups of meteorites remains one of the fundamental problems of cosmochemistry.[31] This problem has become even more important because there are other characteristic signatures, relating to isotopes of other elements such as Ti and Cr (see (Warren 2011) and Figure 3.14), and it is reasonable to think that the understanding of the mechanisms that generated these variations would allow for major advances in our vision of the genesis of the first planetary bodies.

It is nevertheless important to note that variations in isotopic signatures for elements other than oxygen relate to much smaller numbers of nuclei. They are expressed in notation as ε (for ten thousand), and Ti and Cr are both minor elements (~ 0.1% of the rock). It is probably because, unlike oxygen, they result from variations in the initial abundances of presolar grains (grains subsequently destroyed by weathering or metamorphism on the parent bodies, but whose signature was preserved in the bulk rock).

---

rapidly than ^{17}O, which itself diffuses a little more rapidly than ^{18}O). The difference thus created between ^{17}O and ^{16}O abundances is approximately 0.52 times smaller than the one between ^{18}O and ^{16}O.

30 This refers to the effect of selective UV photodissociation from an isotopic point of view. This effect is linked to the absorption by some molecules of a gas (CO) of UV light with a specific wavelength producing their dissociation. The light at this wavelength is progressively attenuated when it travels through a gaseous environment, which leads to a decrease in the rate of photodissociation of the target molecule with distance. When the molecules that have different combinations of isotopes ($C^{16}O$, $C^{17}O$, and $C^{18}O$) are photodissociated differentially by different wavelengths, this effect can provoke isotopic fractionation.

31 Although their origin remains poorly understood, the isotopic signature variations of oxygen are currently used to classify new samples whose nature is difficult to determine.

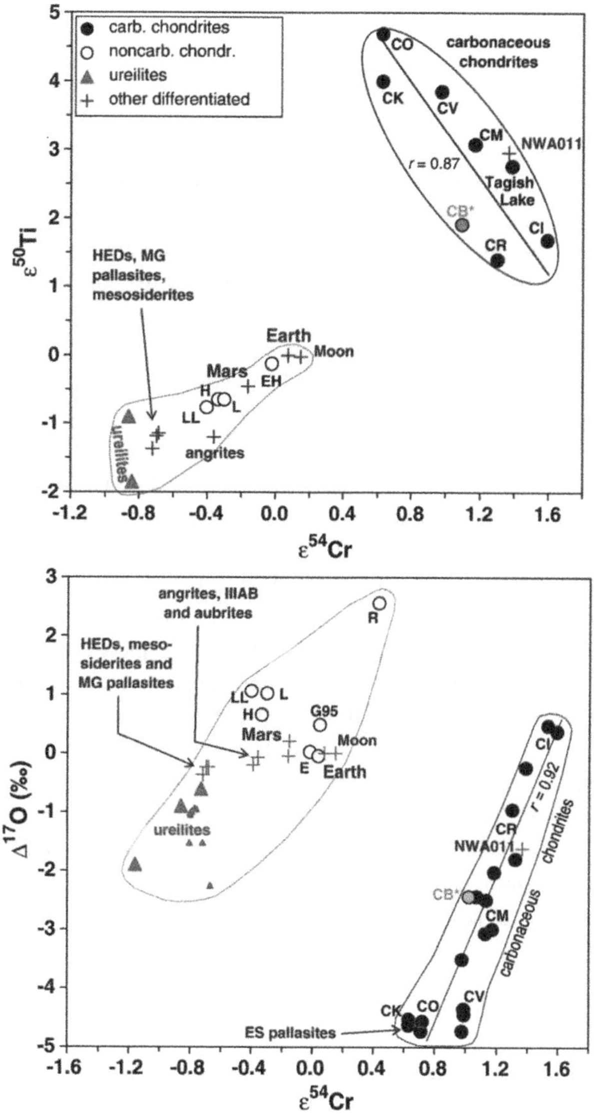

Figure 3.14. *Isotopic signatures in Ti, Cr, and O divide meteorites into two main families: carbonaceous and non-carbonaceous objects; some authors (Kleine et al. 2020) propose that carbonaceous objects formed beyond Jupiter and non-carbonaceous objects formed between Jupiter and the Sun (source: based on (Warren 2011)). For a color version of this figure, see www.iste.co.uk/encrenaz/solar2.zip*

3.5. Differentiated meteorites

Differentiated meteorites come from planetary bodies that underwent sufficiently intense fusion for their materials to have separated into distinct layers (see Figure 3.6). This history is *a priori* very different from that of the chondrites' parent bodies, and we have long considered differentiated meteorites and chondrites to have come from distinct and unrelated bodies. The preceding sections, in particular Figures 3.7 and 3.14, have nevertheless shown a more complex situation, with strong links between some groups of chondrites and differentiated meteorites. Indeed, it is even possible that some chondrites come from partially differentiated bodies, either because the internal heat of these bodies was insufficient to melt them all the way to the surface, or because these bodies had already differentiated when chondritic materials accreted on their surface. This hypothesis was proposed by Weiss and Elkins-Tanton (Weiss and Elkins-Tanton 2013), who showed that some chondrites had retained an intense, unidirectional magnetic field. Such a field can only be explained by cooling of the chondritic rock in the upper layers of a body with a liquid metallic core like Earth's, a large fraction of which was convective. It is hence possible that some chondrites and some differentiated meteorites come from the same body. Nevertheless, these two large families of meteorites still contain very different types of information: while chondrites are privileged witnesses to the genesis of the protoplanetary disk, differentiated meteorites are a precious source of information about the internal evolution of planets and asteroids. In this section, we will aim to describe the main groups of differentiated meteorites and to give some examples of their contribution to the understanding of the geology of planetary bodies. We will not focus on the case of lunar meteorites, although they represent an equivalent mass to the samples brought back by the Apollo and Luna missions (382 kg): indeed, compared to the latter, their contribution to the understanding of the Moon's geology remains modest.

3.5.1. *Fragments of the asteroid Vesta*

HED meteorites are vestiges of ancient magmatic activity that occurred in the crust of a planetary body. Eucrites are rocks with basaltic[32] chemical composition, whose texture is generally like that of volcanic rocks. In other words, the crystals that formed are not visible to the naked eye, which suggests a rapid cooling of the magma after eruption onto the surface. But some eucrites exhibit larger crystals (typically several millimeters), which implies that they crystallized more slowly and

---

32 Their concentration of SiO_2 varies between 45 and 52%, and their concentrations of CaO, FeO, and MgO are very high. Their concentration in K_2O is, on the other hand, very low.

at greater depth. Diogenites are also magmatic rocks, but are "cumulates", which are rocks that formed during the separation by gravity of crystals as large as several centimeters that precipitated while the magma cooled. This process induces chemical fractionation of the rock compared to the magma from which it emerged: the rock is comparatively enriched in Fe and Mg because the host minerals of these chemical species (olivine and/or pyroxenes) are the first to crystallize and accumulate. Eucrites and diogenites thus have very different textures and chemical compositions, but their common origin is not in doubt, as they share other features, such as their Fe/Mn ratio (an indicator of the redox conditions of the parent body) and their isotopic signature in oxygen (see the box in Figure 3.13).

Moreover, howardites are "impact breccias" formed by fragments of eucritic and diogenitic rocks that were crushed and then consolidated, which suggests that these two types of rocks are found near each other on the same parent body. To complete this brief tour of the rocks in the HED family, we must mention one more category: the mesosiderites. These rocks are made up in large part of metal, associated with fragments of eucritic and diogenitic rocks. One hypothesis (Rubin and Mittelfehldt 1992) is that mesosiderites are breccia formed by the impact of a metallic body (a fragment of the core of a previously melted and differentiated asteroid) on the surface of the HED parent body.

HED meteorites are the first rocks of asteroidal origin whose parent body has been identified with a high degree of certainty, on the basis of spectral measurements: they come from the asteroid (4) Vesta (Figure 3.15). It is thus possible to use them to try to reconstruct the geology of this planetary body. Figure 3.15b shows two hypotheses that both explain the link between eucrites and diogenites. However, the chemical composition of certain diogenitic minerals presents features (a very negative anomaly in Eu for pyroxenes) that likely resulted from the contamination of their parental magma by eucritic material. This suggests that the diogenites' parent magmas intruded late into a eucritic crust that was already established, which thus supports the second hypothesis presented in Figure 3.15b. As a result, the diogenites should be younger than the eucrites. Unfortunately, this hypothesis is difficult to verify based on the ages obtained for these two types of rocks. Indeed, if we try to apply radiochronological methods, the results are hard to interpret as a result of the complex thermal history of the rocks resulting from events that occurred at a later stage on Vesta (metamorphism, impacts). The dating method is adequate to determine that the magmas from which eucrites originated crystallized in the first 10 million years of the Solar System, about 4.564 billion years ago, but it does not allow us to determine whether or not the diogenites are significantly younger.

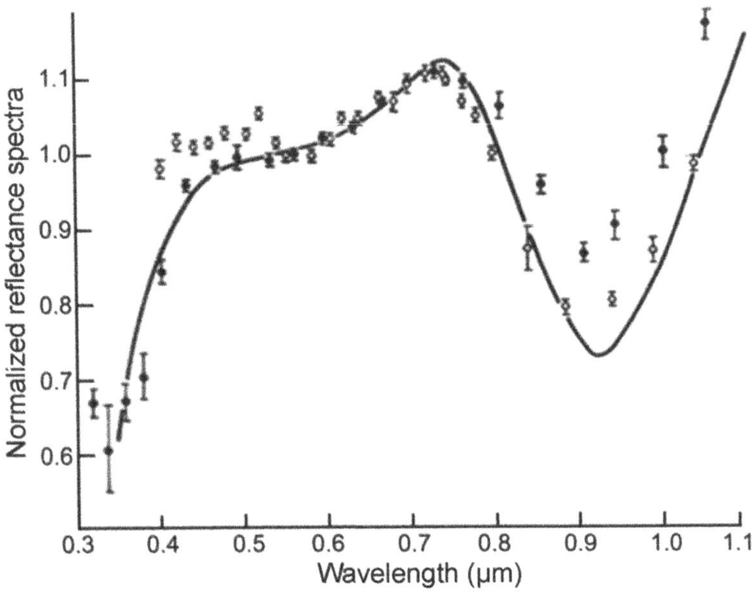

a)

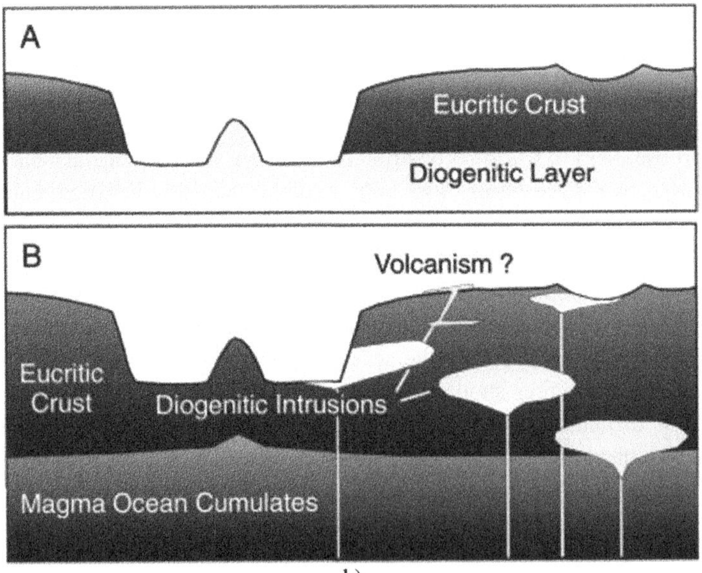

b)

Figure 3.15. *The eucrites and Vesta. For a color version of this figure, see www.iste.co.uk/encrenaz/solar2.zip*

COMMENTS ON FIGURE 3.15. – *(a) Spectrum of visible and infrared light reflected by Vesta (points with error bars) obtained at Cerro Tololo in Chile and at Mount Wilson, compared to a laboratory spectrum (solid line) of the Nuevo Laredo eucrite (source: (McCord* et al. *1970)). This impressive correspondence made it possible, for the first time, to establish a link between a group of meteorites (the HED) and a specific asteroid. Today, this correspondence has been reinforced with additional observations of Vesta (by the Hubble Space Telescope and the Dawn mission) and by improvements in laboratory spectra of HED meteorites. (b) Schematic of the geology of Vesta, through the study of petrology and the geochemistry of HED meteorites. Two hypotheses are proposed here: (A) diogenitic cumulates occur deep under eucritic basalts; (B) diogenitic cumulates occur in the form of intrusions in eucritic basalts. Geochemical data, which show that the magma generating diogenitic cumulates has reacted with the eucritic crust, support hypothesis B. But the Dawn mission was not able to settle the issue completely because of the ejecta from craters that cover Vesta's surface (source: Barrat* et al. *2010)).*

In summary, the study of meteorites from Vesta allows us to propose some hypotheses about the geological evolution of this asteroid: during differentiation, which occurred in the first few million years after the formation of the Solar System, its silicate fraction mostly melted. This led to the formation of a magma ocean on its surface. In cooling, the latter crystallized into large quantities of olivine and pyroxene, minerals that accumulated to form Vesta's deep rocks. Eucritic basalts then formed from the residue of this magmatic ocean. As for the diogenites, they cannot be simply the result of crystallization of a magmatic ocean because their chemical composition indicates a wide diversity of parent magmas. This diversity may be due to a late production of the diogenites' parent magma by secondary fusion of Vesta's deep rocks and/or to their variable level of contamination by the assimilation of eucritic rocks in intrusion in which they occurred.

3.5.2. *Iron meteorites*

Metallic meteorites come from the cores of differentiated asteroids shattered during destructive collisions, in contrast to so-called "constructive" collisions, which allow planetesimals and planet embryos to grow during accretion. They are composed, for the most part, of two metallic minerals with distinct crystalline structures[33] and contain different proportions of Fe and Ni: *kamacite*, which is

33 Kamacite has a body-centered cubic crystal lattice, while taenite has a face-centered cubic crystal lattice.

relatively poor in Ni (< 8%), and *taenite*, which is richer in Ni (> 16%). Other than Fe and Ni, these meteorites contain a number of trace elements (Co, S, Cr, elements of the platinum group such as Ir, etc.). Such elements can be expressed in the form of accessory minerals (sulfates, phosphides, phosphates, chromites, etc.). The relative proportions of kamacite and taenite in the rock are governed by its bulk chemical composition. If the rock contains less Ni (6% or less), it will only contain kamacite. On the other hand, if it is rich in Ni (> 16%), it will be almost completely composed of taenite.

The large majority of iron meteorites contain intermediate quantities of Ni between the two extremes. They are thus made up of the two minerals in variable proportions, which can be brought to light by the two minerals' different reactions to acid etching (Figure 3.16). This results in a pattern made of bright bands (kamacite) that have grown at the expense of a residual mineral (taenite). This is the so-called "Widmannstätten" pattern, named after one of its two discoverers.[34] Iron meteorites that thus contain the two minerals are called *octahedrites* because the bands of kamacite, which developed with orientations defined by the crystal structure of the taenite, are parallel to the sides of octahedra. The width of these bands is linked to the cooling rate of the metal core from which the meteorite came (Goldstein and Short 1967): as a first approximation, the narrower they are, the faster the cooling. But, in practice, the measurement of the width of the bands does not provide a reliable determination of this cooling rate. We now prefer to use a method relying on a comparison of variations in the concentration of Ni measured through the kamacite bands and the residual taenite with the results of computer simulations (Goldstein and Ogilvie 1965).

Indeed, if the diffusion of Ni was instantaneous, the center of the bands and their edges would have the same concentrations. But this is not the case: the center of the taenite bands is less rich in Ni in relation to their edges, and this imperfect diffusion is directly linked to the cooling speed. We thus estimate that the cooling rates for most iron meteorites vary between 100 and 10,000°C/Myr. This means that the thermal history of the cores of asteroids from which we have samples did not last for more than 10 million years.

34 These patterns were discovered in 1808 by Count Alois von Widmannstätten (1753–1849), director of the imperial porcelain museum of Vienna, but we know now that they were described in 1804 by British mineralogist William Thomson (1760–1806).

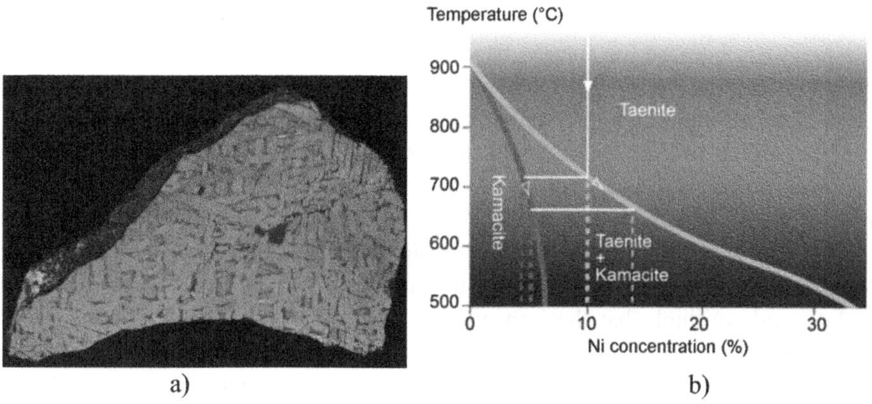

Figure 3.16. *Widmannstätten pattern of iron meteorites. For a color version of this figure, see www.iste.co.uk/encrenaz/solar2.zip*

COMMENTS ON FIGURE 3.16.– *(a) Polished and acid-etched slice of the Paloduro meteorite (sample MNHN no. 3590, 8.2 cm width) (source: (Zanda and Jambon 2012), photo: L.-D. Bayle). Etching makes it possible to reveal the internal structure of the rock, which is composed for the most part of two metallic minerals formed by unmixing in the solid state of a single initial crystal. This unmixing occurs during cooling of a homogeneous Fe-Ni alloy from a high temperature (the alloy itself is the result of crystallization of a metallic liquid). The mineral that forms brighter extended bands is kamacite, which contains relatively little Ni (< 8%). The darker interstitial mineral is taenite, which contains more Ni (> 16%). At high temperatures, taenite is the only stable alloy, and this alloy can contain a wide range of Ni contents. But with cooling, this alloy requires a small amount of Fe, and it will "expel" the excess iron, which forms kamacite crystals. The temperature at which unmixing begins, the relative proportions of kamacite and taenite that result from the process, as well as the composition of this taenite, are all governed by the mixture's initial composition. Diagram (b) illustrates this process for an alloy with an initial content of 10% Ni. The formation of kamacite begins around 700°C, and it thus contains around 4% Ni. With cooling, the taenite becomes progressively enriched with Ni, following the yellow line, while the proportion of kamacite increases, and its composition changes slowly, following the orange line. At the end of the process (~0°C), the rock with 10% Ni is composed first and foremost of kamacite, whose average composition is around 8%, with its composition in taenite at about > 32%. In practice, the slowness of the diffusion at low temperatures causes the resulting mineral compositions to be heterogeneous (source: (Rotaru* et al. *2018)).*

Iron meteorites poor in Ni and composed exclusively of kamacite are called *hexahedrites*, because geometric bands parallel to hexahedra can appear after impact. Meteorites rich in Ni and composed exclusively of taenites are called *ataxites*, because they have no visible structure. Onto this structural classification of iron meteorites a chemical classification relying on the concentration of trace elements, primarily Ga, as well as Ge and Ir, is superimposed. Concentrations of these trace elements are reported in relation to the bulk concentration in Ni on the graphs, whose scales are logarithmic to highlight distinct sets. Fourteen groups, designated by roman numerals and letters such as "IAB", have thus been identified to date, and we think that the iron meteorites that make up each of these groups formed on the same parent body. But more than 15% of iron meteorites cannot be easily classified in any of these groups, and we consider that they probably correspond to more than 50 additional parent bodies. We thus possess in our collections samples of the cores of about sixty asteroids, most of which were destroyed by collisions.[35]

3.5.3. *Pallasites*

Pallasites are undoubtedly the most beautiful meteorites (see Figure 3.5). They are made up of a metallic matrix analogous to iron meteorites in which are set crystals rich in Mg (most often olivine, and sometimes pyroxene), which are green and transparent when the meteorite is fresh. These minerals are the ones that make up the terrestrial mantle – or at least the upper portion. The most obvious explanation of pallasites would be that this set of mantle silicates and metal comes from the boundary between the mantle and the underlying core of an asteroid.[36] But the olivine crystals are often grouped in packets. They seem to have been formed as monomineralic rock, and subsequently separated by liquid metal.[37] It is difficult to understand how metal could be inserted between olivine crystals if the original rock composed of these silicate crystals was placed above the metallic core. A variety of

---

35 It is possible to estimate the time of interplanetary travel of an asteroid fragment from when it detached from the asteroid to when it hit the Earth (Eugster 2003). To do so, we measure the atomic nuclei produced internally by nuclear reactions induced by the cosmic, galactic, and solar radiation to which they are exposed. We refer to this period of time as the "exposure age". For iron meteorites, this age typically varies between a few hundred million to almost 2.3 billion years (Voshage 1984; Eugster 2003), which suggests that the destruction of the parent asteroids of iron meteorites happened a very long time ago. For stony meteorites (chondrites and achondrites), these ages are much shorter, which suggests that these objects are not as resistant to the effects of interplanetary travel.

36 In the case of Earth, we call this zone the "D" region.

37 See the cluster bottom right in the slice of the Esquel meteorite in Figure 3.5.

processes linked to impacts have been mentioned, in particular the possibility of a tectonic readjustment at the core/mantle boundary following a surface impact on the body. But the hypothesis currently considered the most likely is that of fragmentation, following a collision, of a differentiated body in which the outer portion of the core was liquid. The subsequent partial reaccretion of this body associated mantle debris with the residual liquid metal. We must admit that such a succession of events could have occurred fairly frequently in the asteroid belt because, despite the low probability of a sample of a given asteroid arriving on Earth as a meteorite, our collections contain samples of pallasites that come from at least five different parent asteroids, as shown by their isotopic oxygen signatures. We should also note that their concentrations of trace elements indicate that none of these types of pallasites comes from any of the sixty or so asteroids sampled in the form of iron meteorites.

The silicates preserved in the metallic matrices of pallasites are the only mantle debris from ancient asteroids to which we have access to for now. It is, indeed, surprising that we have a number of samples of basaltic planetary crusts and metallic cores but no samples of the mantles of the completely fragmented parent bodies that are the origin of these various meteorites. A reasonable explanation for this is that the large majority of meteorites that represent planetary crusts are the product of a small number of parent bodies that lost fragments relatively recently (primarily Vesta and Mars). This is indicated by their exposure age, which is equal to or less than 120 Myr (Eugster 2003), while those of iron meteorites can reach 2.2 Gyr (see note 35). This suggests that the silicate portions of these asteroids, which fragmented for the most part a few hundred million years ago (Eugster 2003), no longer reach Earth. They were probably not able to survive long trips in the interplanetary space in isolation because of multiple collisions and of space weathering, which results from the combined action of "cosmic gardening"[38] and irradiation by cosmic particles and by the solar wind.

3.5.4. *Fragments of the planet Mars*

Until recently, all of the rocks identified[39] as being of Martian origin were relatively young volcanic (basalt and related rocks), probably from a small number

38 "Cosmic gardening" is the process that leads to the formation of regolith, the powdery material that covers the surfaces of planetary bodies without atmospheres, particularly the Moon. It results from the bombardment of these surfaces by interplanetary dust.

39 Although they cannot be compared to rocks brought back from Mars by space missions, this identification is considered very reliable. It is based on several criteria: (a) Martian rocks constitute a distinctive and individualized group separate from all others from the

of sites (Nyquist *et al.* 2001; Lapen *et al.* 2017). To this, there was only one exception – and not the least important: the pyroxene cumulate ALH 84001. Discovered in 1984, ALH 84001 was first taken for a diogenite, until its Martian origin became known (Mittlefehldt 1994). This rock, older than all other known Martian rocks to that point (4.1 Gyr (Lapen *et al.* 2010)), made the news shortly thereafter, because it contained tubular structures reminiscent of fossilized bacteria, and these structures were interpreted as possibly being fossils of ancient Martian life (McKay *et al.* 1996). This hypothesis has since been refuted numerous times (Treiman 2003; Barrat *et al.* 2011), but the controversy seems to not be completely over, because the first authors were still championing the idea in recent years (McKay *et al.* 2009).

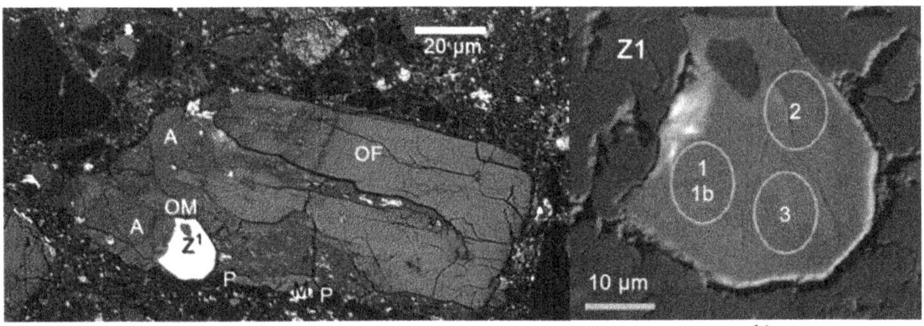

a) b)

Figure 3.17. *(a) View from a scanning electron microscope of a rocky fragment of the Martian impact breccia containing a zircon (Z). The letters indicate the different minerals (OF/OM: orthopyroxenes, A: augite, M: magnetite, P: plagioclase feldspar) (source: (Hewins* et al. *2017)). (b) Close-up view of zircon Z1, showing areas of analysis reported in Figure 3.18 (source: (Humayun* et al. *2013)). For a color version of this figure, see www.iste.co.uk/encrenaz/solar2.zip*

mineralogical, chemical, and isotopic perspectives (yellow ellipse marked "SNC" in the box in Figure 3.13); (b) the crystallization ages of SNC rocks indicate they are all very young (≤ 1.4 Ga with few exceptions), which suggests that they come from a planet, because volcanism was extinguished a long time ago on all asteroids – see section 3.5.2, which shows that the metallic cores of these asteroids had finished cooling 10 Myr after the genesis of the Sun. For dynamic reasons, the planet Mars seems like the best candidate; (c) finally, and this is the strongest argument, all contain traces of gas that are exactly like the Martian atmosphere measured in 1976 by the Viking mission – an atmosphere whose composition is very different from that of the other planets like Earth or Venus.

In 2011, a completely new type of Martian meteorite was discovered and nicknamed *Black Beauty*[40] by the group that was the first to work on it (Agee *et al.* 2013). It is an impact breccia (Humayun *et al.* 2013; Witmann *et al.* 2015; Hewins *et al.* 2017), the first at our disposal from Mars, probably from the ancient high plains that make up the planet's southern hemisphere. Under study while the Curiosity rover began to send back its first *in situ* analyses of more differentiated rocks than SNC meteorites, and while a new reading of the tectonics and topography of Mars was being developed (Bouley *et al.* 2016), this breccia played a crucial role in the emergence of a completely new vision of Mars's crust. Indeed, until then, we had considered Mars's crust as being completely basaltic (in agreement with the chemical and mineralogical nature of SNC meteorites).

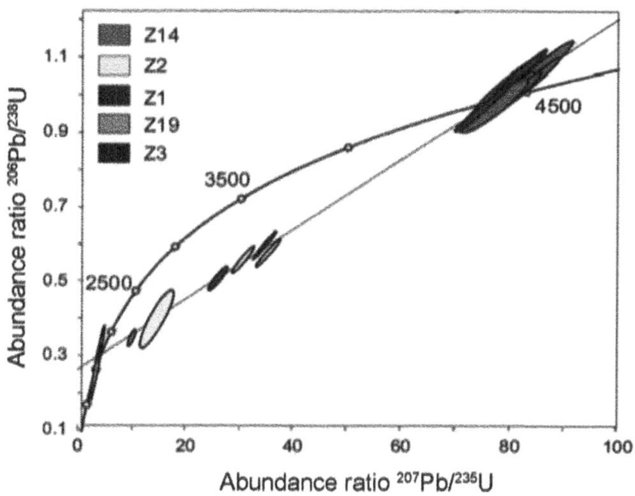

Figure 3.18. *Determination of the age of zircons in the Martian impact breccia NWA 7533 (source: (Humayun* et al. *2013)). For a color version of this figure, see www.iste.co.uk/encrenaz/solar2.zip*

40 Meteorites always bear the name of the place where they fell. When it is in the desert as was the case here, it is given the name of the closest post office, except in dense collection areas (DCA) where many meteorites have been found a long time after their fall as a result of systematic searches. They thus receive a single acronym (in the present case, NWA = North-West Africa), followed by a number. These names are attributed by the Meteoritical Society's Nomenclature Committee (the scholarly international body that brings together scientists and amateurs interested in meteorites), and the use of "alternative" names is strongly discouraged (and even not accepted in most of the community's scientific journals). The various fragments of the impact breccia bear different numbers, even if we know that they are paired. Among them: NWA 7034; NWA 7533, and NWA 7475, etc.

The decay of radioactive isotopes is used as a clock for radiochronometry. It is necessary to determine the quantity of daughter isotopes accumulated in the matter (F), to evaluate their initial concentration (F_0), and to ensure that the environment remains "closed", which is to say that the concentrations of parent (P) or daughter isotopes have not changed during the process through exchange with the surrounding environment. We can thus use the equation: $F = F_0 + P(e^{\lambda t}-1)$, where λ is the parent isotope's decay constant. The half-life of the isotope used should be commensurate with the age of the event one seeks to date. Thus, with a period of 4.469 Gyr, ^{238}U, (which decays into ^{206}Pb) is particularly appropriate for dating the processes that took place in the early Solar System. Another advantage of this clock is that there exists a second U/Pb chronometric pair: ^{235}U-^{207}Pb (period of 704 Myr). The simultaneous use of these two U/Pb pairs makes it possible to eliminate some effects linked to a recent opening of the system such as a loss of Pb (because these two isotopes of the element will be equally affected). The diagram in Figure 3.18 is called "Concordia". The curve represents the variation over time of the ratios $^{206}Pb/^{238}U$ and $^{207}Pb/^{235}U$: all of the samples whose age measured using these two clocks is the same lie on this curve. Such samples are therefore said to be "concordant", and their age is given by their position on the curve. An opening in the system will cause the sample to deviate from the curve: it will move along a line that connects the age of its formation to that of the perturbation. This line along which the sample moves is called a discordia. Many of the zircons in the NWA 7533 breccia fall on the Concordia curve, and their age is known precisely: 4.43 Gyr. But some of them – or even some regions of these grains (for example, the areas of points 1 and 1b of zircon Z1) align on a discordia. This discordia (the same for all samples) intercepts the Concordia at 1.4 Gyr, which indicates a (thermal) event that may have been local or, on the contrary, may have affected a large part of the planet, given that this age corresponds precisely to the crystallization age of nakhlites.

Box 3.1. *How to determine the age of a meteorite*

In contrast, the Martian breccia contains a variety of minerals such as potassium feldspar and zircons (see Figure 3.17) indicating that the rocky fragments that accumulated in this complex rock have, for the most part, crystallized from a magma that "evolved" 4.43 Gyr ago (Humayun *et al.* 2013). Such a magma would be closer to those that generated the granitic rocks of primitive Earth's first continents than to those that formed basalts. These observations agree with the ones from the ChemCam detector on the Curiosity rover, which detected minerals interpreted as potassium feldspars in the vicinity of the Gale crater (Sautter *et al.* 2016). They are also consistent with the revelation of a thick crustal block in Mars's southern hemisphere (*Terra Cimmeria-Sirenum*), on the basis of geophysical models and orbital observations (magnetic and geochemical anomalies, with a high degree of enrichment in Th and K (Bouley *et al.* 2020)). These data, taken together, suggest that Mars's primitive crust was less different from Earth's than we previously thought. This means that we may someday be able to read Earth's past on Mars and, in particular, to better understand the conditions in which life appeared – information that was mostly erased from our planet by the interactions of plate tectonics over geological time.

One could argue that the contribution of the Martian breccia to this paradigm shift is limited, since potassium feldspar, for example, was revealed by ChemCam. However, this would deny the degree to which the study of rocks in laboratories on Earth is complementary to *in situ* observations, particularly in precise age dating – with the ideal being the return of samples, of course.

Indeed, geochemical analysis of rocky fragments and minerals from the Martian breccia has produced rich information about the physicochemical conditions that prevailed on the Martian crust at the time of their formation. (Humayun *et al.* 2013) have thus been able to show that this crust likely had a thickness of about 50 kilometers 4.43 Gyr ago, and that it had previously been contaminated by up to 5% of chondritic material falling on the surface of the planet. To major advances that resulted from the study of the Martian breccia, we must add the revelation of hydrothermal events with sulfur circulation on the surface of Mars (Lorand *et al.* 2018) and the application of traditional techniques of radiochronology to minerals from this breccia: in particular, zircons (see (Humayun *et al.* 2013) and Figure 3.18), but also phosphates and feldspar (Bellucci *et al.* 2015). This has made it possible to date to 4.43 Gyr the formation of this "evolved" crust, which resulted from the rocky fragments of the breccia, an age somewhat older than that of the pyroxene cumulate ALH 84001. This also made it possible to show that the rocks had undergone a major "perturbation" 1.4 Gyr ago, that resulted in a loss of Pb for some of these minerals (see Figure 3.18). This perturbation may also have been due to a local impact or to a larger one on a planetary scale: its age corresponds precisely to that of most of the oldest Martian SNC volcanic rocks to which we have access (nakhlite meteorites).

3.6. Witnesses to the formation and evolution of the Solar System

Meteorites are probes through space and time. Like a palimpsest, each can contain the partially obliterated traces of successive episodes of the internal evolution of planetary bodies. This can make reading them difficult – beyond the technical problems – but also very rich, and for this reading, we rely on a comparison with observations and astronomical models. But we rely above all on the comparison of meteorites that came from parent bodies which underwent different evolutions, for example, which stopped at different evolutionary stages, in the same way we compare the stars in the Herzsprung-Russell diagram.

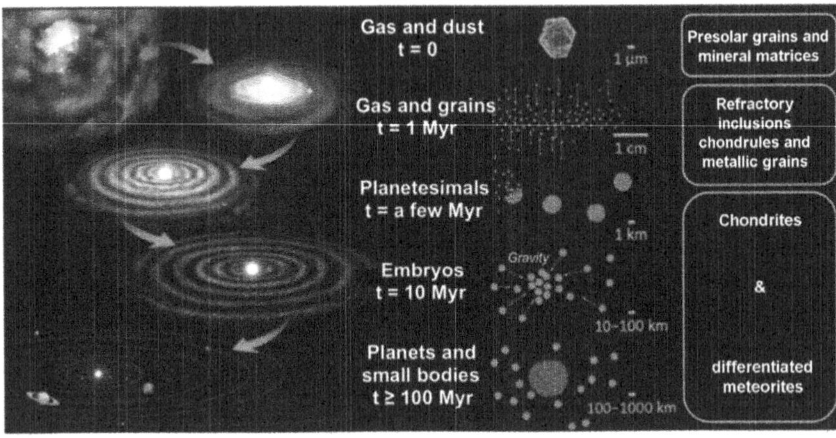

Figure 3.19. *The main steps of the formation of the Solar System from an interstellar gas and dust cloud that collapsed in on itself. For a color version of this figure, see www.iste.co.uk/encrenaz/solar2.zip*

COMMENTS ON FIGURE 3.19. – *The sizes of the solids present in the cloud and then in the disk at each step are indicated. The column on the right lists objects that, in meteorites, carry information from different steps. This view is highly simplified, since in reality we don't know if the genesis of chondrules initiated the growth of the first planetary bodies or if it was, on the contrary, the result of it. In the same way, contrary to what we thought twenty or so years ago, differentiation precedes the genesis of the parent bodies for the chondrites to which we have access. It is possible that Vesta and the other parent asteroids of the differentiated meteorites had a chondritic chemistry and texture at the beginning, but it is impossible to prove it for now. It is nevertheless clear that the chondrites that we find today on Earth come from bodies that accreted late (at a certain distance from the Sun), having accumulated an insufficient quantity of* ^{26}Al *for the heat released by its disintegration to provoke fusion.*

Figure 3.19 recalls the main steps of the formation of the Sun and its planets from an interstellar gas and dust cloud inherited from the interstellar medium that collapsed in on itself. Most of these steps have left their marks on one or more types of meteorites:

– in the matrix, we find grains inherited from the interstellar medium (Bernatowicz *et al.* 1987), including some that condensed in the vicinity of stars that disappeared before the formation of the Sun, as shown by their isotopic compositions;

– the refractory inclusions are considered to be the first solids condensed in the Solar System, in an area that began to cool after all the solids had vaporized – this was traditionally thought to be in the vicinity of the Sun (Grossman 1972) but, to explain the separation between carbonaceous and non-carbonaceous objects (see Figure 3.14), some authors have proposed that the CAI formed beyond Jupiter (Kleine *et al.* 2020), unless they were transported there by winds before accretion (Krot *et al.* 2019);

– chondrules and grains of metal may have been the first millimeter-sized objects formed in the disk after[41] the refractory inclusions and/or farther from the Sun. It is not impossible that they formed through different mechanisms in different groups of chondrites (the sphere being the natural shape of all grains or aggregate of grains that melted in a state of microgravity) and we still don't know if they were the initiator or the consequence of the formation of the first planetary bodies;

– chondrites inform us about the nature and processes that took place on the surface and inside primitive bodies;

– differentiated meteorites are a source of information about the internal structure of planets, including the Earth, and about the geological processes that governed their evolution.

3.7. References

Agee, C.B., Wilson, N.V., McCubbin, F.M. *et al.* (2013). Unique meteorite from early Amazonian Mars: Water-rich basaltic breccia Northwest Africa 7034. *Science*, 339, 780–785.

Alexander, C.M.O'D. (2003). Re-examining the role of chondrules in producing the elemental fractionations in chondrites. *Meteoritics & Planetary Science*, 40, 943–965.

Alexander, C.M.O'D., Nittler, L.R., Davidson, J. *et al.* (2017). Measuring the level of interstellar inheritance in the solar protoplanetary disk. *Meteoritics and Planetary Science*, 52, 1797–1821.

41 Almost all refractory inclusions contain an excess of ^{26}Mg that came from the *in situ* decay of ^{26}Al. This suggests (Russell *et al.* 1996) that they formed over a relatively short time interval, since the half-life of ^{26}Al is only 700,000 years. On the other hand, the formation of chondrules (and grains of metal) began at about the same time, but is apparently due to a repetitive process (Krot *et al.* 2009; Connelly *et al.* 2012; Pape *et al.* 2019) that occurred over a few million years (Bizzarro *et al.* 2004; Connelly *et al.* 2017). The precursor materials of chondrules are not well known. They were less refractory than material in inclusions since they contain the major elements Si, Mg, and Fe.

Barrat, J.-A., Yamaguchi, A., Zanda, B. *et al.* (2010). Relative chronology of crust formation on asteroid Vesta: Insights from the geochemistry of diogenites. *Geochimica and Cosmochimica Acta*, 74, 6218–6231.

Barrat, J.-A., Yamaguchi, A., Bunch, T.E. *et al.* (2011). Possible fluid-rock interactions on differentiated asteroids recorded in eucritic meteorites. *Geochimica and Cosmochimica Acta*, 75, 3939–3852.

Bellucci, J.J., Nemchin, A.A., Whitehouse, M.J. *et al.* (2015). Pb isotopic evidence for an early, enriched, crust on Mars. *Earth and Planetary Science Letters*, 410, 34–41.

Bernatowicz, T., Fraundorf, G., Ming, T. *et al.* (1987). Evidence for interstellar SiC in the Murray carbonaceous meteorite. *Nature*, 330, 24–31.

Bevan, A.W.R. and Axon, H.J. (1980). Metallography and thermal history of the Tieschitz unequilibrated meteorite – Metallic chondrules and the origin of polycrystalline taenite. *Earth and Planetary Science Letters*, 47, 353–360.

Bizzarro, M., Baker, J.A., Haack, H. (2004). Mg isotope evidence for contemporaneous formation of chondrules and refractory elements. *Nature*, 431, 275–278.

Bogard, D.D. and Johnson, P. (1983). Martian gases in an Antarctic meteorite? *Geophysical Research Letters*, 10, 801–803.

Borovicka, J., Spurny, P., Brown, P. *et al.* (2013). The trajectory, structure, and origin of the Chelyabinsk asteroidal impactor. *Nature*, 503, 235–237.

Bouley, S., Baratoux, D., Matsuyama, I., *et al.* (2016). Late Tharsis formation and implication for early Mars. *Nature*, 531, 344–347.

Bouley, S., Keane, J.T., Baratoux, D. *et al.* (2020). A thick crustal block revealed by reconstructions of early Mars highlands. *Nature Geoscience*, 13, 105–109.

Brown, P.G., Assink, J.D., Astiz, L. *et al.* (2013). A 500-kiloton airburst over Chelyabinsk and an enhanced hazard from small impactors. *Nature*, 503, 238–241.

Burke, J.G. (1986). *Cosmic Debris: Meteorites in History*. University of California Press, Berkeley.

Campbell, A.J., Zanda, B., Perron, C. *et al.* (2005). Origin and thermal history of Fe-Ni metal in primitive chondrites. *Chondrites and Protoplanetary Disk, ASP Conferences Series*, 341, 407.

Chapman, C.R. and Salisbury, J.W. (1973). Comparison of meteorite and asteroid spectral reflectivities. *Icarus*, 19, 507–522.

Chladni, E.F.F. (1794). *Über den Ursprung der von Pallas gefundenen und anderer ihr ähnlichen Eisenmassen und über einige damit in Verbindung stehende Naturerscheinungen*. Johann Friedrich Hartknoch, Weimar.

Clayton, R.N. (2003). Oxygen isotopes in the Solar system. *Space Science Reviews*, 106, 19–32.

Clayton, R.N., Grossman, L., Mayeda, T.K. (1973). A component of primitive nuclear composition in carbonaceous meteorites. *Science*, 182, 485–488.

Connelly, J.N., Bizzarro, M., Krot, A.N. *et al.* (2012). The absolute chronology and thermal processing of solids in the protoplanetary disk. *Science*, 338, 651–655.

Connelly, J.N., Bollard, J.D., Bizzarro, M. (2017). Pb-Pb chronometry and the early Solar system. *Geochimica and Cosmochimica Acta*, 201, 345–363.

Connors, M., Hildebrand, A.R., Pilkington, M. *et al.* (1996). Yukatan karst figures and the size of Chicxulub crater. *Geophysical Journal*, 127, F11–F14.

Desch, S.J., Morris, M.A., Connolly Jr., H.C. *et al.* (2012). The importance of experiments: Constraints on chondrule formation models. *Meteoritics & Planetary Science*, 47, 1139–1156.

Duprat, J., Dobrica, E., Engrand, C. *et al.* (2010). Extreme deuterium excesses in ultracarbonaceous micrometeorites from central Antarctic snow. *Science*, 328, 742–745.

Elkins-Tanton, L.T. (2012). Magma oceans in the inner solar system. *Annual Review of Earth and Planetary Sciences*, 40, 113–139.

Eugster, O. (2003). Cosmic-ray exposure ages of meteorites and lunar rocks and their significance. *Chemie der Erde Geochemistry*, 63, 3–30.

Goldstein, J.I. and Ogilvie, R.E. (1965). The growth of the Widmanstätten pattern in metallic meteorites. *Geochimica and Cosmochimica Acta*, 29, 893–920.

Goldstein, J.I. and Short, J.M. (1967). The iron meteorites, their thermal history and parent bodies. *Geochimica and Cosmochimica Acta*, 31, 1733–1770.

Grossman, L. (1972). Condensation in the primitive solar nebula. *Geochimica and Cosmochimica Acta*, 36, 597–619.

Hartmann, W.K. and Davis, D.R. (1975). Satellite-sized planetesimals and lunar origin. *Icarus*, 24, 504–515.

Hewins, R.H., Zanda, B., Humayun, M. *et al.* (2017). Regolith breccia Northwest Africa 7533: Mineralogy and petrology with implications for early Mars. *Meteoritics & Planetary Science*, 52, 89–124.

Humayun, M., Nemchin, A., Zanda, B. *et al.* (2013). Origin and age of the earliest Martian crust from meteorite NWA 7533. *Nature*, 503, 513–516.

Keil, K. and McCoy, T.J. (2018). Acapulcoite-lodranite meteorites: Ultramafic asteroidal partial melt residues. *Chemie des Erde*, 78, 153–203.

Kleine, T., Budde, G., Burkhardt, C. *et al.* (2020). The non-carbonaceous–carbonaceous meteorite dichotomy. *Space Science Reviews*, 216, id.55.

Krot, A.N., Amelin, Y., Bland, P. *et al.* (2009). Origin and chronology of chondritic components: A review. *Geochimica and Cosmochimica Acta*, 73, 4963–4997.

Krot, A.N., Ma, C., Nagashima, K. *et al.* (2019). Mineralogy, petrography, and oxygen isotopic compositions of ultrarefractory inclusions from carbonaceous chondrites. *Geochemistry*, 79, 125–519.

Lapen, T.J., Righter, M., Brandon, A.D. *et al.* (2010). A younger age for ALH84001 and its geochemical link to Shergottite source in Mars. *Science*, 328, 347–351.

Lapen, T.J., Righter, M., Andreasen, R. *et al.* (2017). Two billion years of magmatism recorded from a single Mars meteorite ejection site. *Science Advances*, 3(2), e1 600–922.

Lee, T., Papanastassiou, D.A., Wasserburg, G.J. (1976). Demonstration of ^{26}Mg excess in Allende and evidence for ^{26}Al. *Geophysical Research Letters*, 3, 109–112.

Lewis, R., Bright, D., Steel, E. *et al.* (1987). Presolar diamonds (Cd) in carbonaceous chondrites: Size distribution. *Meteoritics*, 22, 445.

Lodders, K. and Fegley, B. (1998). Presolar SiC grains and their parent stars. *Meteoritic & Planetary Science*, 33, 871–880.

Lorand, J.-P., Hewins, R.H., Humayun, M. *et al.* (2018). Chalcophile-siderophile element systematics of hydrothermal pyrite from Martian regolith breccia NWA 7533. *Geochimica and Cosmochimica Acta*, 241, 134–149.

Marvin, U.B. (1996). History of planetary geoscience. *Encyclopedia of Earth Sciences*, Volume 2. MacMillan Publishing Company, New York.

McCord, T.B., Adams, J.B., Johnson, T.V. (1970). Asteroid Vesta: Spectral reflectivity and compositional implications. *Science*, 168, 1445–1447.

McKay, D.S., Gibson Jr., E.K., Thomas-Kerpta, K.L. *et al.* (1996). Possible relic biogenic activity in Martian meteorite ALH84001. *Science*, 273, 924–929.

McKay, D.S., Thomas-Kepra, K.L., Clernett, S.J. *et al.* (2009). Life on Mars: New evidence from Martian meteorites. *Proceedings of the SPIE*, 7441.

McSween Jr., H.Y. (1994). What we have learned about Mars rom SNC meteorites? *Meteoritics*, 29, 757–779.

McSween Jr., H.Y. (1999). *Meteorites and their Parent Planets.* Cambridge University Press, Cambridge.

Mittlefehldt, D.W. (1994). ALH84001, a cumulate orthopyroxenite member of the Martian meteorite clan. *Meteoritics*, 29, 224–221.

Nittler, L.R. and Ciesla, F. (2016). Astrophysics within extraterrestrial materials. *Annual Reviews of Astronomy & Astrophysics*, 54, 53–93.

Nyquist, L.E., Bogard, D.D., Shih, C.-Y. *et al.* (2001). Ages and geologic histories of Martian meteorites. *Space Science Reviews*, 96, 105–164.

Palme, H. (2001). Chemical and isotopic heterogeneity in protosolar matter. *Philosophical Transactions of the Royal Society of London*, 359, 2061–2075.

Pape, J., Mezger, K., Bouvier, A.-S. *et al.* (2019). Time and duration of chondrule formation: Constraints from ^{26}Al-^{26}Mg ages of individual chondrules. *Geochimica and Cosmochimica Acta*, 244, 416–436.

Rojas, J., Duprat, J., Engrand, C. *et al.* (2019). Micrometeorite mass measurements at Dome C, Antarctica. *50th Lunar and Planetary Science Conference*. Texas.

Rubin, A.E. and Mittelfehldt, D.W. (1992). Mesosiderites: A chronologic and petrologic synthesis. *Meteoritics*, 27, 282.

Sautter, V., Toplis, M.J., Beck, P. *et al.* (2016). Magmatic complexity on early Mars as seen through a combination of orbital, in-situ and meteorite data. *Lithos*, 254–255, 36–52.

Schulte, P., Alegret, L., Arenillas, I. *et al.* (2010). The Chicxulub asteroid impact and mass extinction at the Cretaceous-Paleogene boundary. *Science*, 237, 1214–1218.

Sears, D.W.G. and Dodd, R.T. (1988). Overview and classification of meteorites. In Kerridge, J.F. and Matthews, M.S. (eds). *Meteorites and the Early Solar System*. University of Arizona Press, Tucson.

Shoemaker, E.M. (1959) Impact mechanics at Meteor Crater, Arizona. PhD thesis, Princeton University, Princeton.

Treiman, A.H. (2003). Submicron magnetite grains and carbon compounds in Martian meteorite ALH84001: Inorganic, abiotic formation by shock and thermal metamorphism. *Astrobiology*, 3, 369–392.

Trinquier, A., Birk, J.-L., Allegre, C.J. (2006). The nature of the KT impactor. *Earth and Planetary Science Letters*, 241, 780–788.

Villeneuve, J., Libourel, G., Soulie, C. (2015). Relationship between type I and type II chondrules: Implications on chondrule formation processes. *Geochimica and Cosmochimica Acta*, 160, 277–305.

Voshage, H. (1984). Investigations of cosmic-ray produced nuclides in iron meteorites, 6. The Signer-Nier model and the history of the cosmic radiation. *Earth and Planetary Science Letters*, 71, 181–194.

Walker, D., Stolper, E.M., Hays, J.F. (1979). Basaltic volcanism: The importance of planet size. *10th Lunar and Planetary Science Conference*. Houston, Proceedings, Volume 2. Pergamon Press, New York.

Warren, P.H. (2011). Stable-isotopic anomalies and the accretionary assemblage of the Earth and Mars: A subordinate role for carbonaceous chondrites. *Earth and Planetary Science Letters*, 311, 93–100.

Weiss, B.P. and Elkins-Tanton, L.T. (2013). Differentiated planetesimals and the parent bodies of chondrites. *Annual Reviews of Earth and Planetary Sciences*, 41, 529–560.

Wittmann, A., Korotev, R.L., Jolliff, B.L. *et al.* (2015). Petrography and composition of Martian regolith breccia meteorite Northwest Africa 7475. *Meteoritics & Planetary Science*, 50(Nr 2), 326–352.

Wood, J.A., Dickey, J.S., Marvin, U.B. *et al.* (1970). Lunar anorthosites. *Science*, 167, 602–604.

Yang, J., Goldstein, J., Michael, J.R. *et al.* (2010). Main-group pallasites: Thermal history, relationship to IIIAB iron, and origin. *Geochimica and Cosmochimica Acta*, 74, 4493–4506.

Young, E.D. and Russell, S.S. (1998). Oxygen reservoirs in the early Solar Nebula inferred from an Allende CAI. *Science*, 282, 452–455.

Yurimoto, H. and Kuramoto, K. (2004). Molecular cloud origin or the oxygen isotope heterogeneity in the Solar system. *Science*, 305, 1763–1766.

Zanda, B. (2013). Les Météorites 2/Des météorites primitives au Système Solaire. *Cahiers du Règne Minéral*, 3.

Zanda, B. and Jambon, A. (2012). Météorites 1/Des météorites différenciées aux planètes. *Cahiers du Règne Minéral*, 1.

Zanda, B.B., Bland, P.A., Le Guillou, C. *et al.* (2009). Volatile element distribution in matrix and chondrules of carbonaceous and ordinary chondrites. *40th Lunar and Planetary Science Conference*. The Woodlands Waterway Marriott Hotel and Convention Center, The Woodlands.

Zanda, B., Humayun, M., Lewin, E. *et al.* (2012). Chemical composition of matrix and chondrules in carbonaceous chondrites: Implications for disk transport. *43rd Lunar and Planetary Science Conference*. The Woodlands Waterway Marriott Hotel and Convention Center, The Woodlands.

Zanda, B., Humayun, M., Lewin, E. *et al.* (2018). Mo-W isotopic evidence against chondrule-matrix complementarity. *81th Annual Meeting of the Meteoretical Society*. Moscow.

Zolensky, M., Barrett, R., Browning, L. (1993). Mineralogy and composition of matrix and chondrule rims in carbonaceous chondrites. *Geochimica and Cosmochimica Acta*, 57, 3123–3148.

4

Formation and Dynamic History of the Solar System[1]

Françoise ROQUES
LESIA, Paris Observatory, PSL University, Paris, France

4.1. Introduction

To the naked eye, the planets are differentiated from the stars by their motion – they are *wandering stars* against a background of fixed stars. This motion occupies a single region in the night sky (see Figure 4.1). This region of the sky, the ecliptic, is occupied during the day by the Sun's progress. The discovery that the planets orbit the Sun in almost the same plane and in the same direction gave rise to the idea, as early as the 18th century, that the planets formed in a circumsolar disk. Observational instruments would confirm and shed light on this intuition.

Today, we observe thousands of exoplanets that orbit stars other than the Sun, and their wide diversity raises questions about the place and originality of our Earth.

The bodies in the Solar System follow quasi-elliptical orbits around the Sun or around a planet, and the steps in their formation are fairly well understood (see sections 1.3.2.1 and 2.4–2.6, Volume 1). The scenario developed to explain the formation of the planets in the Solar System is today being tested by the diversity of exoplanetary systems.

1 For general works on this chapter's topic, see (Murray and Dermott 2000; Lissauer and de Pater 2013; Armitage 2017).

The Solar System 2,
coordinated by Thérèse ENCRENAZ and James LEQUEUX. © ISTE Ltd 2021.

Figure 4.1. *Alignment of the planets and the Moon in mid-April 2020, observed from the edge of the Black Sea in Bulgaria (source: Apod/NASA/Mihail Minkov). For a color version of this figure, see www.iste.co.uk/encrenaz/solar2.zip*

In this chapter, the fundamental concepts of the planetary dynamics are reviewed before a description of the major steps of the process of planetary formation and of the questions posed by the formation of exoplanets.

The motion of the bodies in the Solar System is essentially governed by the laws of Newtonian dynamics (section 4.2) applied to the gravitational force. The two-body problem (section 4.3) makes it possible to understand the main orbital properties of planets. To go further, we will consider the three-body problem, which must be simplified to obtain useful results (section 4.4): we consider that the third body is of negligible mass (in the case of satellites) or that the first is much more massive than the two others (in the Solar System, Jupiter and Saturn).

The orbits of a number of planets and satellites are linked through resonances. This mechanism plays an important role by locking orbits or, conversely, by making them unstable (section 4.5). Some orbital configurations are chaotic and play a part in the long-term stability of planetary systems (section 4.6). The planets and the satellites are not isolated, and two important mechanisms occur: tidal effects can seriously perturb the satellite or the planet (section 4.7), and the behavior of satellites is modified by deformations of the planet, whose flattening is due to its rotation (section 4.8).

Besides the large bodies of the planets and satellites, the population of dust exhibits behavior marked by several nongravitational forces (section 4.9).

4.2. Laws of motion of the planets and satellites

The first steps on the path to understanding planetary motion consisted of making precise observations, then finding regularity in these observations. This regularity was found by Kepler, whose three laws (see section 1.2.2, Volume 1) describe the shape of planetary orbits and the velocity of planets along these orbits. Newton would later understand that this motion is due to gravitational force. He was also able to connect a force to the motion created by this force and to show that Kepler's laws represent the motion created by gravitation. This path led to a revolution that changed our vision of the world forever: the same force that causes an apple to fall makes the planets turn.

4.2.1. *Kepler's laws*

At the beginning of the 18th century, Johannes Kepler deduced, from Tycho Brahe's astronomical observations, the laws that bear his name and that describe planetary motion. He found these equations empirically by advancing the hypothesis of a "magnetic vortex" to explain the origin of this motion. We encounter these laws today when we apply Newton's laws of dynamics to the gravitational force applied to two bodies (see section 4.2.2). Kepler's laws correctly describe planetary motion around the Sun because the planets have a negligible mass in comparison to the Sun.

FIRST LAW.– *The trajectory of a planet is an ellipse with the center of the Sun at one of its foci.*

An ellipse is a curve in which, for all points, the sum of the distances to two fixed points (the foci) is constant (Figure 4.2).

The motion of a planet could be described in the following manner:

$$r = \frac{a(1-e^2)}{(1+e\cos f)} \quad [4.1]$$

a statement in which r is the distance of the planet from the center of the Sun, a is the semi-major axis of the ellipse, and e is its eccentricity. f, the true anomaly, is the angle measured in relation to the point of the shortest heliocentric distance, called perihelion (periapsis). The point of the greatest heliocentric distance is called aphelion (apoapsis in the general case).

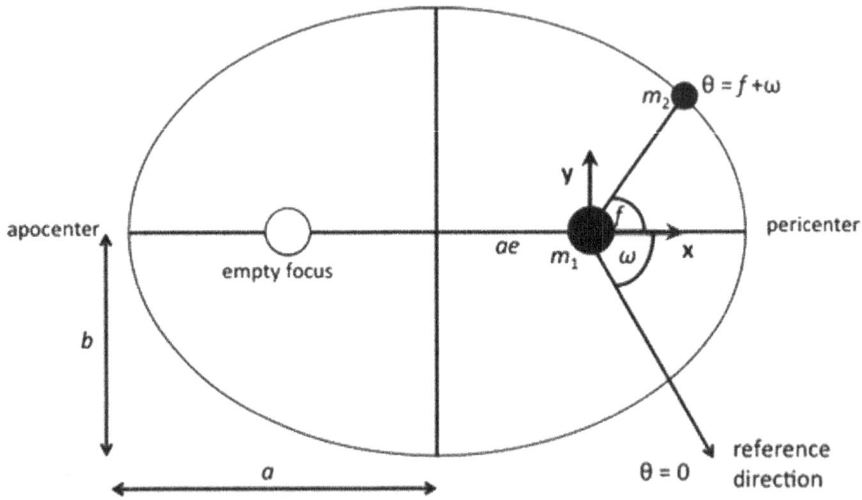

Figure 4.2. *Parameters of the ellipse (source: based on (Shields* et al. *2016))*

SECOND LAW.– *While a planet is in motion, the radius vector joining the center of the Sun and the planet describes equal areas during equal amounts of time. The planet slows as it moves away from the Sun and accelerates when it moves toward the Sun.*

THIRD LAW.– *The ratio of the cube of the semi-major axis of the orbit of a planet to the square of its orbital period around the Sun is the same for all of the planets:*

$$a^3/P^2 = C \qquad [4.2]$$

where P is the orbital period.

If the semi-major axis is expressed in astronomical units (au) and P is expressed in terrestrial years, the constant C is equal to 1. Put another way, for the planets in the Solar System, the period of a planet whose semi-major axis is a au is equal to $a^{3/2}$ years.

Newton's law of universal gravitation would later show that this constant C is proportional to the mass of the (star + planet) system (see section 4.2.3).

4.2.2. *Gravity*

Kepler's laws describe planetary motions, in other words their position and their velocity, but do not explain the cause of this motion. In studying these formulas, Isaac Newton in 1687 discovered gravitational attraction: between two bodies separated by a distance d, there is a force of reciprocal attraction proportional to the mass of the bodies, m_1 and m_2, and to the inverse of the square of the distance between them. Its force is:

$$F = G\, m_1 m_2 / d^2 \qquad [4.3]$$

G is the universal gravitational constant. In the system of units in which distance, time, and mass are expressed in meters, seconds, and kilograms respectively, G has the value $6{,}672\ 10^{-11}\ m^3\, kg^{-1}\, s^{-2}$.

Each of the two bodies is subjected to an equal and opposite force F. If N bodies are present, each will evolve under the action of the gravity from N-1 other bodies.

Newton's law was generalized by Einstein's theory of relativity, which must be applied in the vicinity of very large masses. In the Solar System, the calculation of Mercury's orbit requires taking relativity into account because of the proximity of the Sun.

Gravity is essentially an "astronomical" force. On a human scale, the gravitational attraction of Earth dominates, and the attraction of bodies other than Earth is practically undetectable. Ocean tides, the familiar manifestation of the gravitational attraction of the Moon and the Sun on the Earth, are highly variable from one location to another on Earth because of the propagation of waves in the oceans, which have complex shapes and variable depths.

4.2.3. *Newton's fundamental laws of dynamics*

Newton also formulated the fundamental laws of motion. These laws form the basis of classical mechanics and connect the force applied to a body to the motion of that body.

NEWTON'S FIRST LAW.– *An isolated body, of mass M, maintains a constant velocity* $\mathbf{v}$.

NEWTON'S SECOND LAW.– *If a mass M is subjected to a force* $\mathbf{F}$, *its velocity changes from* $\mathbf{v}$ *to* $\mathbf{v} + d\mathbf{v}$, *such that:* $d\mathbf{v} = \mathbf{F}/M$.

The force modifies the acceleration of the body, and this acceleration is equal to the force divided by the mass of the body.

This law allowed Newton to show that Kepler's laws are a direct consequence of the gravitational force as applied to the system made up of the Sun and a planet whose mass is negligible compared to the Sun. This is the case since the Sun's mass corresponds to 99.8% of the mass of the entire Solar System.

Kepler's third law says that the period of motion of a planet of mass m and a star of mass M around their center of gravity is P, such that:

$$P^2 = \frac{4\pi a^3}{G(m+M)} \quad [4.4]$$

NEWTON'S THIRD LAW.– *When a solid A exerts on a solid B a force* $\mathbf{F}_{A\to B}$, *then the solid B exerts on A the force* $\mathbf{F}_{B\to A}$. *Whether the solids are at rest or in motion, the forces* $\mathbf{F}_{A\to B}$ *and* $\mathbf{F}_{B\to A}$ *are opposed:* $\mathbf{F}_{B\to A} = -\mathbf{F}_{A\to B}$.

Thus, the Earth's gravity attracts the Moon with a force $\mathbf{F}_{E\to M}$ and the Moon attracts the Earth with a force $\mathbf{F}_{M\to E}$, such that: $\mathbf{F}_{M\to E} = -\mathbf{F}_{E\to M}$ (see Figure 4.3).

According to the law of gravity, the magnitude of this force is equal to: $F_{E-M} = F_{M\to E} = \frac{G m_E m_M}{{d_{ME}}^2}$ where m_E and m_M are the mass of the Earth and the Moon and d_{ME} is the distance between Earth and the Moon.

Figure 4.3. *Gravitational forces between Earth and the Moon. For a color version of this figure, see www.iste.co.uk/encrenaz/solar2.zip*

4.2.4. *The orbital elements*

To define the orbital parameters of a body, a reference system is required (X, Y, Z in Figure 4.4). The orbit of a planet is related to a reference plane, the ecliptic, which corresponds to Earth's orbit around the Sun. In the case of satellites, the reference plane is generally the planet's equatorial plane.

The trajectory of the object intersects the reference plan at two points, the ascending node and the descending node. The intersection of the orbital plane and the reference plane is the node line.

The orbit of a planet is determined by five elements (Figure 4.4):

– the semi-major axis a;

– the eccentricity e of the ellipse;

– the inclination i of the orbital plane with respect to the reference plane;

– the longitude Ω of the ascending node (where the orbit intersects the reference plane from south to north), measured from the direction of the vernal point. This is defined by the intersection of the planes of the ecliptic and the terrestrial equator on a fixed date called J2000, corresponding to January 1, 2000 at noon. The vernal point corresponds to the apparent position of the Sun on the vernal equinox;

– the argument ω of the perihelion (angle of the direction of perihelion with the node line).

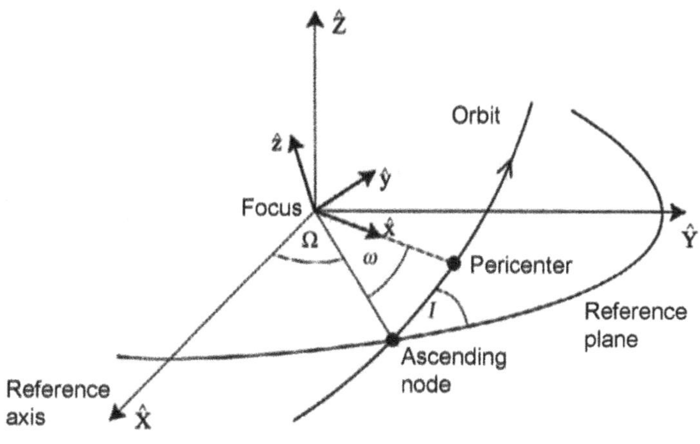

Figure 4.4. *Orbital coordinates in three-dimensional space. Z indicates north (source: based on (Murray and Dermott 2000))*

Furthermore, the position of a point on the orbit also depends on the time t_0 of its passage at perihelion. At a time t, the position of a point is given by the true anomaly, v, an angle measured in the orbital plane from perihelion.

For exoplanets, the reference plane is the plane perpendicular to the target line between Earth and the observed system. The word inclination refers to the angle between the orbital plane and this reference plane. The angle between the orbital plane and the perpendicular of the star's axis of rotation is thus called the spin-orbit angle. The direction of the star's axis of rotation is only known for a small number of systems.

Because of mutual interactions between bodies, planetary motions are not really Keplerian, and the parameters of orbits are not constant over time. An important perturbation is the precession of the nodes: the orbit keeps its shape, but its perpendicular axis pivots in the opposite direction from the motion of the body in orbit (top motion). In the case of planets, these variations are very slow: the node line sweeps the reference plane (ecliptic) in several dozen million years. Another perturbation is the advance of the perihelion: the ellipse turns while remaining in the same plane. The advance of the perihelion of Mercury's orbit is 531"/century. The rate calculated from gravitational equations is 488"/century, a difference discovered by Le Verrier. The equations of relativity, which generalize Newton's equation, are necessary to explain this excess of 43"/century. In the case of comets, the perturbations caused by the proximity of planets can be significant.

4.3. The two-body problem

The two-body problem concerns two bodies subject only to their reciprocal gravitational attraction. This problem is integrable, which means that the bodies' orbits can be calculated exactly.

Each of these two bodies, M_1 and M_2, with mass m_1 and m_2 respectively, is subject to the other body's gravitational attraction. This force has the equal intensity for both bodies:

$$F = G\frac{m_1 m_2}{d^2} \qquad [4.5]$$

and the force is directed toward the other body.

The first step in the problem is to separate the motion of the two bodies' center of gravity and the motion of these two bodies in relation to this center of inertia. As the two bodies are isolated, the motion of the center of inertia is straight and uniform.

Then, we can replace the motion of the two particles in relation to the center of inertia with the motion of a single (fictional) particle of mass $\mu = m_1 m_2/(m_1+m_2)$ subject to a force F.

The solution is an ellipse, a parabola, or a hyperbola (see Figure 4.5), depending on its energy.

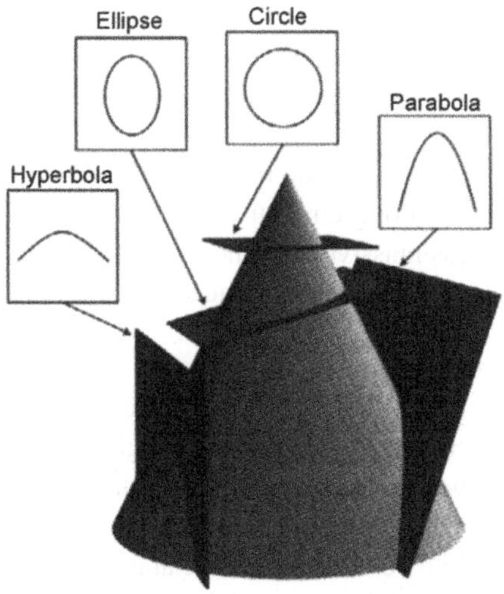

Figure 4.5. *Conic sections (source: based on (Murray and Dermott 2000))*

The reduction of the two-body problem and the solutions to the two-body problem are in (Murray and Dermott 2000).

4.4. The three-body problem

Unlike the two-body problem, the problem posed by a three-body gravitational system is extremely complex in the general case, and its solutions are far from being understood (Chenciner 2007). When there are large differences between the masses of the bodies or when the orbits are almost circular or coplanar, which is most often the case in the Solar System, we can make approximations that simplify the problem and allow for analytical solutions.

When one of the three bodies has a negligible mass in relation to the others (the Sun, a planet, and a small asteroid, a ring particle, an artificial satellite), its effects on the others may be negligible. The simplified problem is known as the *restricted three-body problem*. The small body is called the "test particle". If the two massive bodies also travel along circular orbits, we speak of the *circular restricted three-body problem*.

When one of the three bodies is much more massive than the other two, we speak of the *Hill problem*. Independent of mass, we can consider the case in which the orbits all occur in the same plane, which is the *three-body problem in a plane*.

4.4.1. *Jacobi constant and Lagrange points*

Now we will consider the circular restricted three-body problem. Two bodies with mass m_1 and m_2 travel along circular orbits in relation to their center of mass. The third body is much less massive than the less massive of the other two. The test particle may or may not be in the orbital plane of the two other bodies.

The two massive bodies revolve around O, the two bodies' center of mass. We are in a reference plane (O, x, y) that revolves around the axis Oz at the angular velocity of the first two bodies. This velocity is given by Kepler's third law and depends on the mass of the two bodies and their distance from each other. If we use a system of units in which the distance between the two bodies, the sum of their masses, and the gravitational constant are 1, the angular frequency of the two bodies is 1. We refer to the velocity of the test particle as v.

The first two bodies are fixed on the x axis and their positions are: $\boldsymbol{r_1} = (-\frac{m_2}{m_1+m_2}, 0)$ and: $\boldsymbol{r_2} = (\frac{m_1}{m_1+m_2}, 0)$. The 3rd point is found in $\boldsymbol{r}$. Thus, $|\boldsymbol{r} - \boldsymbol{r_i}\,|$ is the distance between the body i and the test particle. Therefore:

$$C_J = x^2 + y^2 + \frac{2m_1}{|r-r_1|} + \frac{2m_2}{|r-r_2|} - v^2 \qquad [4.6]$$

C_J is a constant of motion, the Jacobi integral. The first two terms are respectively twice the potential centrifugal energy, the 3rd and 4th terms are twice the potential gravitational energy, and the 5th term is twice the kinetic energy. The Jacobi constant is the only integral of motion in the circular restricted three-body problem. It does not provide exact solutions, but it can be used to determine the areas where the particle cannot be.

Thus, we can consider the locations where the velocity of the particle is null. In this case, each value of C_J defines the surfaces (curves in the plane problem) that delimit the areas where the test particle cannot go.

The circular restricted three-body problem has five points of equilibrium, the Lagrange points, L_1, L_2… L_5 (see Figure 4.6). They lie in the orbital plane of the two massive bodies, and the points L_1, L_2 and L_3 are on the line connecting the two massive bodies. The two latter points, L_4 and L_5, each form a triangle with the two massive bodies (see also section 1.2.4, Volume 1).

4.4.2. *Tadpole and horseshoe orbits*

The L_1, L_2 and L_3 Lagrange points are unstable, which means that a test particle placed near one of these points will move away. The L_1 and L_2 points for Earth make it possible to host artificial satellites such as the Gaia astronomical satellite.[2]

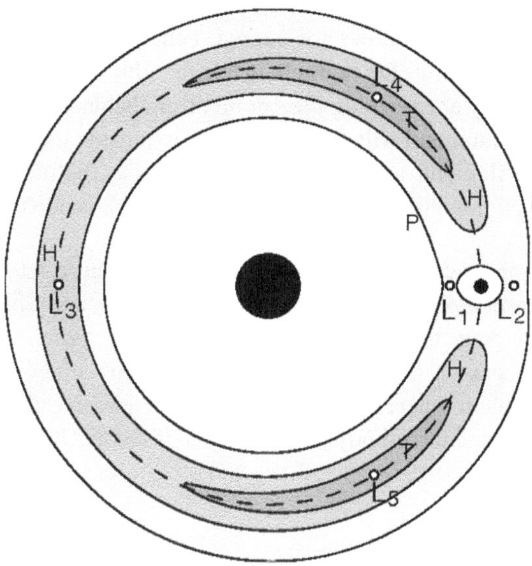

Figure 4.6. *Circular restrained three-body problem (source: based on (Marzari* et al. *2002))*

2 See also: http://www.esa.int/Science_Exploration/Space_Science/Gaia/Gaia_enters_its_operational_orbit.

COMMENTS ON FIGURE 4.6.– *The two bodies are the black disks, and the ratio of the masses $m_1/m_2 = 100$. L_1... L_5 are the Lagrange equilibrium points, and the curves correspond to a null velocity for the three values of the Jacobi constant. The white area around the second body is the Hill sphere (see section 4.4.3) for the secondary body. The dotted line is the circle with a radius equal to the semi-major axis of the orbit of the secondary body. The letters, T, H, and P indicate the areas of tadpole, horseshoe, and passing orbits for the test particle.*

The L_4 and L_5 points are stable equilibrium points as soon as $m_1/m_2 > 25$, which is the case for all known configurations in the Solar System more massive than the Pluto-Charon system. If a test particle is located near one of these positions, it will oscillate around this position along a *tadpole orbit*. The L_4 and L_5 points are important in the Solar System. The Trojan asteroids occupy Jupiter's Lagrange points (see section 3.2.3). A dozen or so bodies oscillate around Neptune's Lagrange points, and Mars also has Trojan bodies. Saturn's satellites Thetis and Dione also have small bodies trapped at their Lagrange points (see section 1.3.3.1).

The three-body problem allows for strange behaviors. If a test particle has an orbit near the satellite's orbit, for example inside its orbit, it will move more quickly and catch up with it. By approaching the satellite from behind, it will exchange some of its angular momentum and will find itself slightly outside its orbit. It will then be slower than the satellite and will fall behind, move away, before approaching from the front. Its interaction with the satellite will bring it back inside the satellite's orbit.

In a reference plane connected to the satellite (see Figure 4.6), the test particle will follow a *horseshoe orbit*. Saturn's two satellites Janus and Epimetheus perform this dance, exchanging their orbits about every four years. Since Janus and Epimetheus have masses on the same order, their study reveals the Hill approximation (problem where a body is much more massive than the two others) rather than the restrained three-body problem, even if the mechanism is the same.

4.4.3. *Hill sphere*

The area of gravitational influence of a planet on its satellite is its Hill sphere, whose radius is:

$$R_H = \left(\frac{m_2}{3(m_1+m_2)}\right)^{1/3} a \qquad [4.7]$$

where m_2 is the mass of the secondary body, m_1 the mass of the primary body, and a is the semi-major axis of the orbit. R_H is in fact the distance of the satellite to its L_1 or L_2 point.

The planetocentric orbits are stable over long periods if they are inside the planet's Hill sphere; natural satellites are inside the Hill sphere. For their part, stable heliocentric orbits are all far outside the planets' Hill spheres. If a body enters a planet's Hill sphere, a comet, for example, it can remain there for a period of time. We then speak of a *temporary satellite*.

4.5. Perturbations and resonances

A resonance appears between two bodies if some of their orbital parameters, mean-motion, precession of the node line, or motion of perihelion change in a commensurable way. Today, we know of a hundred or so satellites, more than 500 comets, and 30,000 asteroids and Kuiper objects, plus the rings. All of these objects show us that the Solar System is highly structured and that all of these bodies obey very complex laws, essentially based on Newtonian gravity and the very subtle effect of resonances.

The *mean-motion resonances* (see also section 1.2.3, Volume 1) are among the most important. They connect bodies whose orbital periods are commensurable (see Figure 4.7). The mean-motion is $n = \frac{2\pi}{P}$, where P is the orbital period. For example, two bodies are in 3:2 resonance if one makes three revolutions around the Sun while the other makes two, as is the case for Neptune and Pluto. We thus have: $2n_{\text{Neptune}} - 3n_{\text{Pluto}} = 0$, or $3P_{\text{Neptune}} - 2P_{\text{Pluto}} = 0$.

In a planetary ring, the areas where the particles are in resonance with a satellite are the site of complex mechanisms such as density waves or undulations. Figures 1.16, 1.17, and 1.22 show a variety of aspects of resonances in the rings of Saturn.

When a small satellite orbits in a ring, the mean-motion resonances accumulate in proximity to the satellite, which has a paradoxical effect. The satellite's gravitation becomes a mechanism of repulsion; the satellite creates an empty space around its orbit and prevents the ring from spreading out. Several of these "ring shepherds" have been observed in planetary rings (see Figure 1.16).

Resonances can connect more than two bodies, as in the 1:2:4 resonance between Jupiter's satellites Io, Europa, and Ganymede.

Resonances can also connect several orbital parameters such as the satellite's mean-motion and the planet's rotational velocity. A spectacular example is the *synchronous rotation* of a number of satellites (the Moon in particular), which exhibit a 1:1 spin/orbit resonance, which means that the orbital period of the satellite is equal to its rotation period, which implies that the same side always faces the planet. Tidal forces deform the two bodies in the shape of a bulge directed toward the other body, but with a delay due to the body's rotation. In the satellite, which is smaller, friction will slow the rotation to a velocity equal to its revolution around the planet. Synchronous rotation reduces friction. The planet is subject to the same effect over a longer time scale. All of the regular satellites in the Solar System are in synchronous rotation with their planet. When the satellite's orbit is eccentric, friction occurs even when it is in synchronous rotation, which heats the satellite's interior. This mechanism is responsible for very active volcanism on Io, the innermost of Jupiter's large satellites (see Figures 1.3 and 1.4).

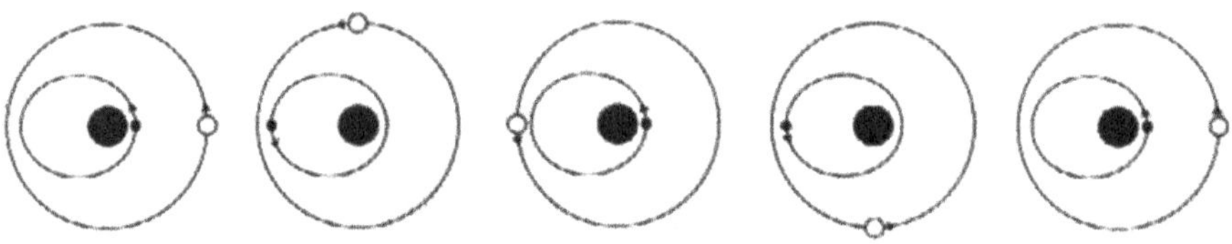

Figure 4.7. *Two bodies in 2:1 resonance; the outer body revolves once around the central body while the inner body revolves twice (source: based on (Murray and Dermott 2000))*

4.6. Stability and chaos in the Solar System

For three centuries, astronomers have sought to establish the stability of the Solar System. Computer-based experiments have begun to allow for a response. The first integrations showed the stability of the system of the outer planets, at least as far as resonant configurations, confirming the calculations of Lagrange and Le Verrier. The problem is more difficult for the terrestrial planets. To simulate the motion of the giant planets, an integration step of 40 days is sufficient, but for a simulation of the Solar System as a whole, a much shorter integration step is necessary to take into account, in particular, the motion of Mercury and the Moon. Calculating the Moon's motion requires taking into account the influence of the Sun and the planets. The first precise analytical theory of the Moon's motion is attributed to Charles Delaunay (1816–1872; published between 1860 and 1867). Today, computers make it possible to perform the calculations quickly and precisely.

The Solar System turned out to be chaotic with a characteristic time of 5 million years. A change of 15 meters in the Earth's position translates into a change of 150 million kilometers 100 million years later. Despite this chaotic behavior, it is possible to explore possible trajectories using computers. Figure 4.8 shows a possible case for the evolution of the planets' orbits. By varying initial conditions, the exploration of configurations shows that the Solar System's inner planets may experience collisions or may be ejected before the death of the Sun.

The simplicity of the law of gravity allowed us to think that the motion of celestial bodies could be predicted with unlimited precision and that it would be sufficient to add more and more terms to the series in order to calculate the orbits over longer and longer periods. But the mathematician Henri Poincaré (1854–1912) showed that the series used in celestial mechanics could diverge after a certain order (the terms in the series stop being smaller and smaller). Several configurations of this type, called chaotic, are present in the Solar System. Thus, projections of planetary motions can hardly extend beyond 100 million years, which is insufficient to study the stability of orbits on the scale of the age of the Solar System.

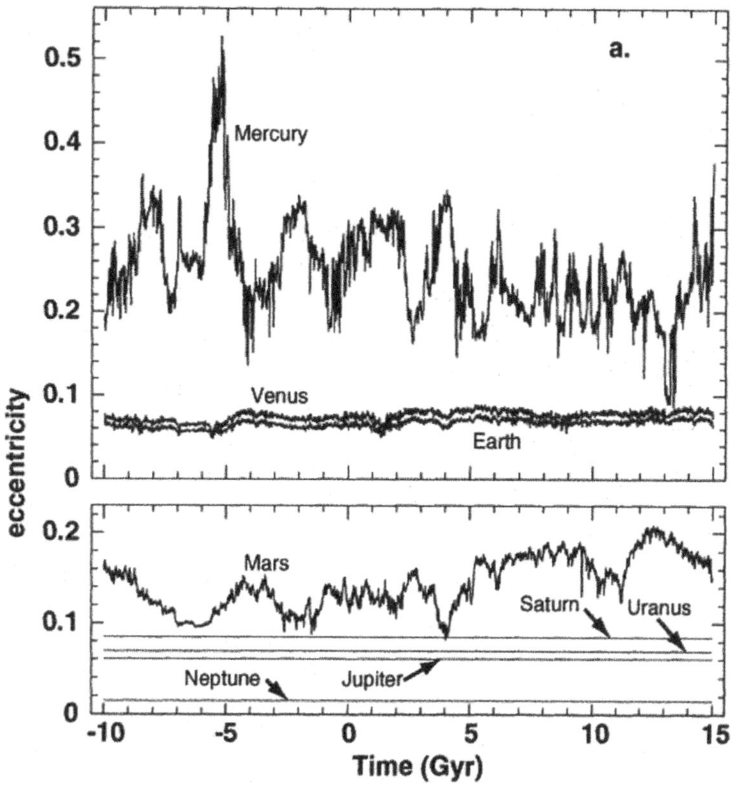

Figure 4.8. *An example of integration of planetary orbits from -10 to 15 billion years. Planetary eccentricities show the stability of the giant planets and the most important variations of the eccentricity of Mercury's orbit (source: based on (Laskar 1994))*

New techniques have been developed for studying the stability of dynamic systems. The study of chaotic areas in the three-body problem show that the orbit of some asteroids is not predictable. Hyperion's axis of rotation (see section 1.3.3.1) has chaotic motion on a time scale of 20 or so days. Pluto's orbit exhibits chaotic behavior on a time scale of 20 million years. The asteroids in the 3:1 Kirkwood division (mean-motion resonance with Jupiter) have an orbit that can abruptly become very eccentric and cross Earth's orbit. Some of the meteorites that land on Earth may come from this area. Studies have shown that the inclination of Earth's axis of rotation would have had chaotic variations without the stabilizing presence of the Moon (Figure 4.9). As for Mars, it has very probably had very strong variations in its obliquity, up to more than 60 degrees (see section 4.3.3.7, Volume 1).

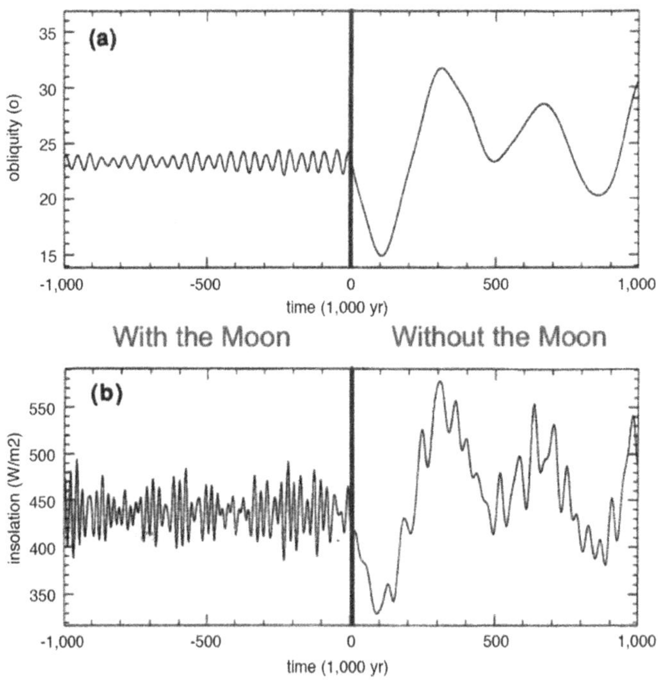

Figure 4.9. *(a) Evolution of the obliquity of the Earth and (b) of sunlight exposure at 65° north latitude, with and without the Moon (source: based on (Laskar* et al. *1993))*

4.7. Orbits in relation to a flattened body

In the previous chapters, objects have been considered to be point objects. The results obtained apply to many situations, even if stars and planets are far from being isolated, because the gravitational field of a spherical body is the same as that of a point with the same mass. But several effects cause the body not to be spherical, and the force field of a non-spherical body is different from an isolated mass.

The gravitational field can be determined fairly precisely by studying the orbit of an artificial satellite orbiting relatively close to the body, or from the rates of precession of a planet's satellites and rings. Knowledge about the gravitational field opens access to information about the internal structure of the body.

Most planets have almost perfect axial symmetry, the primary deviation from sphericity being an equatorial bulge caused by the planet's rotation. It is therefore interesting to study how the force field of an axisymmetric body deviates from the force field of a spherical body.

We often express the gravitational field in the form of potential:

$$\Phi_g(\boldsymbol{r}) = -\int_{\infty}^{r} \frac{\boldsymbol{F_g(r')}}{m} . \mathrm{d}\boldsymbol{r}' \qquad [4.8]$$

We find the gravitational force by taking the gradient of the potential, and the equation for motion is:

$$\frac{\mathrm{d}^2\boldsymbol{r}}{\mathrm{d}t^2} = -\nabla\Phi_g \qquad [4.9]$$

The planet's gravitational potential can be constructed using Legendre polynomials in spherical coordinates, with ϕ the longitude and θ the angle between the planet's axial symmetry and the direction of the particle:

$$\Phi_g(r,\varphi,\theta) = -\frac{Gm}{r}\left[1 - \sum_{n=2}^{\infty} J_n \mathrm{P}_n(\cos\theta)\left(\frac{R}{r}\right)^n\right] \qquad [4.10]$$

The terms $\mathrm{P_n}(\cos\theta)$ are Legendre polynomials:

$$\mathrm{P}_n(x) = \frac{1}{2^n n!}\frac{\mathrm{d}^n}{\mathrm{d}x^n}(x^2-1)^n \qquad [4.11]$$

The gravitational momentum J_n depends on the distribution of the planet's mass. The potential of a body in hydrostatic equilibrium without rotation has a spherical symmetry, and all of the terms J_n are null; the gravitational potential is simplified to: $\Phi_g(r,\phi,\theta) = -\mathrm{G}M/r$. If the distribution of matter in the planet is symmetrical with respect to the planet's equator, the momentum J_n is null for odd numbers of n. This is a very good approximation for giant planets.

If we consider a particle at a distance r from the center of the planet which has an almost circular orbit that is close to the planet's equatorial plane, with mean angular velocity $n(r)$, it will oscillate radially and vertically in relation to its circular orbit with a radial frequency $\kappa(r)$ and a vertical frequency $\mu(r)$.

The orbital, epicyclical, and vertical frequencies are:

$$n^2 = G\frac{m}{r^3}\left[1 + \frac{3}{2}J_2\left(\frac{R}{r}\right)^2 - \frac{15}{8}J_4\left(\frac{R}{r}\right)^4 + \cdots\right] \quad [4.12]$$

$$\kappa^2 = \frac{Gm}{r^3}\left[1 - \frac{3}{2}J_2\left(\frac{R}{r}\right)^2 + \frac{45}{8}J_4\left(\frac{R}{r}\right)^4 + \cdots\right] \quad [4.13]$$

$$\mu^2 = 2n^2 - \kappa^2 \quad [4.14]$$

When the planet is spherical, μ, κ and n are equal. If the planet is flattened, μ is slightly larger than n, κ slightly smaller, and:

$$\frac{d\Omega}{dt} = n - \mu \quad [4.15]$$

$$\frac{d\varpi}{dt} = n - \kappa, \quad [4.16]$$

Ω being the longitude of the ascending node and ϖ the argument of periapsis.

Thus, a flattened planet causes the orbit of its satellites to precess. This precession takes place on a time scale that is typically 10^5 years. The measurement of this precession rate makes it possible to precisely calculate the flattening of the planet's mass (which can differ from the geometric flattening of its surface) and to thus gain access to its internal structure.

On the other hand, the Sun's pull on a planet's equatorial bulge will cause the planet's axis of rotation to precess. It is this effect that will change the position of the north pole on the celestial sphere. The torque due to other planets, which is much weaker than the solar torque, has an important effect since it changes the planet's obliquity. Thus, the obliquity of Mars varies chaotically. Without the Moon, the obliquity of Earth would be very variable, which would have caused important climate instability. Earth's obliquity only varies between 22.5° and 24°, and these quasi-periodic variations, associated with variations in eccentricity and precession of the equinoxes, determine glacial periods (see Figure 4.9). These are the Milankovitch cycles (see section 4.3.3.7, Volume 1).

4.8. Tidal effect

In a satellite of a given volume, the planet's gravitational attraction will be different at different points of the satellite. This differential force is the tidal force (see also section 1.2.7.1, Volume 1). The net force on the satellite will determine the motion of its center of mass, but the tidal force will on one hand deform the satellite and on the other hand produce a torque that will modify its rotation. If the tidal force varies further over time, because the satellite has an eccentric orbit, for example, it can cause the satellite to be heated.

The tidal effect is reciprocal, but the effect of the planet on the satellite is more powerful than the effect of the satellite on the planet.

If the satellite rotates on its axis faster than it orbits the planet, tidal deformation will move to maintain alignment with the planet, but with a certain delay because of friction inside the satellite. For most of the large satellites, the propagation of this deformation slowed the rotation of the satellite down to synchronous rotation: the same side of the satellite will always face the planet. This stabilization takes place on time scales of a few million years.

There are a number of exoplanets close enough to their star for this tidal effect to have brought them into a stable synchronous rotation, which translated into a permanent day and night side. Models show that intermediate regions can be favorable to the presence of liquid water, and thus potentially may be habitable.

In planets with a massive enough satellite, the tidal effect of this satellite can be tangible. In particular, the deformation of the planet creates motion in the fluid parts in relation to the rigid parts, which, on Earth, manifests in our familiar ocean tides (see Box 4.1).

The torque between the satellite and the deformation of the planet slows the rotation of the planet, just as the planet's effect on the satellite's deformation modifies the satellite's rotation, but over much longer time scales.

This torque also slowly changes the radius of the satellite's orbit. Thus, at the same time that Earth's rotation is slowing, the Moon is moving away from Earth at a speed of 3.74 cm/year (see also section 4.2.5.3, Volume 1). The position of equilibrium for a planet-satellite system is double synchronous rotation, where the two bodies face each other permanently. In the Solar System, only Pluto and Charon have reached this state of equilibrium today. The Earth-Moon system should reach it in 50 billion years, which is to say long after the Solar System's lifetime, when the rotation and orbital periods would have been 47 days.

4.8.1. *Tidal deformation*

For any point on the planet with a radius R located at a distance $d >> R$ from a satellite of mass m, the tidal force is the difference between the gravitational attraction of the satellite at that point and at the center of the planet.

For points on the Ox axis, whose origin is the center of the planet and is oriented toward the satellite, the tidal force is:

$$F_m(x) = \frac{Gm}{(d-x)^2} - \frac{Gm}{d^2} \approx \frac{2xGm}{d^3} \quad [4.17]$$

The tidal force is oriented toward the exterior of the planet for points close to the planet-satellite axis. The other points are attracted toward the center of the planet (see Figure 4.10). If the body can be deformed, it will lengthen along the planet-satellite axis. On Earth, the oceans respond more quickly than the continents, creating oceanic tides.

The tidal effect can violently deform a body near a planet or cause it to disintegrate. The limit inside which a satellite cannot survive this tidal effect is the Roche limit (see sections 1.2.7.2, Volume 1 and 4.8.3 in this volume), which determine the area of planetary rings. Inside this limit, solid matter remains to form small bodies that are more or less aggregated. If the matter moves away from the planet beyond this limit, collisions will quickly lead to the formation of a satellite.

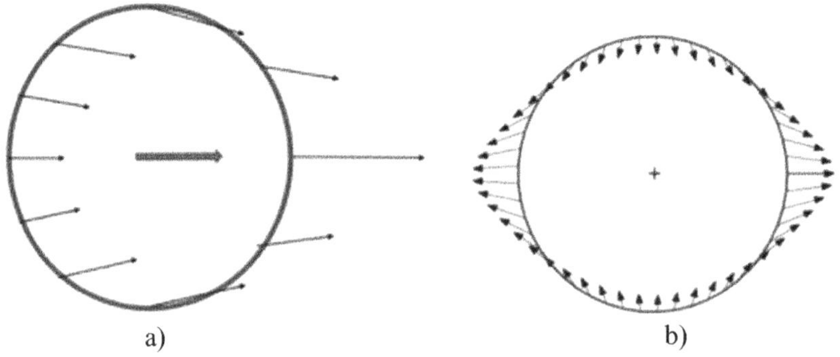

Figure 4.10. *(a) Attraction of a (non displayed) satellite on the surface (black arrows) and the center (red arrow) of the planet; (b) tidal force (black arrows minus the red arrow) on the surface of the planet. For a color version of this figure, see www.iste.co.uk/encrenaz/solar2.zip*

On the Earth's surface, the tidal effect is on the order of 10^7 times weaker than Earth's gravity, and the effect of the Moon is twice as great as that of the Sun.

The Moon's pull on Earth is only balanced at the center of the planet. The side of the Earth closer to the Moon is more attracted (and thus approaches the Moon), and the far side is less attracted (and thus moves away from the Moon) (see Figure 4.10). This double deformation will travel around the Earth in 24 hours to try to remain aligned with the Moon. The presence of continents and oceans makes the tidal effect very complex. On the continents, the tidal force provokes modifications on the order of 20 centimeters. In the absence of continents, the difference in level of the surface of the oceans should be on the order of 70 cm, and the daily rhythm should be semi-diurnal. In the seas and oceans, propagation of the tidal force combines with the volumes of water involved, the depth of the seabed, and the shape of the coast to cause differences in level of up to 17 meters. The tides can be semi-diurnal (in Europe), semi-diurnal with diurnal inequalities (two cycles per day with very different levels), diurnal (in Vietnam), or mixed (diurnal or semi-diurnal depending on the date).

Box 4.1. *Ocean tides on Earth*

4.8.2. *Tidal torque*

Tidal dissipation creates secular rotation and orbital rates for planets and satellites. For an isolated planet-satellite system, the total angular momentum is conserved, but there can be a transfer of momentum between the rotation and the orbital period.

A planet and its satellite are each deformed by tidal effect due to the other body. Let us suppose that the satellite has reached synchronous rotation, and so its deformation is directed exactly at the planet, with possible small oscillations due to its eccentricity. However, the planet has not reached synchronous rotation, and its deformation, that is to say its tidal force, is not directed toward the satellite because the planet is not fluid. The delay will depend on the planet's rotational speed and the orbital speed of the satellite (directly related to its distance). If the planet and the satellite rotate in the same direction and the planet rotates more quickly than the satellite, the deformation will be "in front" on the satellite. The torque between the planet's deformation and the satellite will slow the planet's rotation and accelerate the rotation of the satellite, which will move away. If, on the other hand, the satellite is close enough for its orbital velocity to be greater than the planet's rotation (Phobos, for example), the tidal torque will slow the satellite, which will fall to the planet (see section 4.3.4, Volume 1).

Tidal torque depends on the size of the deformation and the angle between the directions of the deformation and the satellite. Each of these two factors is proportional to m_s/d^3, where m_s is the mass of the satellite and d is its distance from the planet. Thus, the torque between the satellite and the deformation it generates in the planet varies with m_s^2/d^6.

Earth is subjected to tidal torque on the part of the Moon, as well as the Sun.

The slowing of Earth is measurable using shell fossils, which indicate that the year was about 400 days long 350 million years ago. The dates of eclipses also bear the trace of this change over the last two millennia, and the precision of atomic clocks also makes it possible to detect directly the length of day.

The Pluto-Charon system has evolved more quickly and has reached its state of equilibrium of double synchronous rotation.

The tidal torque is significant between the Sun and Mercury, which is locked in a state of spin-orbit equilibrium: Mercury rotates three times while it orbits twice around the Sun.

4.8.3. *Roche limit*

In a planet's gravitational field, a satellite is subjected to tidal force, which works against the forces of cohesion that result from internal gravitational forces in the satellite and nongravitational intramolecular forces, or forces of mechanical cohesion. When the satellite approaches the planet around which it orbits, the differential effect increases: mechanical distortions, or tides, arise; there is a limit distance beyond which cohesion is no long assured. This limit L_R, called the Roche limit, can be expressed as a function of the radius R_p of the planet and the densities ρ_p and ρ of the planet and the satellite: for this, we consider that the satellite is made up of two particles of mass m and radius r of density ρ_s and that the differential force on the planet is equal to the force of attraction of one particle on another (see Figure 4.11):

$$L_R = 2R_p \left(\frac{\rho_p}{\rho}\right)^{\frac{1}{3}} \quad [4.18]$$

A precise calculation gives:

$$R = 2{,}456R_p \left(\frac{\rho_p}{\rho}\right)^{\frac{1}{3}} \quad [4.19]$$

We find these calculations at https://en.wikipedia.org/wiki/Roche_limit.

When an object approaches a planet at a distance lower than L_R, it fragments and gives rise to a ring.

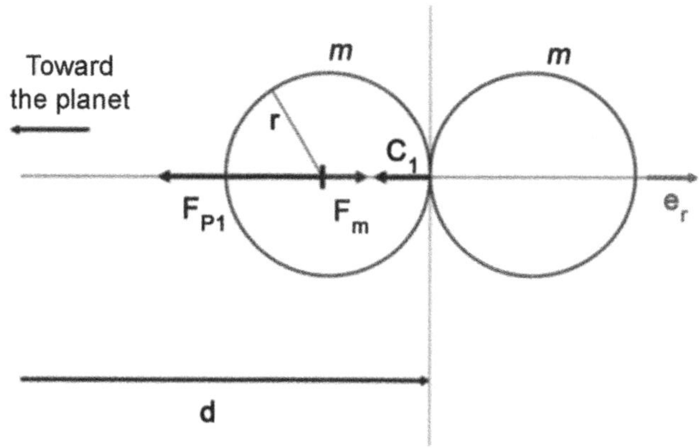

Figure 4.11. *Assessment of forces acting on a mass m at a frame linked to the planet (supposed to be Galilean): d is the distance to the planet. F_{P1} is the planet's gravitational force of attraction, F_m is that of the second sphere of mass m, C_1 is the contact force between the two spheres. For a color version of this figure, see www.iste.co.uk/encrenaz/solar2.zip*

4.9. Nongravitational forces and orbits of small bodies

Contrary to what was done in the preceding paragraphs which deal with gravitational interactions between the Sun, the planets, and the satellites, we will now consider the nongravitational forces, which are solar radiation pressure, solar wind, and friction on gases. Gravity acts on a body in its entirety, while these forces only act on the surface; they therefore affect small particles more strongly because they have a larger surface-to-volume ratio.

4.9.1. *Radiation pressure (micrometer-sized grains)*

Solar radiation exercises a repulsive force on all bodies in the Solar System. This force is equal to:

$$\boldsymbol{F}_{rad} = \frac{L_{\odot} A}{4\pi c d_{\odot}^2} Q_{rp} \boldsymbol{r} \qquad [4.20]$$

$L_{\odot}$ is the Sun's luminosity, $d_{\odot}$ is the distance of the particle from the Sun, c is the speed of light, A is the cross-section of the particle, Q_{rp} is the radiation pressure force, a dimensionless parameter equal to 1 for a perfectly absorbent particle, and is << 1 for particles smaller than the wavelength. r is a unit vector directed from the Sun toward the particle.

For particles larger than the wavelength, the pressure force increases with the cross-section, that is, for a spherical particle of radius r, as r^2. As the gravity on the body increases with mass, the ratio of the radiation pressure over the gravitational force varies with $1/r$. For a grain in heliocentric orbit, this dimensionless ratio, β, is represented in Figure 4.12 for a variety of compositions: $\beta = 5{,}7\ 10^{-5} \frac{Q_{rp}}{\rho r}$ where r is in cm and ρ in g/cm^3.

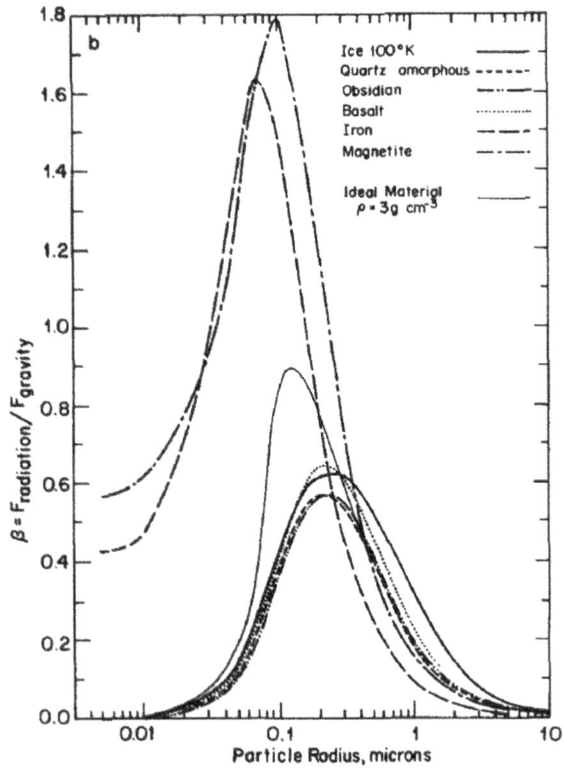

Figure 4.12. *Ratio between radiation pressure and gravitational force as a function of particle size for different compositions of particles (source: based on (Burns* et al. *1979))*

β is independent of the distance from the Sun, and its effect is important for micrometer and submicrometer-sized particles. These particles are subject to the gravity of a Sun with apparent mass $M_{\odot}(1-\beta)$. The particles for which $\beta > 1$ are pushed away more strongly than they are attracted by the Sun, and they will be ejected from the Solar System. For dust released by a body in orbit around the Sun, it is sufficient that $\beta > 0.5$ for it to escape the Solar System.

This effect is particularly important for comets. Ions align in an antisolar direction because they follow the solar wind, and dust will have an antisolar trajectory, although one that is more curved because, when leaving the comet on a Keplerian orbit, they are subject to radiation pressure which will cause them to deviate as they move away from the comet (see Figure 2.21).

4.9.2. *Poynting-Robertson effect (small macroscopic particles)*

Radiation pressure is a force directed in the antisolar direction. An additional effect is due to the fact that the re-emission of the solar flux is isotropic in a reference plane related to the particle. Re-emission is thus greater in front of the motion. This translates into a slowing force that has the opposite direction to the particle's velocity. The force expressed above becomes:

$$\boldsymbol{F}_{\text{rad}} = \frac{L_{\odot} A Q_{rp}}{4\pi c d_{\odot}^{2}}\left[\left(1-\frac{2v_r}{c}\right)\boldsymbol{r}-\frac{v_\theta}{c}\boldsymbol{\theta}\right] \quad [4.21]$$

The first part corresponds to the radiation pressure, seen above. The second and third part, with the two components of velocity, v_r and v_θ, are the Poynting-Robertson component.

In the Solar System, these forces will rapidly eject micrometer particles toward the exterior, and they will cause centimeter-sized particles to spiral toward the Sun more slowly. The typical time scale of this decrease in years is: $t_{\text{PR}} \approx 400\ \frac{r(\text{au})^2}{\beta}$, where r(au) is the distance from the Sun expressed in au.

Zodiacal light, a bright band visible in the ecliptic, is composed of particles of 20 to 200 microns, whose lifetime is on the order of 10^5 years on the level of the Earth. It is resupplied by collisions between asteroids and by particles that fall from the Kuiper disk.

4.9.3. *The Yarkovsky Effect (meter to kilometer-sized particles)*

On a rotating body heated by the Sun, the hemisphere corresponding to evening will be warmer than the hemisphere emerging from nighttime. The difference in radiation between the two hemispheres manifests in the Yarkovsky effect in the daytime. This effect will slow/reduce the body's orbit if its rotation is retrograde and will accelerate/increase its orbit if its rotation is prograde. A similar effect occurs between the spring/summer hemisphere and the autumn/winter hemisphere.

This effect is particularly sensitive for bodies between a meter and a dozen kilometers in size. It was demonstrated for the first time by the difference in the purely gravitational orbital model for the (6489) Golevka asteroid. This near-Earth asteroid has a radius of 300 m.

4.9.4. *Yorp torque (asymmetric bodies)*

If the body is asymmetrical, the composition of the absorption, reflection, and re-emission of solar flux manifests as Yarkovsky-O'Keefe-Radzievskii-Paddack torque, otherwise known as the Yorp effect. This torque is sensitive for bodies that are not ellipsoids or that exhibit variations in albedo and can cause acceleration or slowing of rotation. The torque increases with the inverse of the square of the object's size. We believe that it especially affects the rotation of asteroids less than 20 km, and the variation in the rotational velocity of an asteroid of 5 km occurs on a time scale of a hundred million years. It is possible that some binary asteroids are created by matter ripped from an asteroid after a rotation accelerated by the Yorp effect.

4.9.5. *Friction from solar particles (submicrometer dust)*

Dust smaller than a micron is subject to slowing by particles carried by the solar wind. This effect is on the order of 20% of the Poynting-Robertson effect for particles of obsidian, and even more so for the particles smaller than 0.1 μm that spiral toward the Sun.

4.9.6. *Friction in gas*

Interplanetary space is currently almost empty of gas, but the effect of friction was significant in the protoplanetary disk. It also plays a role for particles in planetary rings that can end up entering the planet's atmosphere. Particles are slowed by the gas. The force is:

$$F_d = -\frac{C_d A \rho_g v^2}{2}$$

where C_d is the coefficient of friction, on the order of 1, A is the cross-section of the body, ρ_g is the density of the gas, and v is the velocity of the particle in relation to the gas. For a particle on a circular orbit in the planet's static atmosphere, this slowing can make the particle lose energy, which will fall and thus paradoxically increase its velocity. This friction is the primary cause of the fall of artificial satellites to Earth.

In the Solar System's protoplanetary disk, the pressure gradient slowed the velocity of the gas in relation to the Keplerian velocity. The parameters of the disk make it possible to estimate that the velocity of the gas was 0.5% below Keplerian velocity. Small particles are not very sensitive to this slowing because they are coupled with the gas. Kilometer-sized bodies are not very perturbed by this either because of their mass. The effect has maximum effect on particles that encounter, during their orbital period, a mass of gas on the same order as their own. The drift of a meter-sized body in the vicinity of the terrestrial planets can reach 10^6 km per year and thus the body can disappear in about a hundred years! It was thus necessary for the body to grow very quickly from a centimeter to a kilometer in the protoplanetary disk. This stage in the scenario of planetary formation is still the subject of discussion (see section 2.6, Volume 1).

4.10. Formation of planetary systems

The title of this section is too ambitious. It will be limited to showing the scenario of the formation of the Solar System, which is now fairly well understood, even if some stages still need to be made more precise. As for exoplanetary systems, they show a wide diversity, and an understanding of their formation is still to come.

In the Solar System, a similar scenario, with different spatio-temporal conditions, explains the difference between terrestrial planets, gas giants, and icy giants. The formation of small bodies also depends heavily on their location, but the conditions are very different depending on whether we are talking about asteroids, Kuiper objects, Oort cloud objects, or satellites and rings. The question of physico-chemical processes explaining how essentially solid bodies could form from an essentially gaseous protoplanetary disk is a question that is not yet completely resolved. The dynamic processes are more well established: we know today that, while most bodies in the Solar System have stable orbits, many objects have been disturbed and are no longer where they formed.

Figure 4.13 summarizes some of the potential initial steps in the formation of the Solar System (see also section 1.2.6, Volume 1).

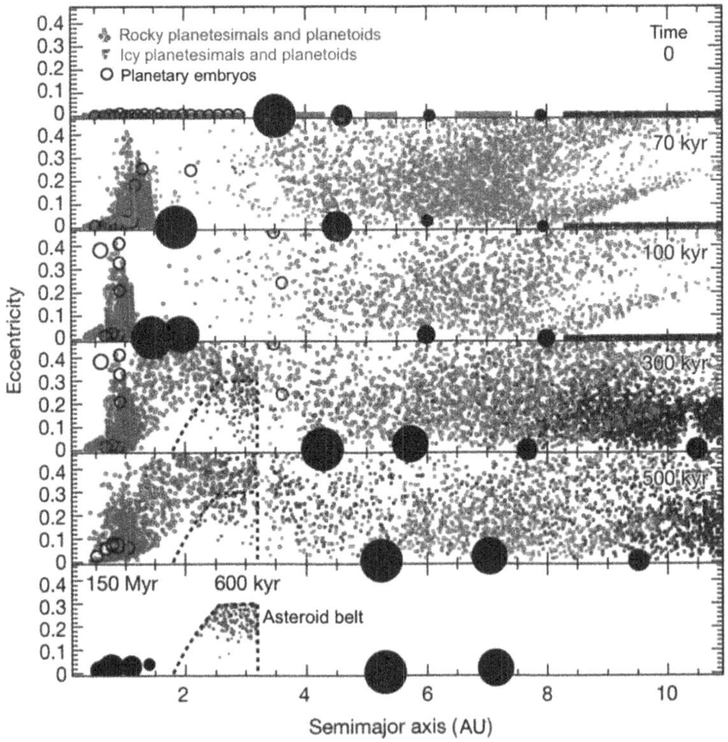

Figure 4.13. *Possible scenario for the initial evolution of the Solar System in the Nice model. For a color version of this figure, see www.iste.co.uk/encrenaz/solar2.zip*

COMMENTS ON FIGURE 4.13.– *The planets are represented by circles whose size depends on their mass during the different stages. The distribution of gas in the protoplanetary disk is not shown. Planetesimals and planetoids are represented by red dots when they are rocky and blue dots when they are icy. The simulation begins when Jupiter is completely formed. It migrates toward the Sun and is then captured by Saturn with which it is in 3:2 resonance. Then they both migrated outward. The lower diagram illustrates the situation reached after 150 million years (source: based on (Walsh* et al. *2011)).*

4.10.1. *A disk of planetoids*

The planets form in circumstellar disks (see section 2.4, Volume 1) that contain a halo of gas and a disk of dust. Several processes will take place in the disk: the accretion of dust into small bodies a kilometer in size, planetesimals, then the increase in size of the largest bodies and the evacuation of gas. These first steps, when the gas is still present, are described in section 2.5 of the first volume.

Once the gas has dissipated, the bodies' motion is governed by gravity and collisions. We thus find a disk of planetesimals where the largest objects begin to attract all the smallest bodies that are passing by. A body of radius R will increase its mass from the objects with which it collides. Because of gravity, its cross-section of collision, instead of being its geometric surface (proportional to R^2) will increase by a factor $F_g=(1+V_e^2/V^2)$, where V_e is the escape velocity of the body and V is the relative velocity of the encounter. As V_e is proportional to the radius R of the body, the increase velocity of the mass M, dM/dt, can be proportional to R^4, which favors the growth of the larger body. Models show that this acceleration in accretion is the most likely evolution of a disk of planetesimals.

When the area where bodies may be attracted is empty, the body stops growing. The large body that results is called a *protoplanet*. The size of a protoplanet depends on the matter available in the disk, which depends particularly on the distance from the star. Near the star, protoplanets are at most as large as the Moon because only metals and rocks exist in solid form. Beyond what we call the *ice line*, lower temperatures allow for condensation of ice, including water ice. The amount of solid matter available, which is much larger, allows for the formation of bodies equal to a dozen terrestrial masses. The length of this stage also increases with distance from the star, between a few tens of thousands to a few million years. But the majority of the solid matter in the disk is still in the form of planetesimals.

However, some conditions can lead to regular growth of bodies that do not lead to the formation of planets. For large enough values of V over V_e, F_g approaches the order of 1, and dR/dt can become independent of R. The ratio between the largest masses and the small bodies can decrease rapidly. The growth of small bodies can overtake that of the largest. Orderly growth leads to a disk of small bodies, which can reach a state of equilibrium and have a very long lifetime.

4.10.2. *Formation of terrestrial planets*

In the inner Solar System, each planet is the result of planetesimals that condensed in a relatively narrow ring of the protoplanetary disk. Mutual perturbations will increase the eccentricity of the orbit of these protoplanets, and catastrophic collisions will increase the size of the largest protoplanets until planets are formed. Computer simulations of this process can lead, in an unpredictable manner, to systems with features that are similar to terrestrial planets. Earth could have formed, for example, from a protoplanet like Mars + 10 like the Moon + about a hundred like Iapetus, or from 5 Mars + 3 Moon + 12 Iapetus. However, several simulations resulted in systems with fewer planets and planets on more eccentric orbits. This may be because these models do not take into account the residual small bodies in which the planets are immersed. Now, once formed, the planets have continued to be subject to the bombardments of planetesimals coming, in part, from the outer Solar System, which may have recircularized their orbits.

The impacts of planetesimals bring mass and energy to the planet. The planet is also heated by radioactive elements with short lifetimes like aluminum 26 during the first million years of planetary life, then by potassium, thorium, and uranium during the first billion years. These sources of heat can soften the planet. The heaviest elements will then fall to the center, and the interior of the planet will differentiate. This can also happen in planetesimals and planetoids if the quantity of radioactive elements is sufficient.

In addition to energy from planetesimals and radioactivity, other sources of heat contribute to the heating of the still-forming planet: the planet's potential energy when it contracts in response to an increase in pressure or when it differentiates, as well as exothermic chemical processes. The planet radiates this heat into space. In the planet, the heat propagates by conduction and, if it is partially melted, by convection. Conduction happens relatively slowly, and convection only concerns areas that are melted enough to allow for matter to move. Also, as a first approximation, when the planet forms, its temperature is the result of equilibrium between the energy deposited on the surface, radioactive decay, and heat radiating into space. If the planets are bombarded by bodies larger than a hundred or so kilometers, the impacts bring enough energy to completely differentiate a planet.

It also forms an atmosphere. The proto-atmosphere can become optically thick and can trap the heat of the impacts. This effect, which causes the planet's temperature to increase, becomes significant when the planet's mass is greater than a tenth of terrestrial mass. The temperature can then exceed 1,600 K, and heavy components will fall to the center of the planet, leading to complete differentiation.

The current atmosphere of terrestrial planets, composed of heavy elements ($Z \geq 3$), is probably the result of outgassing of planetesimals falling on planets. Several mechanisms are involved during these falls: small planetesimals disintegrate in the atmosphere, leaving behind their energy and their matter. Bodies larger than a few dozen meters will impact the ground after having been eroded by the atmosphere, which will have a temporarily high temperature. If the impactor is even larger, it can cause some of the atmosphere to escape. If the size of the impactor is equal to or greater than the scale height of the atmosphere, a large part of the atmosphere will escape. But the impactor will also bring its volatile components to the atmosphere. For bodies on the order of 100 km that contain 1% of volatiles, there will be equilibrium between the loss of atmosphere and the contribution of volatiles for a mass of atmosphere by surface element equal to the current terrestrial oceans. Terrestrial planets may have lost their atmospheres several times during their period of accretion.

Light elements can also escape from the atmosphere through Jeans instability. A molecule of gas has an agitation velocity that depends on its temperature, and if this velocity is greater than the planet's escape velocity, the gas will leave the atmosphere. Thus, the hotter and lighter the gas, the more easily it escapes. And the smaller the planet, the lower its escape velocity, and the more easily it loses its atmosphere (see also section 1.3.2.2, Volume 1). If hydrogen and helium escape easily from the atmosphere of terrestrial planets, heavy gases were also able to escape their proto-atmospheres when they were hot.

The current atmospheres of terrestrial planets probably formed at the end of the accretion period after outgassing of the hot planet and the contribution from the last planetesimals. The end of the process of terrestrial planetary formation occurred in a gas-free environment. Isotopic analyses indicate that volatile elements and water present today on Earth come from planetesimals that were propelled from the outer Solar System toward the center of the Solar System by planetary perturbations. This brings us now to focus on what happened in the outer Solar System, where the protoplanets that formed were much more massive (see also section 1.3.2.1, Volume 1).

4.10.3. *Formation of Jupiter*

The formation of Jupiter is an important characteristic of the Solar System. Next to Jupiter, the asteroid belt has a very low mass (a thousandth of the mass of Earth) and Mars is also a small planet (a tenth of the mass of Earth).

The composition gradient in the asteroid belt, which is the result of a process of heating that took place very early in the Sun's lifetime, has not affected the condensation of planetesimals since then. This shows stability in the asteroid belt since its formation. Moreover, meteorites that have landed on Earth from the inner part of the asteroid belt have made it possible to date the formation of asteroids. These considerations, as well as the significant proportion of gas (hydrogen and helium) on Jupiter, lead us to think that this planet formed very early in the history of the Solar System.

The process of the formation of gaseous planets from planetesimals requires the formation of a planetary core large enough to accrete gas before the nebula dissipated (10^6 years). A core on the order of 10 terrestrial masses is required to attract the surrounding gas (see Figure 4.14). We think that Jupiter found itself at a different location from the Sun when it formed, just beyond the ice line, because this is the place where the most matter was available, as we will see. At this distance, the temperature of the nebula was lower than the condensation temperature for water ice. Outside this limit, solid matter thus includes not only rocks, but also ice, which significantly increases the nebula's surface density and thus the speed at which these bodies form. Furthermore, volatile materials coming from the inner Solar System also condense at this distance. The surface density was able to reach a sufficient value to explain a rapid formation of Jupiter's core.

Models show that, once it begins, the fall of gas accelerates until a mass on the order of Jupiter's mass is accreted. From that moment, the planet will create an empty space around itself, pushing matter from the disk away. A similar mechanism takes place in the rings of Saturn where the presence of a small satellite translates into an empty space in the ring. The planet will then cool and contract.

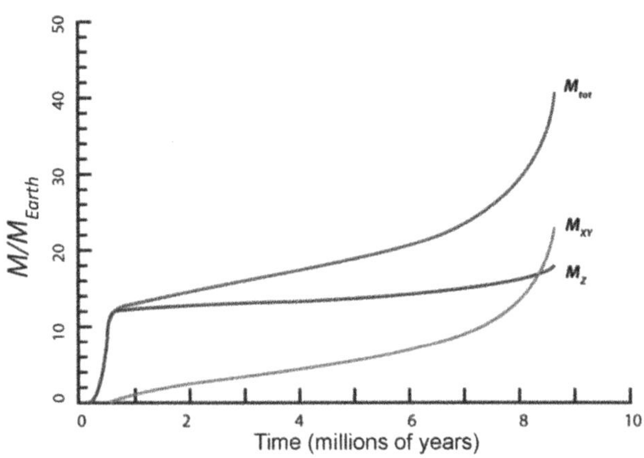

Figure 4.14. *Process of the formation of Jupiter. The red curve is the mass of solid matter, the green curve is the mass of gas, and the blue curve is the total mass of the planet (source: based on (Pollack* et al. *1996)). For a color version of this figure, see www.iste.co.uk/encrenaz/solar2.zip*

4.10.4. *Formation of giant planets by core accretion*

The first attempts to explain the formation of giant planets saw the opposition of the model of formation by core accretion and the model of collapse through instability of a gas disk, a process similar to the one that forms the stars. This second model was abandoned in the end to explain the planets in the Solar System, which are now well explained by the core accretion model. But it may remain valid for the formation of large exoplanets that are very distant from their star (see section 4.10.5).

While the proportion of heavy elements (chemical elements other than hydrogen and helium) in Jupiter is around four times the solar value, it is 10 for Saturn and on the order of 50 for Uranus and Neptune. Thus, the giant planets accreted solid matter more efficiently than gas. The D/H ratio for Jupiter and Saturn is on the order of $2\ 10^{-5}$, which is the value of the interstellar medium, while it is higher for Uranus and Neptune, which is explained by enrichment in deuterium by icy bodies.

The difference stems from the fact that the time scales for the formation of a solid core are longer at a larger distance from the Sun. Protoplanets must have reached a sufficient mass to accrete gas before the solar wind and radiation emptied the protoplanetary disk.

Jupiter formed before the dissipation of gas. The proto-Saturn, which is farther from the Sun than Jupiter, took more time to reach the critical mass of 10 terrestrial masses and did not begin to accrete gas until after the nebula had already partially dissipated. Uranus and Neptune formed even later. Since attempts to model their formation in their current location requires time lengths longer than the age of the Solar System, they must have formed much closer to the Sun in the protoplanetary disk before migrating outward.

It is likely that the processes of solid core and gas accretion were not completely disconnected. In the process described in Figure 4.13, the dissipation of the nebula took place after Jupiter and Saturn had reached the phase when gas accretion accelerated, while Uranus and Neptune were in the phase of simultaneous accretion of gas and solids. This process stopped when the gas nebula dissipated. The fall of gas can also be stopped by the creation of an empty space around the orbit of the planet, by the mechanism of resonance.

The planets then interacted gravitationally with the residual disk of planetesimals and the other giant planets, and migrated out of the locations where they formed through the exchange of angular momentum. The presence of Pluto and numerous other Kuiper objects in 3:2 resonance with Neptune suggests a process of mutual perturbations where Jupiter migrated slightly toward the inner Solar System and where Saturn, Uranus, and Neptune moved away from the Sun. Then, Neptune captured Pluto and other Kuiper objects called plutinos, forcing them to migrate as well, which explains the significant eccentricity of their orbits.

Global models have been proposed to explain all of the properties, position, and composition of the planets and small bodies. The Nice model (see also section 1.2.6, Volume 1) is based on the idea that the giant planets had to travel through resonances to explain the eccentricity and inclination of their orbits. Since these planets contain gas (hydrogen and helium), their formation took place before the gaseous component of the protoplanetary disk was dispersed. Their formation thus had to lead to planets with circular orbits in the same plane. The planets and planetesimals remaining after their formation migrated in the gaseous environment and ended up perturbed and/or locked by mutual mean-motion resonances. A scenario is acceptable if it leads to a system that agrees with observations, four giant planets with fairly similar orbits to current giant planets, while the asteroids and the terrestrial planets are preserved. One of the potential scenarios is illustrated in Figure 4.13 above. In another scenario, a fifth giant planet was ejected in the game of ping-pong played by the planets since their formation. Figure 4.15 shows this latter scenario, in which Jupiter migrated slightly inward, while Saturn and Uranus migrated slightly outward and Neptune migrated somewhat more outward: a fifth planet had perturbed the other planets before being ejected by Jupiter.

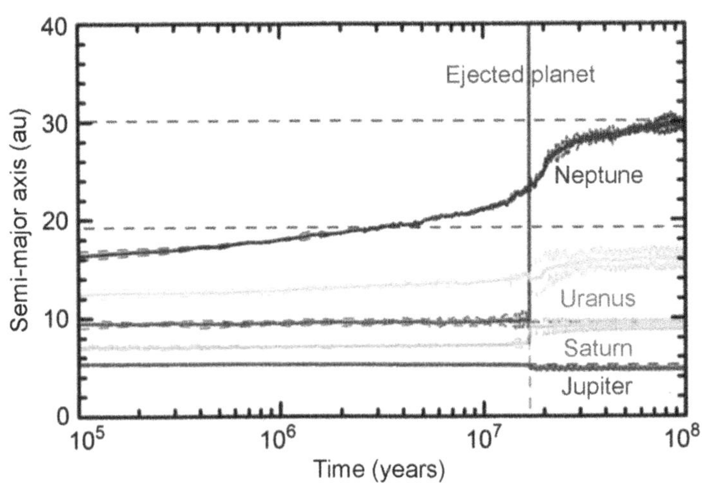

Figure 4.15. *Possible scenario for the evolution of giant planets formed in a disk of planetesimals. For a color version of this figure, see www.iste.co.uk/encrenaz/solar2.zip*

COMMENTS ON FIGURE 4.15.– *Five planets initially in 3:2, 4:3, 2:1, 3:2 resonance in a disk of mass equal to 20 terrestrial masses. The curves in color show the semi-major axes, periapses, and apoapses of the planets, while the dotted lines show the semi-major axes of the current planets. At about 60 million years, a planet is ejected, and the interactions between the planets and planetesimals reduce the eccentricities of the planets (source: based on (Nesvorný and Morbidelli 2012)).*

4.10.5. *Formation by disk instability*

Even if, as we've seen, some solutions exist to resolve the problems encountered by the "solid core" scenario to form Saturn "in time" or to form Uranus and Neptune, some researchers have conceived more radical solutions: let us return to the model of formation by local collapse in a disk.

This model of instability provides a possible explanation for the exoplanets that are very distant from their star, at locations where the density of solid matter would not have permitted the formation of a core massive enough to begin to attract gas.

For the Solar System, this scenario might describe formation by gravitational instability in the gaseous protoplanetary disk. This disk is effectively not homogeneous, and inevitably, locally high densities can exist. In principle, the thermal pressure of the gas prevents gravitational collapse of these high densities, and, moreover, differential Keplerian rotation tends to disperse them rapidly. However, if the disk is sufficiently dense and cold, these lumps can become gravitationally unstable.

This scenario was first proposed in the 1970s to explain the formation of all planets, but was quickly abandoned for terrestrial planets because it is very difficult to develop instabilities in the inner disk because of the high temperature and the strong shear caused by differential rotation of the disk. In the outer disk, however, the conditions are more favorable because thermal pressure is lower and the shear is weaker. Computer simulations have thus shown that a protoplanetary disk can effectively develop local instabilities. The advantage of formation by instability is that it is in principle extremely fast, on the order of a few hundred years in the area around Jupiter's orbit.

However, a major unknown remains: for initial instabilities to develop and form planets, they must be able to cool rapidly as they contract. However, no simulation has yet shown that this was possible, and the instability scenario does not seem able to form the giant planets of the Solar System. It has however experienced a resurgence of interest with the recent discovery of giant extrasolar planets orbiting at large distances, sometimes more than 100 au, from their star. *A priori*, it is completely impossible for such planets to form in the "solid core" scenario, while the instability model becomes more effective in these outer regions. An alternative is, of course, that these planets formed from a solid core closer to the center and then migrated outward.

This mode of formation could also explain the existence of isolated objects with low mass discovered in areas of stellar formation. The function of mass (the number of objects as a function of mass) of objects in the areas of stellar formation show that the mechanism of stellar formation could also generate bodies that are much less massive than stars, brown dwarfs, for example. These bodies with too little mass to initiate thermonuclear reactions can have masses similar to the planets. It is very difficult or even impossible to detect them, except at the beginning of their lives when, still hot, they emit a sufficient amount of infrared radiation. If they are near a star, it will then be difficult or even impossible to differentiate them from a "real" planet formed by core accretion. Numerous objects that are between 13 and 60 times the mass of Jupiter have been discovered near stars, isolated or in pairs, forming a still-mysterious family: What are they made of? How were they formed? Are they super-planets or mini-brown dwarfs? The two families are probably intermingled in this range of mass (see also sections 2.8.4 and 2.8.5, Volume 1).

4.10.6. *Disappearance of the gas*

The first stages of the process of planetary formation occur in a disk, whose mass is mostly in the form of gas, primarily hydrogen. Grains and growing planetesimals interact strongly with the gas, and we have seen that this gas is undoubtedly essential for enabling the accretion of smaller grains.

While the gas planets had to have formed before the disappearance of gas, the terrestrial planets, which formed more slowly, completed their formation in a gas-free disk. The observation of young stars shows that primordial protoplanetary disks disperse on time scales between 1 and 10 million years, with the average lifetime being probably on the order of 3 million years for a solar-type star (see Figure 4.16). This places dispersion of the disk during the final phase of mutual interactions between large embryos.

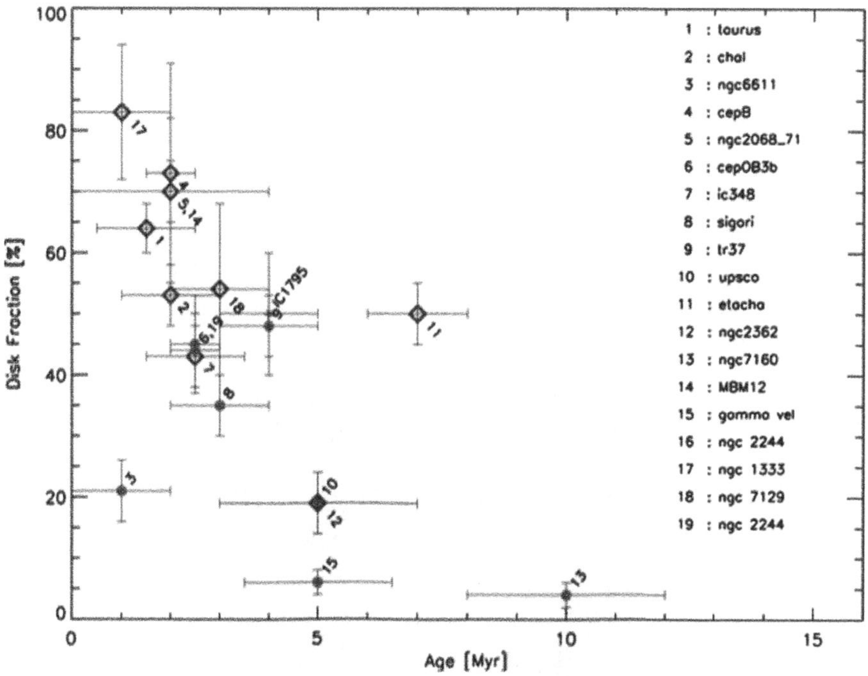

Figure 4.16. *Proportion of open star clusters surrounded by massive protoplanetary disks in different star clusters with different ages (source: based on (Roccatagliata et al. 2011)). For a color version of this figure, see www.iste.co.uk/encrenaz/solar2.zip*

We have yet to explain why and how the disk dispersed. There are several possible mechanisms for this, such as, for example, the stellar wind from a star in the T-Tauri phase (see also section 2.3.3, Volume 1), or the viscous accretion of the disk by the star. The most likely cause seems however to be the effect of photo-evaporation due to ultraviolet radiation from the young star, which, together with the viscosity of the disk, is able to very quickly disperse the disk of hydrogen gas after having "cut" it in half. Photo-evaporation results from the effect of ultraviolet (UV) radiation from the star on gas molecules. The interaction of UV photons will effectively heat the gas, and if its thermal agitation velocity exceeds the local Keplerian velocity, the gas will be ejected. Since the Keplerian velocity is proportional to $a^{-0.5}$, where a is the semi-major axis of the orbit, while the energy carried by a UV photon does not decrease with distance from the star, molecules in outer regions will be the most easily ejected during photo-gas interaction. But the flux of photons, and thus the rate of interaction with the gas, will diminish with a.

We can thus calculate that there is a critical radius, called r_g, beyond which the energy deposited by photo-evaporation exceeds the orbital energy:

$$r_g \sim 1{,}4 \frac{M/M_S}{T/10^4\text{K}} \text{ua} \qquad [4.22]$$

where T is the temperature of the gas after interaction with a photon, M is the mass of the star, and M_S is the mass of the Sun. Under these conditions, because of the rotation of the gas, the critical radius for dispersion is instead equal to $r_{cr} \sim 0.2r_g$. For typical UV radiation, we have $T \sim 10^4$ K. For a solar-type star and an MMSN-type disk (*minimal mass solar nebula*), we then obtain: $r_{cr} = 2$ au.

Figure 4.17. *Image of the disk of the young star DoAr 44 with surrounding dust (orange) and gas (blue). The smallest cavity in the gas reveals the presence of a young planet cleaning the disk. The bar at the bottom corresponds to 60 au (source: Alma (ESO/NAOJ/NRAO)). For a color version of this figure, see www.iste.co.uk/encrenaz/solar2.zip*

However, while the disk is very dense, at the distance r_{cr}, the flux of matter toward the inner disk due to viscosity is greater than the flux of matter ejected by photo-evaporation. But the mass of the disk decreased by accretion to the star, and after some time, the gaseous matter beyond r_{cr} will be ejected from the system before it is able to cross the r_{cr} limit to make itself "safe". An opening is thus created in the disk, which ends up being cut in two. The inner part of the disk is protected from photo-evaporation, but is so small (<2 au) that it will very quickly be accreted to the star (while, before, the flux of matter spiraling toward the star was compensated by the matter coming from farther away in the disk). The outer part of the disk will, itself, be progressively dispersed from inside to out. Computer models indicate that the truncation of the disk occurs at the level of r_{cr} on a scale of a few million years, while the dispersion that follows is much faster, perhaps a few 10^5 years. The Alma observatory allows us to access the internal structure of young disks (see Figure 4.17).

4.10.7. *Catastrophic collisions*

In the final phase of planetary formation, a few hundred very large bodies provoked catastrophic collisions that modified the composition and/or condition of the surface of several planets and satellites.

The collision of a large body with the Earth may have led to the formation of the Moon from material torn from Earth's mantle.

Mercury's extremely high density indicates that the proportion of metals in relation to silicates is much higher than in other terrestrial planets. This feature is well explained by the impact of a massive body that ripped away the planet's initial silicate mantle. Collisions by large bodies could also explain bodies with atypical axes of rotation such as Uranus, which tilted under the impact of a body with mass equal to Earth. Venus's inverted axis of rotation was, in the early days, explained by a collision with a Mars-type body. However, models of the planet's dynamic evolution showed that friction between the mantle and the core and especially tides due to a very dense atmosphere could have inverted the planet's rotation or caused its axis to tilt (see section 4.3, Volume 1). Images from Miranda suggest that this satellite of Uranus was fractured by an impact, then reformed. Studies of these phenomena show that this period lasted between 10 and 100 million years after the birth of the Sun.

4.10.7.1. *Formation of the Moon*

Isotopic dating indicates that the Moon formed between 30 and 200 million years after the Earth. This places the Moon's formation near the end of the agitated phase of collisions between planetary embryos (see also section 4.2.5.3, Volume 1). Recent models postulate that the Moon's formation was due to a giant impact between the proto-Earth and another protoplanet, called "Theia", which may have been the size of Mars. In this model, Theia crashed into Earth at a high velocity and partially destroyed it; a cloud of extremely hot debris, essentially formed by Theia's mantle, began to orbit around the proto-Earth, with the Moon subsequently accreting from this ring of debris as it cooled. Such an impact would explain several of the Moon's unusual features:

– the fact that the Moon is very low in iron compared to Earth, and also very low in volatile elements (H_2O, nitrogen, CO_2, etc.);

– the chronology of its formation;

– the very high angular momentum of the Earth-Moon pair.

Recent analyses of lunar samples have however brought to light a major problem with this theory: the fact that the surface of the Moon has the same isotopic composition as Earth for the elements O, Ti, Cr, W, and K. This is not possible if the Moon is essentially made up of Theian matter, whose isotopic composition has *a priori* little chance of resembling Earth's, because it probably formed elsewhere in the Solar System. To try to resolve this paradox, several recent models have been proposed. It is possible, for example, that the disk of post-impact debris was so hot and dense that an isotopic equilibrium with terrestrial composition was able to be produced. It is also equally possible that Earth and Theia had a common origin. Another possibility: a more energetic impact could have torn more matter from Earth, causing the disk of pre-Lunar debris to be dominated by terrestrial matter. Finally, if the rotation of the proto-Earth was initially extremely fast, the impact with Theia could have torn an enormous amount of matter from the terrestrial mantle to form the pre-Lunar disk. All of these theories have their advantages and disadvantages, but we should note that none of them raises questions about the fact that the Moon formed from the Earth-Theia impact.

4.10.7.2. *The case of Mercury*

The particularly high metal/rock ratio on Mercury was first explained by its proximity to the Sun. Models of condensation based on a high temperature in the nebula close to the Sun predicted material rich in metal at 0.3 au. Another model proposed an evaporation of the silicate mantle, connected to strong solar activity. Another hypothesis considered the impact of a body with 20% of Mercury's mass on a planet that was already differentiated, which had two times its current mass. The matter that was torn away, primarily silicates from the mantle, fell toward the Sun or was ejected by the Poynting-Robertson effect (see section 4.9.2). This latter model correctly predicts the current composition of the planet and integrates well into the general model of planetary formation. The energy of this collision would have destroyed a slightly smaller planet: Mercury may thus be considered as a survivor of the protoplanets present in the inner Solar System.

4.10.8. *Small bodies*

The planetesimals that remain between the orbits of Mars and Jupiter are those that survived perturbations due to Jupiter, which prevented them from coming together into a single body of planetary size. Perturbations from Jupiter effectively dispersed most of them and then stabilized the orbits of a minority of the objects. This belt of small bodies, the asteroids, continued to evolve as a result of rare mutual collisions (see section 2.2.1).

The planets were subjected to bombardment by the planetesimals that remained, which were dispersed by the giant planets. Some came to bombard the planets of the inner Solar System. These planetesimals, made of ice, thus could have brought water to Earth. The planetesimals ejected out of the Solar System formed the Oort cloud (see section 1.3.4.4, Volume 1). Those that were outside the planets were not disturbed and form the Kuiper belt (see section 1.3.4.3, Volume 1). The study of craters shows the progressive decrease in these bombardments over time. These events did not however completely end, as shown by the fall in 1994 of the comet Shoemaker-Levy 9 on Jupiter (see Box 5.1, Volume 1).

In the outer Solar System, the giant planets all have complex systems of rings and satellites. The rings are located inside the Roche limit (see section 4.8.3), where the tidal effect of the planet prevents the accretion of the particles of the rings into satellites. The Roche limit is the site of complex phenomena that cause rings and satellites to interact on time scales much shorter than the age of the Solar System.

There are two kinds of satellites. Regular satellites have circular, direct orbits in the planet's equatorial plane. Irregular satellites are characterized by eccentric, inclined, and/or retrograde orbits. They often exist outside the systems of regular satellites. There are many of these around the giant planets. New observational techniques make it possible to regularly increase the number of these small satellites, which are captured bodies that probably came from the Kuiper disk. The satellites of Mars, Phobos and Deimos, are probably captured asteroids.

The orbits of regular satellites lead us to think that they formed in a circumplanetary disk. This disk may be the remnant of the nebula from which the planet formed. It is more likely the result of the capture of a body by the planet. The captured body was broken when it crossed the Roche limit. Collisions led fragments to form a disk in the planet's equatorial plane, and the part of the disk outside the Roche limit transformed very quickly into satellites.

4.10.9. *Planetary migration*

We have known since the 1970s that a planet immersed in a disk of gas interacts gravitationally with it, and that this interaction can be strong enough to cause the planet to migrate toward the star or away from it. Curiously, this mechanism has initially been largely ignored in models of planetary formation, undoubtedly because the first versions of these models did not really need migration. The situation changed radically in 1995 with the discovery of the first exoplanets, and particularly the very massive *hot Jupiters* that are very close to their star (see sections 2.8.4 and 2.8.5, Volume 1), which could never have formed in the standard model (see Figure 4.13). Besides, problems encountered by the solid core scenarios to form Uranus and Neptune *in situ* also made the possibility of migration attractive (see section 4.10.4). Today, in 2020, all models of planetary formation take into account migration, which is an essential mechanism to explain some features of the Solar System as well as a number of exoplanets.

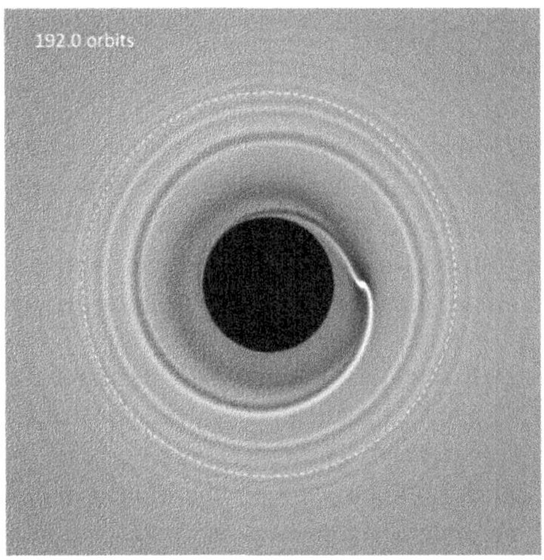

Figure 4.18. *Type 1 migration of a planet of terrestrial mass in a protoplanetary disk. We see that the planet creates grooves (density waves) in the disk, but is not massive enough to be able to empty the region around itself. The density waves lead to migration (see the text) (source: Frédéric Masset, CEA). For a color version of this figure, see www.iste.co.uk/encrenaz/solar2.zip*

However, it is necessary to distinguish between two distinct mechanisms of migration: the migration of protoplanets via interaction with the primordial gaseous disk, and the one that occurred later, of planets with a residual disk of planetesimals.

4.10.9.1. *Type I migration*

When a protoplanet attains a mass comparable to Earth's while the gas disk is still present, it begins to interact dynamically with the disk (see Figure 4.18). More specifically, the planet interacts with density waves that it creates in the gas disk. For "standard" density profiles, the result of these interactions is a loss of angular momentum for the planet, prompting its migration inward. This migration can thus be very rapid and cause the planet to fall into the star in only a few 10^4 years! This mechanism has a major problem: it's too efficient! How can we explain that the terrestrial planets in our Solar System were not thus swallowed up by the Sun? There are several solutions to this paradox. The first is that the Sun did swallow a number of terrestrial protoplanets, and that the ones we see today come from farther away in the disk and migrated to occupy the spaces left empty. The corollary is then of course that there was more solid matter at the beginning in the disk than what we see today. Another solution is that the mechanism of accretion that formed the terrestrial planets was not completed until after the dispersion of the gas disk. In this case, there is little migration, because only planetoids smaller than terrestrial mass were formed when the primordial disk was still present.

4.10.9.2. *Type II migration*

For giant protoplanets larger than 10 terrestrial masses, the situation is different. These objects effectively create a groove in the gas disk and empty the region around them. Once this empty space is created, the planet's radial evolution follows that of the disk. Now, since this disk is slowly spiraling and approaching the star, because of its viscosity, the planet will also migrate at the same pace. This regular migration is *a priori* slower than type I migration, and it occurs on a time scale comparable to the disk's viscous time scale. It seems inevitable for all growing giant planets. And this raises a question: How can we explain the presence of giants in the Solar System as well as all the giant exoplanets observed at several au from their star? How could they have escaped this final destiny? The response here may be that there is strength in numbers.

Indeed, if a planet always migrates inward, two planets together may be able to stop this migration, or even to reverse it, on the condition that the openings created by the two planets overlap and that the inner planet is 2–4 times more massive than the other. This may be what happened to the Jupiter-Saturn pair during a complex migratory process called the *Grand Tack* in reference to the pushing back of boats around a buoy during a race (see (Masset and Snellgrove 2001), Figure 4.13, and section 1.2.6, Volume 1).[3]

Type I and type II migrations stopped after the dispersion of the gas disk. The planets can still continue to migrate, but now it is through interaction with the planetesimals and planetoids that were not involved in planetary accretion. The interactions of the planets with this disk of planetesimals may be complex, because unlike with the gas, they can be perturbed on very eccentric orbits, "rebound" from one planet to another, or even be ejected from the system (Figure 4.19). We think that this is the kind of process of interaction between planetesimals and planets that is at the origin of the current structure of the outer Solar System. It is in particular this planetary game of billiards that may have placed Uranus and Neptune in their current position at 20 and 30 au from the Sun, while these planets undoubtedly formed much closer to the Sun. This could resolve the problems encountered by the solid core scenario to form giant planets far away from their star.

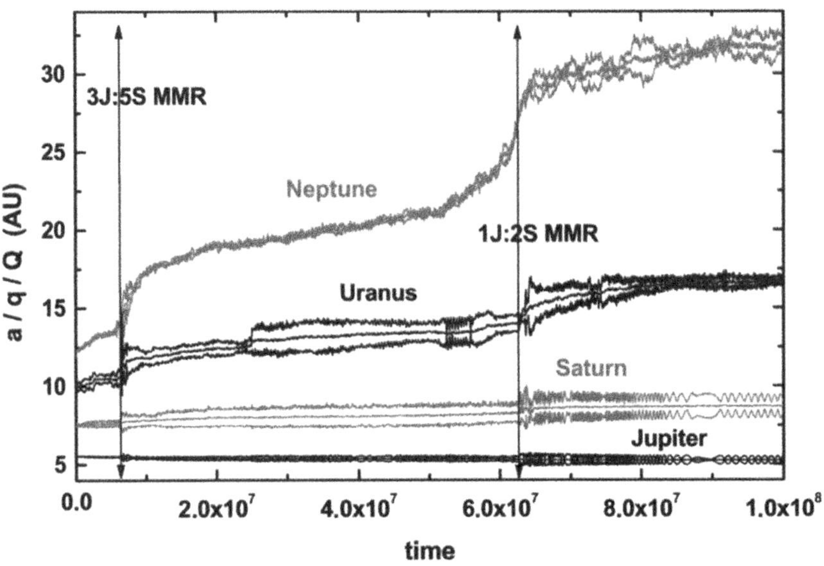

Figure 4.19. *Evolution of giant planets in the presence of a disk of planetesimals of 50 terrestrial masses. The extremes in variations of the semi-major axes are indicated, as well as the Jupiter-Saturn resonance (source: based on (Morbidelli* et al. *2007))*

3 See also Raymond, S. and Nesvorny, D (2020).

4.10.10. *Fate of the small bodies*

All of the matter that formed the terrestrial planets is found, at a given moment, in bodies of a few kilometers to a few hundred kilometers in size. The question concerning the families of small bodies still present in the current Solar System is to know why they did not evolve into larger bodies and how they arrived at their current positions.

4.10.10.1. *The asteroids in the main belt*

The Solar System does not only consist of planets – far from it. It is also inhabited by innumerable small bodies, most of which are grouped into two vast structures. The most well known is undoubtedly the asteroid belt located between Mars and Jupiter (see section 3.2.1). This region contains thousands of bodies whose total mass does not exceed a thousandth of the mass of Earth. It is three to four orders of magnitude below what we would have expected for a planet that would have formed in that location. Why did the mass remain dispersed in so many bodies? Why are the orbits of the asteroids so eccentric and inclined? Why do they have such a diversity of composition? Why are a number of small asteroids differentiated, although neither heat, nor accretion during their formation, nor radioactivity could have heated them sufficiently? Moreover, the distribution of spectral types of asteroids as a function of the semi-major axes shows that there was some mixing of the planetesimals in this region. The heliocentric distribution of classes of asteroids respects the decreasing temperature of the primordial disk, with materials that condensed at high temperatures closer to the Sun and materials that condensed at lower temperatures outside the asteroid belt. All of these elements show that they are not debris from an ancient planet that was destroyed, as was thought before we knew more about the properties of these objects. It is instead a region where a planet never could have formed, very probably because of perturbations from Jupiter.

A very important point is that the total mass of this belt is very low in relation to what we might expect in a protoplanetary disk having formed terrestrial and giant planets. If we consider, for example, the minimal mass solar nebula (MMSN[4]), we consider that there should have been initially 1,000 times more solid matter in the region of the asteroids than there is today. The great challenge for all models of formation is thus to explain how 99.9% of the mass of the asteroid belt could have disappeared. There is no consensus yet how the belt formed. What is certain is that it is impossible that the dynamic perturbations of Jupiter (at least, the current Jupiter) could have by themselves ejected so much matter from between 2 and 4 au.

4 See (Armitage 2017).

One solution may be that the perturbations of Jupiter acted indirectly by exciting large planetary embryos that formed in the belt. These embryos, once placed on eccentric orbits, could themselves have perturbed and ejected the majority of the primitive asteroids before disappearing, either by falling into Jupiter or by being ejected. Another scenario supposes that, at the beginning of the protoplanetary disk, these four giant planets formed closer to the Sun than their current positions and then migrated in the disk. Jupiter would have first migrated inward (type II migration) and ejected all or part of the bodies present in the current asteroid region.

But once Saturn had almost rejoined it, the interactions between the two giant planets caused them to migrate outward, a fate that Uranus and Neptune also shared. In the dynamic chaos that followed, some planetesimals from outer regions could have been injected into the < 4 au region. The asteroid belt would thus in the end be populated by a mix of bodies that formed there, the majority, and others that came from different regions of the Solar System, but its total mass remained very low. This is the scenario presented in Figure 4.13.

4.10.10.2. *The Kuiper disk and the Oort cloud*

The populations of objects in the Kuiper disk and the Oort cloud have different orbital properties (see section 3.1.2). Even if some objects in the Kuiper disk were clearly perturbed by planets, a part of the Kuiper disk is made up of objects on orbits that are circular and minimally inclined on the ecliptic. This can only be explained if these objects formed where they are currently located. On the contrary, comets in the Oort cloud, which are on very eccentric and inclined orbits, cannot form in this configuration. However, we know they contain a significant portion of volatile elements, and this proves that the matter they are composed of condensed in the outer Solar System.

Computer models show that Oort objects formed in the vicinity of the giant planets, at 3–30 au from the Sun, before being ejected by gravitational perturbations from these planets. When they reach 10,000 au, galactic gravitational perturbations then take the lead to recircularize and disperse their orbits, to end up with the current situation in the Oort cloud, which has a very slightly flattened ellipsoidal shape. It is possible that the Oort cloud also contains objects that came from other stars, ejected then captured in circumsolar orbits. Models also show that 10 to 1,000 terrestrial masses of planetesimals had to be ejected from the vicinity of the planets to achieve the current system.

When we consider the distribution of the matter that originated the planets, we find an abrupt interruption beyond Neptune. Now, the presence of Kuiper objects that formed in their current position shows that the absence of a planet beyond Neptune (apart from the potential tenth planet that would have been a captured or strongly perturbed object) is not explained by the lack of planetesimals. Another explanation is required. Scenarios involving extensive planetary migration could be an answer, but they need to be refined.

4.10.10.3. *Planetary satellites and small bodies*

The satellites in the Solar System exhibit a wide diversity of properties that are explained by different mechanisms of formation, ranging from ancient formation *in situ* to recent captures.

Around the giant planets, there are three kinds of satellites: the innermost bodies are small, near or in the rings themselves: they probably share a common origin with the rings (see sections 1.3.3.2 and 1.5). At distances between a few and a few dozen planetary radii, we find larger satellites on more regular orbits. Their regularity is the sign of formation in a circumplanetary disk, similar to planetary formation. Finally, all of the giant planets' systems of satellites also have satellites that are distant from the planet, which are often small and have eccentric, inclined, or even retrograde orbits. These objects, initially on circumsolar orbits, were probably captured by the planet.

Regular satellites have almost circular orbits, mostly inside the Hill sphere. Their properties are clearly explained by formation in a disk in the equatorial plane of the planet. This disk can be formed at the same time as the planet. It can also come from a later capture of circumsolar matter. In this disk, the satellites are formed by core accretion. Jupiter's Galilean satellites exhibit decreasing density as distance from the planet increases, a sign of formation in a disk that was "heated" by the planet.

Rings can only last inside their planet's Roche limit, which prevents them from accreting into satellites. Dust rings like Jupiter's are recent, and their particles are constantly being renewed. The rings of Uranus and Neptune are also very likely the result of captured bodies that were fractured as they crossed the planet's Roche limit (see sections 1.5.5 and 1.5.6). But the question of knowing if Saturn's massive rings were formed at the same time as their planet or were also the result of posterior capture is still open (see section 1.5.4).

The discovery of rings around an asteroid-type body like Chariklo and a dwarf planet like Haumea shows that many planetesimals must have had rings that subsequently disappeared (or are too small to be detected). It is possible that this is connected to the large number of binary objects in the Kuiper belt.

There is no observational trace of rings around the terrestrial planets, but theories based on computer simulations explain the orbits of Mars's two small satellites by successive cycles of formation and destruction of the rings and satellites.

4.10.11. *Exoplanetary formation*

Are the properties of the thousands of exoplanets that have been discovered (see section 2.8, Volume 1) compatible with the formation scenario described for planets in the Solar System?

Some systems of exoplanets support the mechanism of planetary formation by accretion in a protoplanetary disk: the Kepler-90a system has a structure that is similar to the Solar System, although more compact, with the orbit of Kepler-90h being at a radius of 1 au (see Figure 4.20). This system of eight planets around a solar-type star, coplanar, whose inner planets are terrestrial and whose outer planets are gaseous, is well explained by a classic formation in a disk, followed by a migration of the system toward the star.

However, a number of other systems exhibit properties that are incompatible with the processes of planetary formation in the Solar System. For example, some exoplanets orbit at distances much greater than the sizes of protoplanetary disks. Thus, Ross 458(Ab)c is a planet equal to 11 Jovian masses detected at more than 1,000 astronomical units from its star, which corresponds to an orbital period of more than 50,000 years! We have tried to explain this by formation near the star followed by outward migration, but the models don't work. Thus, despite its low mass, this planet probably formed like a star, by gravitational collapse of a molecular cloud. Another strange case is planet GJ 3512b. This giant planet, which is in a very eccentric orbit around a star ten times less massive than the Sun, poses another problem: it is very difficult to explain the formation of a planet in a disk around a star that is barely 250 times more massive. An alternative theory of planetary formation based on instabilities in a massive circumstellar disk could explain this type of planet. But the details of this mechanism still need to be refined.

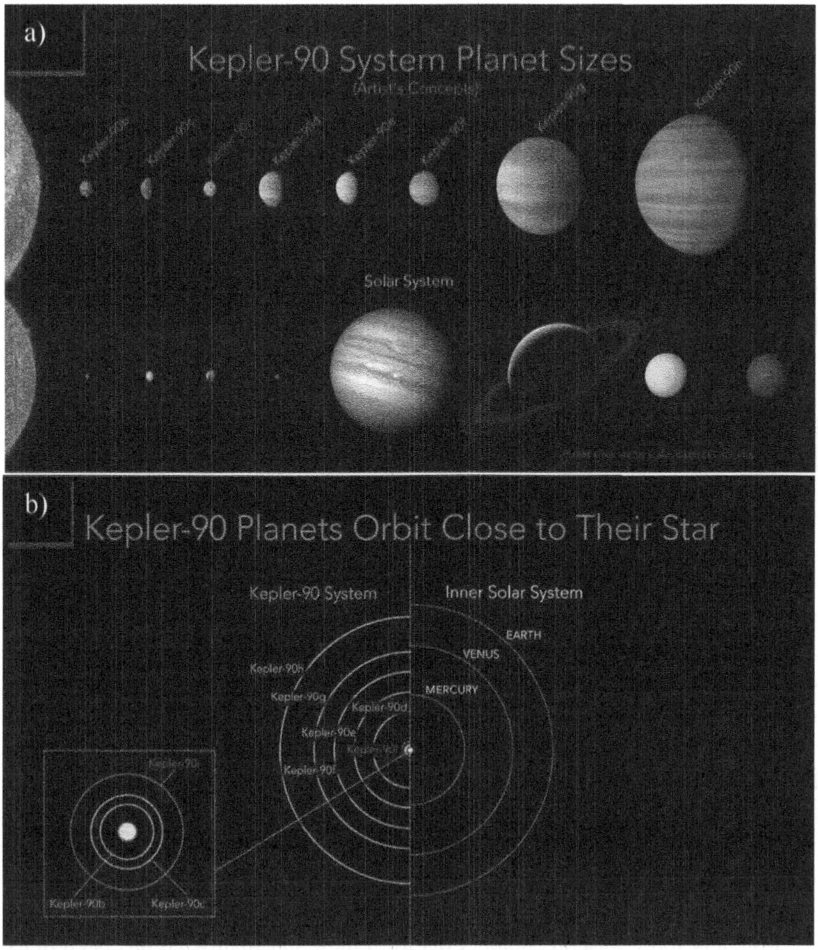

Figure 4.20. *(a) Comparison of the sizes of planets in the Kepler-90 system and the Solar System. The sizes of the planets are to scale, but the distances are not. The drawings of the surfaces are conceptual. (b) Comparison of orbits of the Kepler-90 system and the Solar System (source: NASA, Wikimedia Commons). For a color version of this figure, see www.iste.co.uk/encrenaz/solar2.zip*

Contrary to the giant planets in the Solar System, which are all more than 5 au from the Sun, a large number of gaseous exoplanets have been discovered very close to their star, thus well inside the ice line, far from the icy region where gaseous planetary cores can form (see section 2.8.5, Volume 1 and section 4.10.4 in this volume). Moreover, in the large family of these *hot Jupiters* with periods less than 200 days, there is an accumulation of *very hot Jupiters* that orbit their star in less than 10 days, which corresponds to planets 10 times closer to their star than Earth is to the Sun! If we consider that they formed, like the giants in the Solar System, in the outer icy section of the disk, we would have to explain how they then migrated inward in the system. Several types of mechanisms have been conceived: an interaction between the planet and the protoplanetary disk, an interaction with another planet (or even a star if the central star is binary), or the tidal effect of the central star. These mechanisms are fairly well understood, but they do not explain at all why the planet's migration did not end with it falling into the star. Two mechanisms have been proposed to explain why migration in a disk would so often stop near 0.1 au. The planet's fall can be stopped when it begins to be subjected to the star's tidal effect. The star's perturbation can also oppose the viscous torque of the disk and confine an inner edge of the disk, as do shepherd satellites for planetary rings (see section 1.3.3.2). Planetary migration then stops once it arrives in the empty region. The presence of terrestrial planets can also confine the inner edge of the disk. But none of these mechanisms apply to hot Jupiters.

Some hot Jupiters may have been subjected to perturbation by a planet (or a star) by the so-called "Kozai" mechanism, developed to explain the origin of some irregular satellites in the Solar System, where there was an exchange between the eccentricity of a body and the inclination of the other body. The models show that this mechanism may play a role in the evolution of the planetary orbits of Upsilon Andromeda c and d. This same mechanism may explain the high eccentricity of exoplanet HD 80 606b, which could have been perturbed by the stellar companion of HD 80 606.

A number of giant planets are found near their star, but this may at least partially be observational bias: detection by gravitational microlenses, whose sensitivity does not depend on the distance between the planet and the star, indicate that there are proportionately more gaseous planets far from their star than what has been detected by the methods of transit and radial velocity (which require following the planet for several orbits). There may be a number of giant exoplanets beyond the ice line that remain to be discovered.

Given the large number of very agitated systems, it is not at all impossible that a planet formed in a protoplanetary disk and was then ejected from the system. This may be the case for objects floating far from any star, like PSO J318.5-22 (six Jovian masses). Another possible explanation is that some of these floating planets, also called brown sub-dwarfs, are formed in the same way as the stars, by collapse of a molecular cloud. Only the detection of a solid core would make it possible to tell the difference, but this observation is almost impossible.

We see that the existence of most exoplanets cannot be explained by the same scenario as the Solar System. Theoreticians have not lacked imagination to explain the various configurations that have been observed. Sometimes for this purpose, they dig through the mechanisms that were proposed for the Solar System but were later rejected. Several of these models could explain the formation of some extrasolar systems, as if all of the mechanisms that have emerged from the theoreticians' imaginations had to find their place somewhere in the sky. But each new discovery seems to contribute even greater diversity into the bestiary of exoplanets as well as new questions about their formation.

4.11. References

Armitage, P.J. (2017). Lecture notes on the formation and early evolution of planetary systems [Online]. Available at: https://arxiv.org/pdf/astro-ph/0701485.pdf.

Burns, J.A., Lamy, P.L., Soter, S. (1979). Radiation forces on small particles in the Solar system. *Icarus*, 40, 1–48.

Chenciner, A. (2007). Three body problem. *Scholarpedia*, 2(10) [Online]. Available at: http://www.scholarpedia.org/article/Three_body_problem.

Laskar, J. (1994). Large-scale chaos in the Solar system. *Astronomy and Astrophysics*, 287, L9–L12.

Laskar, J., Joutel, F., Boudin, F. (1993). Orbital, precessional, and insolation quantities for the Earth from -20 Myr to +10 Myr. *Astronomy and Astrophysics*, 270, 522–533.

Lissauer, J. and de Pater, I. (2013). *Fundamental Planetary Science*. Cambridge University Press, Cambridge.

Marzari, F., Scholl, H., Lagerkvist, C.-I. (2002). *Origin and evolution of Trojan asteroids* [Online]. Available at: https://cutt.ly/GyEmys6.

Masset, F. and Snellgrove, M. (2001). Reversing type II migration: Resonance trapping of a lighter giant protoplanet. *Monthly Notices of the Royal Astronomical Society*, 320, L55–L59.

Morbidelli, A., Tsiganis, K., Crida, A. *et al.* (2007). Dynamics of the giant planets of the Solar System in the gaseous protoplanetary disk and their relationship to the current orbital architecture. *Astronomical Journal*, 134, 1790–1798.

Murray, C.D. and Dermott, S.F. (2000). *Solar System Dynamics*. Cambridge University Press, Cambridge.

Nesvorný, D. and Morbidelli, A. (2012). Statistical study of the early Solar system's instability with four, five, and six giant planets. *Astronomical Journal*, 144(117).

Pollack, J.B., Hubickyj, O., Bodenheimer, P. *et al.* (1996). Formation of the giant planets by concurrent accretion of solids and gas. *Icarus*, 124, 62–85.

Raymond, S. and Nesvorny, D (2020). Origin and dynamical evolution of the asteroid belt [Online]. Available at: https://arxiv.org/abs/2012.07932.

Roccatagliata, V., Bouwman, J., Henning, T. *et al.* (2011). Disk evolution in OB associations: Deep SPITZER/IRAC observations of IC 1795. *Astrophysical Journal*, 733(113).

Shields, A.L., Barnes, R., Agol, E. *et al.* (2016). The effects of orbital configuration on the possible climates and habitability of Kepler 62-f. *Astrobiology*, 16, 433–464.

Walsh, K.W., Morbidelli, A., Raymond, S.N. *et al.* (2011). A low mass for Mars from Jupiter's early gas-driven migration. *Nature*, 475, 206–209.

5

Origin of Life and Extraterrestrial Life

James LEQUEUX
LERMA, Paris Observatory, PSL University, Paris, France

5.1. Definition of life

We only know of one example of life: life on Earth. All of our concepts are necessarily influenced by this example. But terrestrial life may only be a special case. There may be life elsewhere in the Solar System and on exoplanets, and it may be different; could we be only an element in a larger, unknown set? In this domain, as in many others, it is important to be skeptical of an overly anthropocentric vision.

That said, we believe that life results from the action of the laws of physics and chemistry on appropriate components when the environment is ready for it. It will thus be the natural result of the evolution of certain chemical systems under favorable conditions. The observation of bacteria and even multicellular organisms around hydrothermal springs at the bottom of the ocean or near volcanoes, and in other unexpected environments such as sub-glacial lakes, shows that these conditions can be extremely varied. It nevertheless appears that the presence of liquid water and a source of energy, as well as nutrients, are necessary conditions for the appearance and maintenance of life. For that matter, all living beings on Earth contain water.

The definition of life is most often based on the principal functions of a living being: the possibility of reproduction, the ability to use energy in the environment, differentiation from the surrounding environment, and creation of an organizational

The Solar System 2,
coordinated by Thérèse ENCRENAZ and James LEQUEUX. © ISTE Ltd 2021.

structure. However, this definition is insufficient. In their formation, growth, and multiplication, crystals meet all of these criteria, and no one would consider them to be alive. We can also use a biochemical definition, such as the presence of DNA. But are we sure that living beings elsewhere than on Earth use DNA for their genetic code, or could they use another kind of molecule? Whatever the case, living organisms must contain genetic material.

Most exobiologists define life by the possibility of Darwinian evolution. In addition to the previous criteria, this implies the possibility of mutations and natural selection. Organisms must also be mortal, or else evolution would be impossible. Monocellular algae corresponds to all of the criteria we have enumerated, and they represent the simplest form of life on Earth. These forms are already complex and elaborate, with walls, proteins, DNA, and potentially a nucleus. Giant viruses can also be considered to be alive, but they can only exist as parasites.

5.2. The appearance of life on Earth

5.2.1. *Physicochemical conditions*

It is reasonable to suppose that the first living organisms on Earth were brought here by comets or meteorites, and that others arrived here at different times: this is the *panspermia* theory developed by physicist Hermann von Helmholtz (1821–1894). But this theory only shifts the problem, because life had to have appeared elsewhere, on Venus or on Mars, from which we occasionally receive a fragment expelled by the fall of an asteroid or a large meteorite. Life could have indeed existed on Mars when there was liquid water on the planet, for a period of time about 1 billion years after its formation. As it turns out, and as we will see, life appeared on Earth during this period. Venus could also have harbored life from 1 to 3 billion years after its formation, before becoming too hot. Thus, the idea that life came from Mars or Venus cannot be rejected, but it is still more likely that it arose on Earth.

When the Earth formed 4.6 billion years ago, it was liquid, so hot that water vapor and other volatile molecules were released through evaporation. Among the first solid elements, zircon crystals appeared on the terrestrial surface after about 150 million years. Since they can only crystallize in the presence of liquid water, this means that water was already present during this era on the cooled Earth. Where did it come from? We think that some planetoids that became part of Earth could have contained water and other volatile elements if they formed at a great distance from the proto-Sun and migrated into the area of rocky planetary formation (see section 4.10.2). Water could also have been brought after Earth's formation by bombardment from comets and meteorites. The oceans were thus present very early in our globe's evolution. The Late Heavy Bombardment, which occurred around 800 million years after the formation of Earth (see section 4.10), devastated Earth's surface, caused the oceans to partially boil away, and may have destroyed any form of life that appeared previously. But it seems that Earth's water and atmosphere were able to survive this event in some form.

What was this atmosphere made of? It could not be interstellar gas because this gas was expelled very early in the region where Earth formed, only leaving behind the solid refractory elements that make up our globe (see section 2.5, Volume 1). This has been confirmed by the analysis of isotopes of rare gases in the current terrestrial atmosphere, whose ratios are different from those in the Sun, which reflect the ratios of primitive interstellar gas. The atmosphere thus formed from the water, carbon dioxide (CO_2), and nitrogen in some of the planetoids or that was brought here later by the fall of meteorites and, less importantly, of comets.

After the Late Heavy Bombardment, a considerable greenhouse effect particularly due to CO_2 initially brought the temperature on the terrestrial surface to a value on the order of 200°C, even though the young Sun was then almost 30% less luminous than it is today. As for the oceans, their temperature has decreased over the last 3.5 billion years from around 65°C to the current temperature (see Figure 5.1). As the first forms of life observed date from this era, they could only have appeared in the oceans.

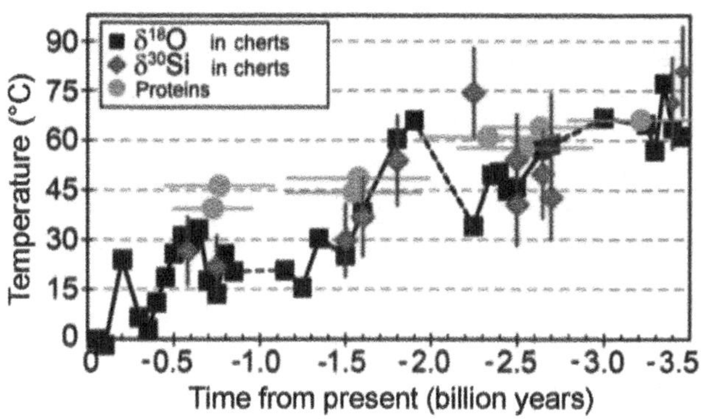

Figure 5.1. *Temperature decrease in the oceans since 3.5 billion years ago, shortly after the Late Heavy Bombardment. For a color version of this figure, see www.iste.co.uk/encrenaz/solar2.zip*

COMMENTS ON FIGURE 5.1.– *The present time is on the left on the x-axis. The temperature was evaluated based on isotopic ratios* $^{18}O/^{16}O$ *and* $^{30}Si/^{28}Si$ *in chert, which is an ancient siliceous sediment. The variations in these ratios are expressed by:* $\delta^{18}O = [(^{18}O/^{16}O)/(^{18}O/^{16}O)_{present} - 1] \times 1{,}000$*, and likewise for silicon. The isotopic ratio* $^{18}O/^{16}O$ *measured in deposits of organic matter in rocks by the decay of proteins is also indicated, confirming the other results (source: based on Tartese* et al. *2017).*

The decrease in Earth's temperature was the result of the reduction in the greenhouse effect, which itself was due to the decrease in the quantity of CO_2 in the atmosphere, with the convective motion of the terrestrial mantle having carried away a large amount of CO_2 in the form of carbonates in basalts produced by volcanism. For that matter, the Earth found itself completely frozen 2.3 billion years ago, and we find traces of this glacial phenomenon in different regions of the world. This event did not affect the temperature of the oceans, either because it did not last for very long or because it only affected land above sea level.

Subsequently, the planet experienced alternating glacial and interglacial periods, along with violent phenomena that could have caused mass extinctions such as the one that wiped out the dinosaurs 66 million years ago, but life always managed to survive and develop.

5.2.2. *The first forms of life*

The oldest confirmed forms of life date back 3.7 billion years; these are structures analogous to current stromatolites, with shells, filaments, rod cells, etc., which are considered microfossils with biological activity. Some were capable of photosynthesis, but without releasing oxygen. These organisms seem to have used several kinds of metabolisms, a sign of an already-evolved form of life: thus, life must have appeared even earlier in Earth's history. Other traces even older than 3.7 billion years have indeed been discovered, but their nature is controversial.

The first bacteria were methanogenic bacteria, which draw energy from the production of methane, which also contributed to the greenhouse effect. Cyanobacteria (monocellular blue algae) appeared 700 million years later, and they produced oxygen from photosynthesis: this oxygen not only destroyed the methanogenic bacteria, but methane itself, which decreased the greenhouse effect and contributed to the lowering of terrestrial surface temperature. We see that the terrestrial atmosphere and the temperature of Earth were already affected by the effects of life, which is still true today.

The first forms of life appeared in an atmosphere dominated by carbon dioxide and in the presence of water. The necessary organic molecules may have been synthesized from a mixture of water, methane, and ammonia vapor, as in the Miller-Urey experiment, but it was a very different reductive environment in the atmosphere at the time (for a description of this kind of laboratory experiments, see (Ponnamperuma 1964), which is incidentally a really visionary article). However, bacteria may have been born in a locally reductive environment. Other possibilities exist: an example is given by hydrothermal sources on the ridges in the Atlantic and the Pacific Oceans, where chemical reactions produce prebiotic molecules from hydrogen sulfide SH_2, dissolved CO_2, and mineral nitrogen substances: they shelter large colonies of bacteria. Despite very different conditions, prebiotic molecules are always carbon-based. Some have imagined that they could be silicon-based, which is tetravalent like carbon, but silicon is much less abundant than carbon, its chemistry is much less rich, and its compounds are often unstable: among 200 molecules detected in the interstellar medium, we only know of a few molecules containing silicon, while most of them contain carbon. The idea should be abandoned.

A completely serious possibility is the contribution of prebiotic molecules by asteroids and especially comets (Altwegg *et al.* 2016). Recent studies based on the chemical composition of the comet 67P/Churyumov-Gerasimenko observed by the European Rosetta probe suggest that while the comets did not bring all water to Earth, they did supply a large quantity of prebiotic material, especially during the Late Heavy Bombardment. The mass of this material could have been equivalent to the total biomass present today on our globe (around 2 10^{12} tons) (Marty *et al.* 2016). What is it made of? We have detected in comet 67P no less than 58 different molecules, primarily organic, including an amino acid, glycine, phosphorus monoxide PO, an indispensable product for DNA synthesis, and particularly chemically active products such as hydrogen cyanide HCN and formaldehyde H_2CO. The comet also contains a large quantity of refractory organic material, which may primarily be polyoxymethylene, a macromolecule that results from the polymerization of formaldehyde, and a similar quantity of mineral matter, primarily silicates. Some meteorites, the carbonaceous chondrites, contain material that is apparently similar to cometary matter but has been reheated, which may have participated in the seeding of Earth. Recent experiments on irradiated ice produced amino acids, sugars, and other complex organic molecules: see for example (Fresneau *et al.* 2017).

Meteorites contain numerous amino acids, but we think that all except glycine formed through aqueous alterations, both in the meteorite and in laboratory analysis. We should also note that among the 80 different amino acids found in meteorites, there are only 8 of the 20 necessary for terrestrial life.

Living matter is characterized by two fundamental molecules whose interaction produces life: proteins and nucleic acids. Although these molecules are very complex, they are made up of only a few chemical components. The proteins in terrestrial living matter, whether animal or vegetable, result from the combination of only 20 amino acids. Deoxyribonucleic acid (DNA), which contains genetic information and directs the production of proteins, is made of sugar, D-deoxyribose, four nitrogenous bases, and a phosphate base. We conclude from this that all living organisms must have a common chemical ancestor: a purely chemical form of evolution thus necessarily preceded biological evolution. This idea dates back to Charles Darwin (1809–1882) himself, who wrote in a letter to a friend:

> But if (and oh what a big if) we could conceive in some warm little pond with all sorts of ammonia and phosphoric salts, light, heat, electricity etc. present, that a protein compound was chemically formed, ready to undergo still more complex changes...

But it is in fact much older, although formulated in a much less scientific form, obviously. Indeed, Diodorus Siculus (around 90–0 BCE), in volume 1 of his *Universal History*, which is supposed to have transmitted the ideas of Anaxagorus (around 500–428 BCE), but which was probably influenced by Lucretius (around 98–55 BCE), wrote (translation by Charles Henry Oldfather (1887–1954)):

> The Earth separated ... into the sea and the land; which was like potter's clay and entirely soft. But as the sun's fire shone upon the land, it first of all became firm, and then, since its surface was in ferment because of the warmth, portions of the wet swelled up in masses in many places, and in these, pustules covered with delicate membranes made their appearance... By night the living things forthwith received their nourishment from the mist that fell from the enveloping air, and by day were made solid by the intense heat; and finally, when the embryos had obtained their full development and the membranes had been thoroughly heated and broken open, there was produced every form of animal life.

The synthesis of proteins from amino acids began with peptides formed by the combination of two of these molecules (see Figure 5.2). These peptides easily form chains of polypeptides. Proteins are polymers of such peptide sequences, where different amino acids are distributed in a defined order for each protein. For example, egg white protein corresponds to a polypeptide chain of 129 molecules belonging to these 20 amino acids. We should note that no peptide has been found in the interstellar medium, comets, or meteorites, which implies that peptide synthesis occurs under specific circumstances: the simplest possibility for providing the necessary energy is a reaction that takes place on a structure of clay, which was effectively produced in the laboratory in 1970.

Nucleic acids are constructed from four molecules called nucleotides, which contain nitrogen cycles, purine or pyrimidine. They are prone to combining in long chains connected by phosphate groups (see Figure 5.3). Atomic phosphorus, nitrogen, and some phosphorylated and nitrogenated molecules necessary for nucleotide synthesis have been detected in comets and the interstellar medium. Some nucleotides have been synthesized by the effect of ultraviolet radiation on ice recovered from interstellar grains (Ruf *et al.* 2019).

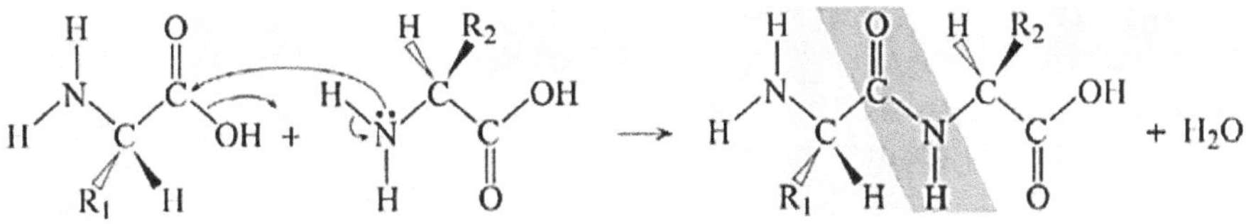

Figure 5.2. *Formation of a peptide link between two amino acids. R_1 and R_2 are characteristic radicals of each amino acid (for example, H for glycine, CH_2OH for serine). The arrows on the left indicate the reaction's detailed mechanism. The amide group –CO–NH– in gray is the peptide link that connects the two sections (source: based on (Luft 2014))*

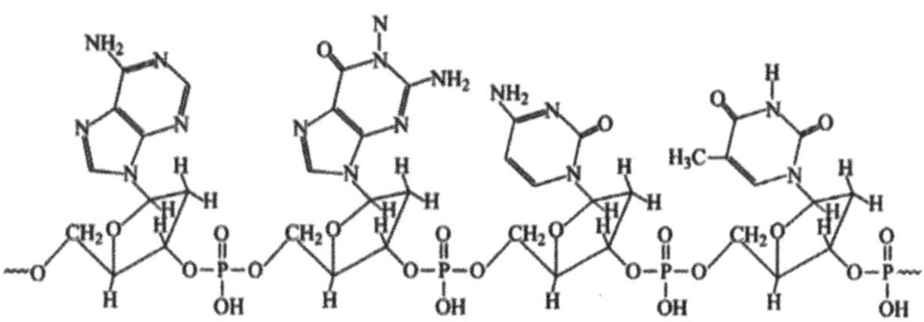

Figure 5.3. *Example of a structural fragment of DNA formed by four nucleotides connected by phosphate groups; carbon atoms are not generally represented to simplify the presentation (source: based on (Luft 2014))*

Deoxyribonucleic acid or DNA is formed from strands made of sequences of polynucleotide chains coiled into a double helix. Given a sequence of nucleotides such as –A–G–C–T– in one of the strands, the same sequence will be found in the reverse order –T–C–G–A– in the other strand, in order to ensure the largest number of hydrogen bonds to connect the two strands and to reduce steric repulsive forces as much as possible. Some nucleotides, particularly adenine and guanine, were able to be synthesized in experiments that aim to reconstitute the conditions for the appearance of life (Ponnamperuma 1964).

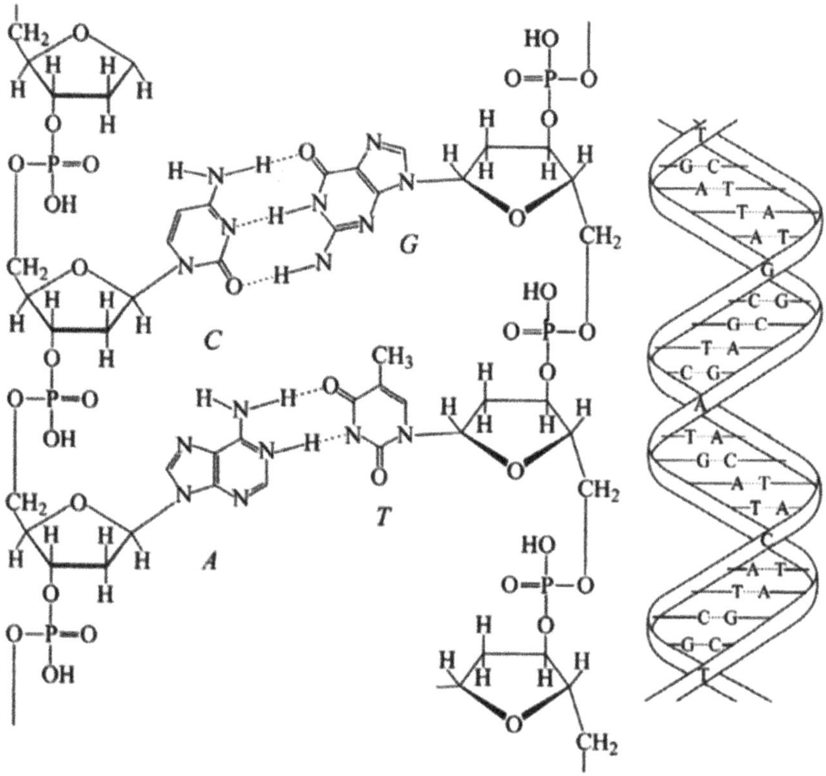

Figure 5.4. *An example of a fragment of the DNA double helix*

COMMENTS ON FIGURE 5.4.– *Carbon atoms are not generally shown to simplify the presentation. We see the four nucleotides from Figure 5.3 (A, G, C, and T). The two helices are connected by N-C bonds on the right and hydrogen bonds on the left (dotted lines). On the right is a simplified view showing an example of the placement of nucleotides (source: based on (Luft 2014)).*

Some sections of DNA contain genetic information: these are genes. This information is encoded in the order in which the nucleotides are laid out. It will itself define the order of amino acids in the protein constructed from the gene. For this, the gene's genetic information is copied by a different nucleic acid, ribonucleic acid or RNA, which then encodes the protein. RNA exhibits a structure similar to DNA, with a ribose-phosphate skeleton and bonds to which nitrogenous bases are attached: adenine, guanine, and cytosine are present as in DNA, while uracil replaces thymine. RNA is able to ensure both the genetic information and the catalysis, which DNA cannot do; on the other hand, this molecule is made up of a single strand combining short sequences of nucleotides, while DNA, with its double helix, which contains millions or billions of nucleotides, is capable of identical self-replication. It is today generally accepted that an "RNA world" preceded the DNA world. What is the advantage of DNA in relation to RNA? The stability of the double helix, much longer than strands of RNA, enables the constitution of large genomes, and also the possibility for evolution thanks to errors that occur during replication. However, we are in the dark with regard to the initial formation of these complex macromolecules, for which only the first steps have been reproduced in laboratory experiments, the formation of amino acids and nucleotides. It is the same for the formation of chlorophyll, hemoglobin, etc.: that is the problem of the origin of life.

5.2.3. *The formation of living cells*

In addition to proteins and DNA, living cells like algae or monocellular bacteria have an envelope that separates them from the outside environment, through which they make exchanges of energy or products, particularly nutrients. This envelope is made of lipids, which in the case of living matter are very often phosphoglycerides (see Figure 5.5). Lipids are amphiphiles; in other words, one of their extremities (the head) possesses a dipole moment that attaches to water molecules, which also have one. The other end, which does not have a dipole moment, is hydrophobic. In the presence of liquid water, lipids stick to each other on the surface perpendicularly such that the hydrophilic heads are in contact with the water, while the hydrophobic tails remain in the air. They thus form a monomolecular film on the surface of the water.

Lipid monolayers can spontaneously assemble into dual structures that are flat, then curved, as shown in Figure 5.6: they thus separate an outer and an inner environment, as for a living cell, and constitute filters capable of allowing some molecules to pass while preventing others. Laboratory experiments on amino acids extracted from the Murchison meteorite, which contain almost 80 of them, saw more or less spherical vesicles similar to liposomes develop rapidly in water. We are not far from cell walls!

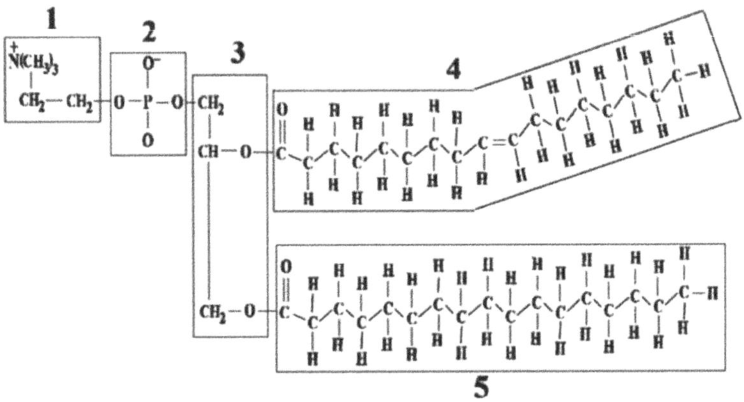

Figure 5.5. *Formula for a typical phosphoglyceride. Group 1 is choline, group 2 is phosphate, and group 3 is glycerol; together, they constitute the hydrophilic head, which possesses a dipole moment thanks to + and - charges. Tails 4 and 5 are fatty acids, unsaturated and saturated, respectively, which are hydrophobic (source: Wikimedia Commons/Danntzikg)*

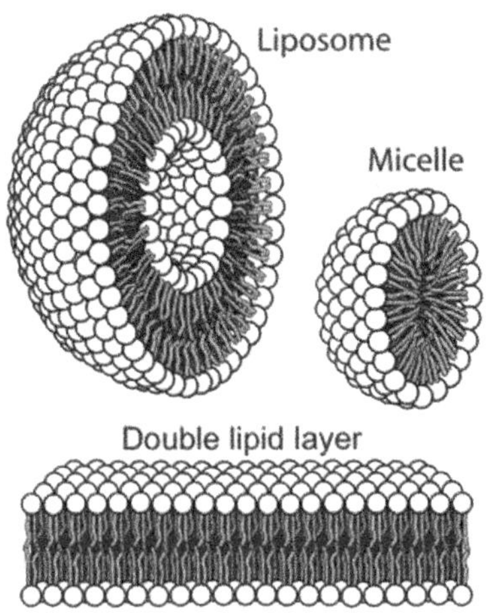

Figure 5.6. *Structure of lipid layers. For a color version of this figure, see www.iste.co.uk/encrenaz/solar2.zip*

COMMENTS ON FIGURE 5.6.– *The hydrophilic heads of molecules are represented by white spheres, and their hydrophobic tails are represented by yellow filaments. Monolayers can close up on themselves, forming micelles, which occur frequently in colloidal solutions. Double layers can also close up on themselves, forming liposomes that can be produced artificially (source: Wikimedia Commons/ LadyofHats).*

We also need to understand not only how real cell walls can be formed, but also how they can trap macromolecules like proteins or DNA. Experiments have shown that if a mixture of vesicles and those macromolecules is subjected to cycles of hydration and dehydration, they can become trapped because of links that form to replace those left open by the lack of water. This can occur in swamps that become periodically dry. We then have to understand the origin of metabolism, which ensures the survival of cells. We have a long way to go.

Once the first bacteria were born, evolution constructed from them living beings that were progressively more complex, according to the laws of natural selection. To attempt to understand the mechanisms that led to the extraordinary diversity of forms of life, it is interesting to try to determine the characteristics of our *Last Universal Common Ancestor*, Luca.

Several teams of biologists have reconstructed a cell containing the minimum number of genes (albeit several hundred) that are capable of transmitting genetic information by RNA synthesis based on DNA and its translation via protein synthesis based on RNA, as well as the energy and structural maintenance of the cell. Luca is at least 3.5 million years old, which was undoubtedly several hundred million years after the appearance of the first forms of life. In the tree of life, Luca appeared just before the great bifurcation between bacteria and Archaea, then between Archaea and Eukaryotes, multicellular organisms that subsequently evolved into mushrooms, plants, and animals (see Figure 5.7).

This took a long time: 2.3 billion years elapsed between the appearance of the first bacteria and the first multicellular organisms, and another 650 million years before the explosion of life in the Cambrian era. In particular, chlorophyll had to be synthesized by plants, which perform photosynthesis by absorbing CO_2 and producing oxygen, which currently makes up a large part of the atmosphere: this took place 2.5 billion years after the formation of the Earth, while bacteria appeared 700 million years earlier. On Earth, evolution has been marked by variations in climate, continental drift, several periods during which a large number of species went extinct, the origin of which is still under debate, and more recently by human causes. Finally, the increase in the Sun's temperature will mean that Earth will become difficult to live on in 1.5 billion years, and even sooner if humans continue to contribute to global warming.

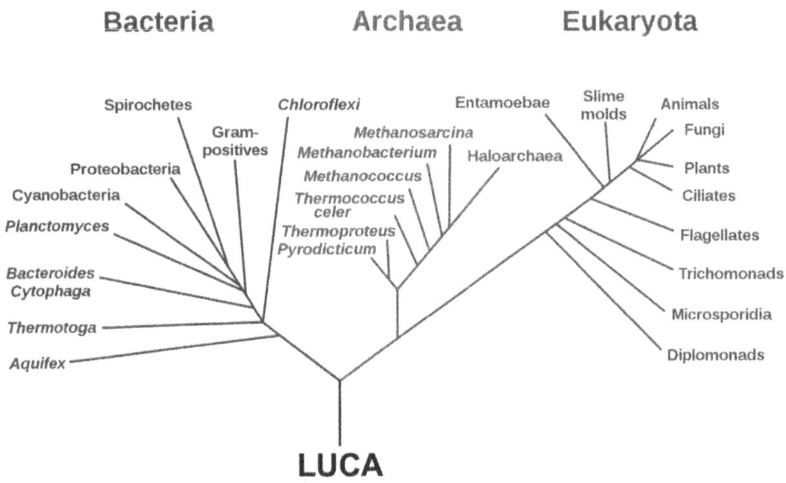

Figure 5.7. *Diagram representing Carl Woese's tree of life (1928–2012) (source: based on Wikimedia Commons). For a color version of this figure, see www.iste.co.uk/encrenaz/solar2.zip*

5.3. Life elsewhere in the Solar System

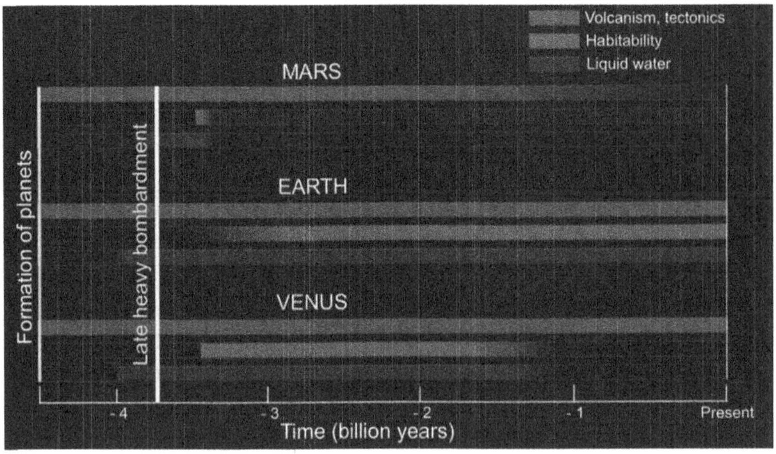

Figure 5.8. *Possibilities of life on Mars, Earth, and Venus. For a color version of this figure, see www.iste.co.uk/encrenaz/solar2.zip*

COMMENTS ON FIGURE 5.8.– *The presence of liquid water (in blue) is mandatory for life. Volcanism (in red), like the plate tectonics present on Earth, influences the atmosphere and forms continents, which facilitates life on planets (in green). The data on Venus are uncertain, and current conditions prohibit any form of life on the surface (source: based on (Darby Dyar* et al. *2019).*

5.3.1. *Mars*

Two planets other than Earth are or were in the habitable zone: Mars and Venus (Darby Dyar *et al.* 2019). Mars is currently a very dry celestial body, where water only seems to exist in the form of ice, but there are a number of pieces of evidence of liquid water in an already distant past, obtained by the successive probes and *rovers* that have explored the planet: geological signatures of fluvial networks, deposits of clay, enrichment in deuterium, fossil ice, etc. (see sections 4.3.2.3 and 4.3.3.7, Volume 1). Liquid water was abundant until 3.5 billion years ago, then it became intermittent until disappearing completely 3 billion years ago. Thus, if life appeared on Mars, it was probably at the end of this period. It would have had to spread quickly, without having the time to reach evolved forms (see Figure 5.8).

To look for places on Mars that would have been suitable for life, we can apply the usual criteria: presence of liquid water, presence of nutrients (C, H, N, O, S), low acidity, low salinity, presence of iron and sulfur in different oxidation states. It is the union of these positive criteria that allowed scientists with the Curiosity mission to conclude that the Yellowknife Bay site could have constituted a "habitable environment" 3.5 billion years ago. The Curiosity rover thus sought organic molecules in this region: chlorobenzene and dichloroalkanes were indeed found, but the organic matter that reacted with the Martian chlorine to form these molecules may have been of external origin (comets, meteorites, or interplanetary dust). Curiosity also discovered a methane emission that lasted several months, which added to weaker continuous outgassing. The presence of methane was completely unexpected, given the highly oxidating characteristic of the Martian atmosphere, which is dominated by carbon dioxide with some oxygen as a result of its photodissociation by ultraviolet solar radiation. On Earth, methane is largely of organic origin. On Mars, its origin may however be currently abiotic (for example, through outgassing of methane clathrates formed underground on the planet at some point in the past). The discovery of Martian methane thus has important implications in terms of exobiology, but its very existence is still highly controversial.

The presence of liquid water on Mars 3.5 billion years ago raises a problem: at that time, the Sun was almost 30% less luminous than today, and Mars should have been much colder: why did it have liquid water? Although no satisfactory response to this paradox currently exists, we generally believe that the overall cold and dry climate on Mars could have been interspersed with hotter and more humid periods that were probably provoked by volcanic episodes. This is not *a priori* the most favorable scenario for the emergence of life, since longevity is one of the criteria for its appearance and maintenance. We can nevertheless note that, as in the case of Earth, a glacial period is not necessarily fatal if liquid water can continue to exist under the ice during cold periods, thus allowing for the potential survival of micro-organisms.

5.3.2. *Venus*

The case of Venus is very different than Mars. Four billion years ago, the Sun was less luminous than today, and the temperature of the surface of Venus should have been about 280 K, given an albedo of 0.3 similar to Earth today. Now, the high D/H ratio on Venus compared to the interstellar medium and the terrestrial ratio proves that water was abundant in the Venusian atmosphere (see section 4.3.3.7, Volume 1). Thus, the conditions for a liquid water ocean may have existed on Venus 3 or 4 billion years ago. In 2016, researchers analyzed the potential lifetime of this ocean by using a global climate model similar to the one used to predict terrestrial climate change (see section 1.3.2, Volume 1 and section 6.3.2.2 (Way *et al.* 2016)). By assuming an ocean equal to a tenth of current terrestrial oceans with the current topography of Venus and a constant rotation period, they showed that the ocean could have existed for more than two billion years. However, they also showed that ocean stability depended vitally on the topography and rotational velocity of the planet. According to this model, an increase in the rotational velocity would lead to a rapid increase in surface temperature.

What happened next? Solar flux increased little by little, leading to water vaporization, then dissociation through ultraviolet radiation, and then release of the hydrogen atoms. The absence of water may have prevented the establishment of plate tectonics similar to Earth's, where the subduction of the plates requires hydrated materials. The absence of plate tectonics may have also prevented the recycling of carbon dioxide in rocks and thus regulation of the greenhouse effect as happened on Earth. The greenhouse effect led to the infernal conditions that exist today on Venus.

If a liquid ocean could have been present on the surface of Venus for several billion years, could life have appeared and developed? Nothing allows us to exclude the possibility *a priori*, but nothing allows us to confirm the hypothesis either, because all potential traces of life must have been erased by resurfacing caused by intense volcanism. Indeed, the ground on Venus is young, only a few hundred million years old.

We can nevertheless wonder if life could have appeared in Venus's clouds, where temperature (230 to 250 K) and pressure (0.04 to 0.5 bar) conditions would not be incompatible with microbial life. We have observed that Venus's clouds absorb ultraviolet solar radiation at a specific wavelength, but the nature of the absorber has not been determined unambiguously. Could it be bacteria? For the moment, this is only speculation. However, we note that there are terrestrial micro-organisms surviving in environments dominated by carbon dioxide, rich in sulfuric acid, and containing iron. We could imagine, in Venus's clouds, a metabolism based on iron and sulfur, in which oxidation of these compounds would be paired with a reduction in carbon dioxide in the absence of oxygen (Limaye *et al.* 2018). Life could have migrated from the surface to these clouds as the planet's temperature rose, adapting to the new conditions. Once again, this hypothesis is purely speculative at this point.

5.3.3. *Satellites of the giant planets*

To date, we have identified four satellites around giant planets that could harbor primitive forms of life: Europa and possibly Ganymede and Callisto, satellites of Jupiter, and Enceladus, a satellite of Saturn (see section 1.1).

On Europa (see section 1.2.1.2) images from Voyager 1 revealed an icy surface, covered with fractures interwoven into a complex network, that may conceal a viscous or even liquid ocean (see Figure 5.9). Tidal deformation due to Jupiter and its other satellites, tides whose amplitude reaches 30 or so meters high, produce these fractures and the heat that melts the deep ice encasing the satellite. The ocean of Europa is probably salty: indeed, Europa exhibits a magnetic field induced by the one on Jupiter, which implies the presence of an ionized and conductive environment. The water in this ocean could be in direct contact with the satellite's siliceous core, thus enabling the development of a complex chemistry that could lead to life, even under the apparently inclement temperature conditions that are reminiscent of subglacial lakes in the Antarctic, where colonies of bacteria nevertheless thrive.

Figure 5.9. *The surface of Europa is covered with cracks, as shown in this image obtained by the Galileo probe in the visible and near infrared spectra. The north pole is on the right. The light areas are ice, while the reddish ones are other compounds that have been added to the ice (source: NASA/JPL-Caltech/SETI Institute). For a color version of this figure, see www.iste.co.uk/encrenaz/solar2.zip*

Geysers, which were glimpsed by the Hubble Space Telescope and confirmed by Galileo, seem to erupt on Europa (cryovolcanism). These geysers are made up of water, fine particles of ice, and water vapor, and thus manifest the presence of water under this surface, heated by tidal forces.

Ganymede and Callisto may also harbor a liquid water ocean under their surface (see sections 1.2.1.3 and 1.2.1.4). However, as they are farther away from Jupiter than Europa, the tidal effects induced by the planet are weaker and, according to theoretical models, this ocean could be trapped between two layers of ice, which would reduce the potential for development of life. As for Io, the satellite closest to Jupiter, it is the site of such volcanism that life is probably not possible there.

Like Europa, Enceladus (see section 1.3.2), a satellite of Saturn, exhibits cryovolcanism, clearly observed by the Cassini probe, which demonstrated the existence of an ocean, probably salty, under the ice (see Figures 5.10 and 5.11). These geysers come from this underlying ocean, heated by the tides due to Dione, another of Saturn's satellites. More recently, hydrogen has been discovered in these jets. This is yet another environment that could be suitable for the appearance of life. Other celestial bodies that could also exhibit cryovolcanism are Triton (see section 1.2.4), a satellite of Neptune, and the dwarf planet Ceres (see section 2.4.1). Saturn's satellite Titan (see section 1.3.1) also exhibits cryovolcanism, but of methane, and probably cannot harbor life.

Figure 5.10. *Enceladus and two fragments of the rings of Saturn, seen backlit by the Cassini probe. Jets emerge in the area of Enceladus's south pole: these are cryovolcanic geysers. The small satellite Pandora, below and to the right on Enceladus, is lit by light reflected by Saturn (source: NASA/JPL-Caltech/Space Science Institute)*

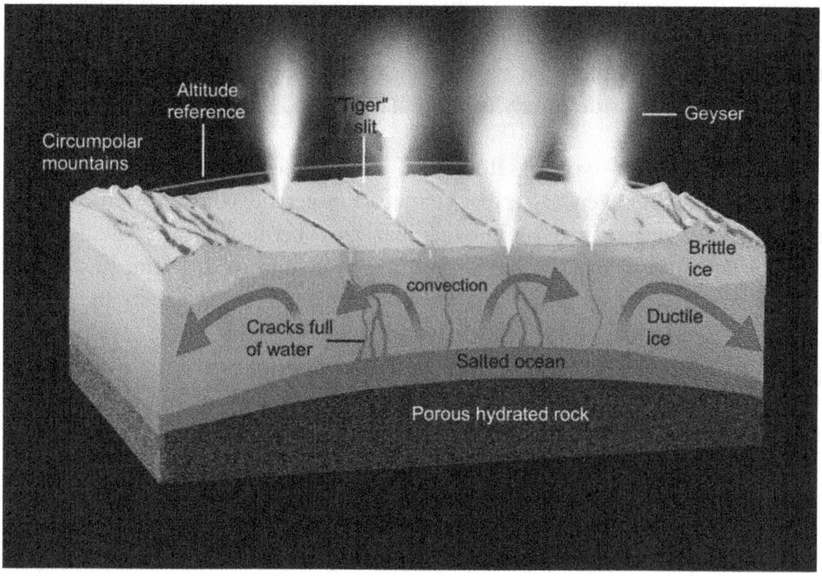

Figure 5.11. *Cryovolcanism on the surface of Enceladus. Heated saltwater rises through the cracks (fissures in the ice called "tiger stripes" because they form parallel lines) and emerge in the form of geysers (source: based on an artist image; NASA/JPL-Caltech/Space Science Institute). For a color version of this figure, see www.iste.co.uk/encrenaz/solar2.zip*

5.4. How can life be detected on exoplanets?

Beyond the Solar System, it would be crucial to detect life on other planets (see section 2.7.7, Volume 1). We can of course wait for a potential incoming signal that would be broadcast by an advanced civilization: we will talk about this later. It would be sufficiently interesting to detect signs of more primitive life.

We know that on Earth, atmospheric oxygen is the result of chlorophyll photosynthesis by cyanobacteria and plants. If photosynthesis were to disappear suddenly, the oxygen would also disappear rapidly, through combustion or by oxidizing a variety of minerals: it thus must be replenished. Chlorophyll will be very difficult to detect, but if we find oxygen on an exoplanet, there is a good chance that the planet harbors living beings capable of photosynthesis. And there are good reasons to think that, as on Earth, oxygen is indispensable for maintaining more advanced beings that have greater energy requirements, because only oxidation-reduction reactions involving oxygen can supply enough energy. If there are living beings on the planet, its atmosphere should also contain CO_2 and methane, but the presence of these elements, by the way very widespread, is no proof of life. It is true that methane in the terrestrial atmosphere seems to be largely of biological origin, but the detection of methane on an exoplanet does not prove anything in the absence of other indicators. It is definitely oxygen that we need to look for.

Unfortunately, the oxygen atom can barely exist in the presence of other substances, and dioxygen O_2 is a symmetric molecule whose spectral lines and bands are very weak. The only band that may be usable for detection on an exoplanet is found in the red range with a wavelength of 760 nm, in a generally crowded spectral range, but it has been observed from space in the terrestrial atmosphere (Sagan *et al.* 1993). It seems preferable to try to detect ozone O_3, which is produced by the effect of ultraviolet radiation from the star on O_2: although it is much less abundant; its spectral bands are stronger. In particular, a band of ozone at a wavelength of 9.6 μm lends itself to detection. Of course, it exists in the terrestrial atmospheric spectrum (see Figure 4.5, Volume 1). The detection of O_2 and O_3 on exoplanets thus should be possible from space, above the atmosphere.

The only practical possibility for detecting the band of O_3 at 9.6 μm on exoplanets is to observe their transits from space (see sections 2.7.3 and 2.7.6, Volume 1). No satellite is capable of doing this currently, and we will have to wait for the launch of the JWST, the successor to the Hubble Space Telescope.

5.5. Communicating with other civilizations?

For more than a hundred years, we have constantly broadcast signals that travel through the terrestrial ionosphere and betray our technically advanced civilization. Is it the case for other planets, and can we capture signals coming from them, whether emitted involuntarily or with the goal of communicating with us? This is the purpose of research efforts like SETI (*Search for Extra-Terrestrial Intelligence*), which have been active since 1960 with often very powerful means, but which have not yet produced any results. We can on the other hand send messages toward the Universe, as was done in 1974 toward the M13 globular star cluster, which is located at around 6,400 parsecs and contains around 300,000 stars. The antenna that was used was the Arecibo radio telescope, and the strength of the emission was several megawatts, so it could have been easily detected from any point in the Milky Way. The message was composed of 1,679 bits distributed over 73 lines of 23 characters each (these are prime numbers, which would, it was assumed, make it easier to decode the message). The problem is that it will take 42,000 years to obtain any potential response! Moreover, deciphering the message is far from trivial, and we have to hope that the extraterrestrials will be more intelligent than we are.

In the same spirit, in 1972 and 1973, NASA installed a plaque on each of its Pioneer probes that may be easier to decipher than the Arecibo message. This was repeated with the two Voyager probes, which left the Solar System for an infinite voyage through space, a message in the form of a videodisc (with a pictographic user manual) intended for potential extraterrestrial civilizations, in the unlikely case that they manage to retrieve and decipher it. In retrospect, these attempts seem naive, but their motivation was more philosophical than scientific, and they didn't cost much.

5.6. References

Altwegg, K., Balsiger, H., Bar-nun, A. *et al.* (2016). Prebiotic chemicals – amino acid and phosphorus – in the coma of comet 67P/Churyumov-Gerasimenko. *Science Advances*, 2.

Darby Dyar, M., Smrekar, S.E., Kane, S.R. (2019). The exoplanet next door. *Scientific American*, 320(2), 53–59.

Encrenaz, T., Lequeux, J., Casoli, F. (2021). *Planets and Life*. EDP Sciences, Les Ulis.

Fresneau, A., Abou Mrad, N., D'Hendecourt, L.L.S. *et al.* (2017). Cometary materials originating from interstellar ices: Clues from laboratory experiments. *Astrophysical Journal*, 837, 168.

Limaye, S.S., Mogul, R., Smith, D.J. *et al.* (2018). Venus' spectral signatures and the potential for life in the clouds. *Astrobiology*, 18, 1181–1198.

Luft, R. (2014). *Biosphère et chimie*. EDP Sciences, Les Ulis.

Marty, B., Avice, G., Sano, Y. *et al.* (2016). Origin of volatile elements (H, C, N, noble gases) on Earth and Mars in light of recent results from the ROSETTA cometary mission. *Earth and Planetary Science Letters*, 441, 91–102.

Ponnamperuma, C. (1964). The origin of life in the Universe. *Space Science Education Conference* [Online]. Available at: http://archive.org/details//nasa_techdoc_19650011877.

Ruf, A., Lange, J., Eddhif, B. *et al.* (2019). The challenging detection of nucleobases from pre-accretional astrophysical analogs. *Astrophysical Journal Letters*, 887, L31.

Sagan, C., Thompson, W.R., Carlson, R. *et al.* (1993). A search for life on Earth from the Galileo spacecraft. *Nature*, 365, 715–721.

Tartese, R., Chaussidon, M., Gurenko, A. *et al.* (2017). Warm Achean oceans reconstructed from oxygen isotope composition of early-life remnants. *Geochemical Perspectives Letters*, 3, 55–65.

Way, M.J., Del Genio, A.D., Kiang, N.Y. *et al.* (2016). Was Venus the first habitable world of our solar system? *Geophysical Research Letters*, 43, 8376–8383.

6

Methods for Studying the Solar System

Thérèse ENCRENAZ[1], Marcello FULCHIGNONI[1], Laurent LAMY[1], Françoise ROQUES[1] and James LEQUEUX[2]

[1] *LESIA, Paris Observatory, PSL University, Paris, France*

[2] *LERMA, Paris Observatory, PSL University, Paris, France*

6.1. History

In this introduction, we present a summary of the first assessments, before the space age, of the geometric and physical parameters of the Solar System (the methods and types of instruments are indicated in parentheses):

– Around 230 BCE: Eratosthenes (276–194 BCE) determines the dimensions of the Earth (zenith distance of the Sun at the meridian on the summer solstice measured at two different latitudes for which the distance on the ground is known, given that the Earth is assumed to be spherical).

– Around 140 BCE: Hipparchus (190–120 BCE) discovers the precession of the equinoxes (comparison of the positions of stars at different times). He determines the diameter of the Moon and the distance of the Earth to the Moon (relative angular dimensions of the Earth's shadow and the Moon during a total lunar eclipse, given that the Sun was assumed to be extremely far away).

– Around 900: Al-Battani (858–929) discovers the slow advance of the apogee of Earth's orbit.

– 1533: Gemma Frisius (1508–1555) invents triangulation to measure large distances on Earth.

The Solar System 2,
coordinated by Thérèse ENCRENAZ and James LEQUEUX.

– 1543: publication of *De revolutionibus orbium cœlestium* by Copernicus (1473–1543).

– 1577: Tycho Brahe (1546–1601) shows that the Great Comet that appeared that year was at least six times more distant than the Moon (simultaneous measurements of its position in relation to the stars from his observatory in Denmark and in Prague).

– 1609: Johannes Kepler (1561–1730), in his *Astronomia Nova*, shows that the orbit of Mars is elliptical and states his first two laws (use of logarithms).

– 1610: Galileo (1564–1622) publishes in his *Sidereus Nuncius* his discovery of mountains on the Moon and the four main satellites of Jupiter, then, in 1613, publishes the discovery of the phases of Venus (refracting telescope).

– 1619: Kepler states his third law in his *Harmonices Mundi*.

– 1655: Christiaan Huygens (1629–1695) discovers Titan.

– 1659: Huygens explains the phenomena of Saturn's rings.

– 1666: Jean-Dominique Cassini (1625–1712) discovers the rotation of Jupiter and specifies the rotation of Mars, which was suspected by Huygens.

– 1671: Cassini, Jean Picard (1620–1682), and Ole Rømer (1644–1710) measure a difference in terrestrial longitudes with precision for the first time (use of eclipses of satellites of Jupiter as a time marker in different places).

– 1672: Cassini and Jean Richer (1630–1696) determine the parallax of Mars, from which the distance from Earth to the Sun can be calculated, using Kepler's 3rd law (simultaneous measurements of the position of Mars with respect to the stars in Paris and Cayenne). Their result was confirmed by Flamsteed (1646–1719).

– 1687: Isaac Newton (1642–1727) publishes his *Philosophiæ Naturalis Principia Mathematica*, which contains the law of gravity, the physical demonstration of Kepler's laws, and the calculation of the parabolic orbit of comets.

– 1716: Edmond Halley (1656–1742) discovers the periodicity of the comet that bears his name.

– 1727: James Bradley (1693–1762) discovers the aberration of starlight (annual displacement of γ Draconis observed with a zenith sector).

– 1747: Bradley discovers the nutation of Earth's axis (same instrument).

– 1780: Horace Bénédict de Saussure (1740–1799) hypothesizes the existence of the greenhouse effect in the Earth's atmosphere.

– 1781: William Herschel (1738–1822) discovers Uranus (reflecting telescope).

– 1798–1824: Pierre-Simon Laplace (1749–1827) publishes his *Mécanique Céleste* (treatise in 5 volumes and 3 supplements).

– 1801: Giuseppe Piazzi (1746–1826) discovers Ceres, the first asteroid.

– 1811: François Arago (1786–1855) observes the polarization of the light reflected by the Moon, then later by the coma of comets, establishing their solid nature (polarimeter).

– 1824: Joseph Fourier (1768–1830) shows that the heat of the Earth's surface comes essentially from the Sun, and provides the order of magnitude of the thermal flux coming from inside the Earth. He includes the greenhouse effect in the Earth's atmosphere.

– 1846: Urban Le Verrier (1811–1877) discovers Neptune "through calculation".

– 1859: Le Verrier discovers the abnormal advance of Mercury's perihelion, explained in 1915 by General Relativity.

– 1864–1868: Gian Battista Donati (1826–1873), Angelo Secchi (1818–1878), and William Huggins (1824–1910) obtain the first spectra of planets and comets, which reveal the presence of carbon compounds in the latter.

– 1866: Daniel Kirkwood (1814–1895) discovers gaps in the distribution of semi-major axes of the orbit of asteroids.

– 1873: Laurence Parsons (1840–1908) measures the temperature of the surface of the Moon (thermocouple, by comparison with black bodies at different temperatures).

– 1889: Henri Poincaré (1854–1912) begins to raise questions about the long-term stability of the Solar System, introducing the idea of chaos.

– 1907: Vesto Slipher (1875–1969) shows the presence of strong molecular bands in the spectra of giant planets. Caused by methane and ammonia, they were identified in 1932 by Rupert Wildt (1905–1976).

– 1912: Alfred Wegener (1880–1930) discovers the continental drifts on Earth.

– 1924: Edison Pettit (1889–1962) and Seth B. Nicholson (1963) measure the temperature of the clouds of Venus (mid-infrared thermocouple).

– 1924: Bernard Lyot (1897–1952) shows that the surface of the Moon and Mercury is covered with volcanic ash, while Mars is instead covered with sand (polarization).

– 1926: William Coblenz (1873–1962), Carl Otto Lampland (1873–1951), and Donald Menzel (1901–1976) show that the giant planets are very cold (mid-infrared).

– 1955: Bernard Burke (1928–2018) and Kenneth Franklin (1923–2007) discover radio radiation in decametric waves coming from Jupiter, revealing its magnetic field (radio telescope).

– 1958: Cornell Mayer (1921–2005), Timothy McCullough (1910–2004), and Russell Sloanaker show that the surface of Venus is very hot (radio telescope).

6.2. Observational techniques

6.2.1. *Remote sensing*

6.2.1.1. *Photometry and imagery*

From the 17th to the beginning of the 20th century, astronomical images were based on drawings. During the 20th century, these were replaced by photographic plates. Then, from the 1960s, the emergence of new technologies opened the way to other generations of receptors in the visible range. The first of them was the photomultiplier, made up of a tube containing several electrodes acting as charge amplifiers. The photomultiplier did not enable image acquisition, but allowed for the photometric measurement of variable objects with a high degree of precision. In planetology, it was used successfully to measure light curves of small objects in the Solar System (asteroids and comets in particular), allowing their rotation period to be determined. It also enabled the observation of stellar occultations during transit by an object in the Solar System (planet, satellite, asteroid; see section 6.2.1.3). Microchannel plates made it possible to perform two-dimensional photometry, but were little used in planetology.

Charge-transfer receptors were introduced in astronomy during the 1980s. They make use of the photoelectric effect in solids and offer a number of advantages: bi-dimensionality, sensitivity, linearity, dynamics, miniaturization, reliability, digitization of data. The most common of these systems is the Charge Coupled Device or CCD, made up of a capacitor containing an insulator between a metal electrode and a semiconductor that is sensitive to photoelectric effects. The application of a difference of potential modifies the distribution of the charges in the semiconductor, creating a potential sink capable of storing charges in the capacitor. The receptor is made up of a bi-dimensional framework of pixels (as many as a million or more), each containing such a capacitor. The flow of data is considerable (on the order of a Gbit for a large image), which requires appropriate computing resources for processing the mass of data.

In the infrared range, beginning in the 1980s, hybrid CMOS (Complementary Metal Oxide Semiconductor) receptors have been used, with the goal of producing image receptors beyond the limits of silicon's photoconductivity. Using the photovoltaic effect, these large-format matrices are made of photodiodes sensitive to the wavelength being observed, with a reading circuit implemented on a silicon base. Since the signals to be detected are very weak (less than an electron-volt), the matrices have to be cooled to low temperatures (below 77 K) using cryogenic liquids (N_2 or He) in order to reduce the dark current (primarily of thermal origin) as much as possible. CCD and CID cameras in planetology were first used to capture images of planetary disks. Figure 6.1 gives an example of the quality of an image obtained in the visible range from the disk of the planet Uranus by the Hubble Space Telescope.

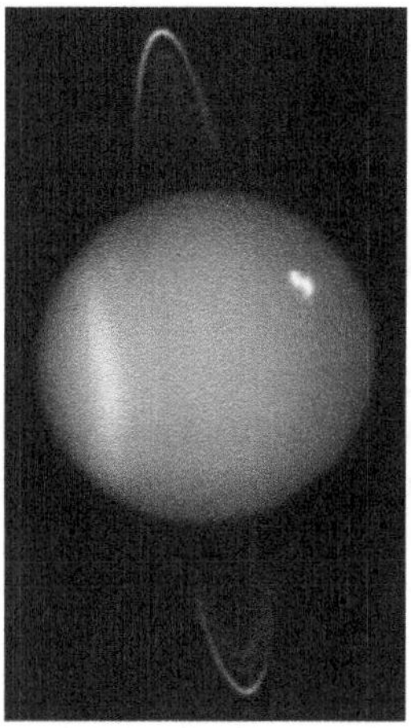

Figure 6.1. *Uranus imaged by the Hubble Space Telescope's CCD camera in 2003. The pole of the ecliptic is at the top of the figure, and the planet is near the equinox (passage in 2006). The planet's blue color can be attributed to methane gas; the white areas can be attributed to methane cirrus clouds at high altitude (source: NASA). For a color version of this figure, see www.iste.co.uk/encrenaz/solar2.zip*

6.2.1.2. *Spectroscopy and polarimetry*

As with all objects in the Solar System, the information we have about the physical chemistry of a body's surface or atmosphere comes from spectroscopic observation of its electromagnetic spectrum, from the ultraviolet to radio waves.

Mineralogical signatures are primarily located in the near infrared. Those of neutral molecules are found in the infrared (vibration-rotation bands) and the millimeter and submillimeter range (rotation bands). Ultraviolet radiation, which has more energy, informs us about the products of molecular dissociation (radicals, atoms, and ions) in comets and planetary stratospheres (see section 5.1.4, Volume 1).

6.2.1.2.1. Spectroscopy

The electromagnetic spectra of objects in the Solar System are characterized by a reflected solar component that culminates at 0.5 μm (maximum of the solar spectrum, corresponding to that of a black body to 5700 K) and a thermal component that dominates beyond a few micrometers (see section 1.3.1, Volume 1). In the reflected component, spectral signatures of atmospheric components are observed in absorption in front of the solar continuum. In the case of the thermal component, the flux emitted at a given frequency ν is given by the following equation (see section 5.1.4, Volume 1):

$$\Phi(\nu) = \int_0^\infty B(\nu)e^{-\tau(z,\nu)}\, d\tau(z,\nu) \qquad [6.1]$$

an equation in which $\tau(z, \nu)$ is the optical depth above the altitude z:

$$\tau(z,\nu) = \int_z^\infty \alpha(z_i,\nu)N(z_i,\nu)dz_i \qquad [6.2]$$

where α is the absorption coefficient of the absorbing molecule at the frequency ν and N is its density. If several atmospheric absorbers are present, the quantity $\tau(z, \nu)$ is the sum of the contributions from different molecules. The quantity $\alpha(z_i, \nu)$ depends on the properties of the molecules present in the atmosphere and is derived from laboratory studies (Hanel *et al.* 2003).

A parameter currently used to measure Φ is the brightness temperature $T_B(\nu)$, which is the temperature of a black body that radiates the same flux at frequency ν. According to the Barbier-Eddington relation, in the case of observation at the nadir, $T_B(\nu)$ is, as a first approximation, the temperature of the atmospheric layer for which the optical depth $\tau(\nu)$ is equal to 1. In the case where the flux is emitted by an entire planetary disk, $T_B(\nu)$ is, as a first approximation, the temperature of the atmospheric layer at which the optical depth is equal to 2/3 (see section 5.1.4, Volume 1).

At the center of a molecular line, the optical depth is larger than at the edges; the level $\tau = 1$ (or 2/3) is achieved at a higher altitude than for neighboring frequencies. We can see that, in the planetary spectrum, the line can appear in emission or absorption, based on the temperature gradient in the region where the line forms (see Figure 6.2). In the troposphere (see Chapter 1, Volume 1), temperature decreases as altitude increases; the line thus appears in absorption. On the other hand, in the stratosphere, temperature increases with altitude, and the line appears in emission (see section 5.1.4, Volume 1).

Another mechanism of emission that manifests primarily in the UV and visible ranges is fluorescence. When an atmospheric component (ion, atom, radical, or molecule) passes from the fundamental state to an excited state through absorption of a solar photon (UV, visible, or sometimes near IR) or due to the effect of collisions by other particles, it can either fall directly back to the fundamental state by emission of a photon of the same wavelength, or it can fall to the fundamental state by cascading through a series of intermediate energy levels. In the first case, there is resonant fluorescence: we observe in the spectrum of an object an emission line of the same wavelength as the transition under consideration. In the second case, we observe a spectrum of fluorescent lines at wavelengths longer than that of the source of the excitation; these wavelengths are such that the sum of the corresponding transition energies is equal to the transition energy that produced the excitation.

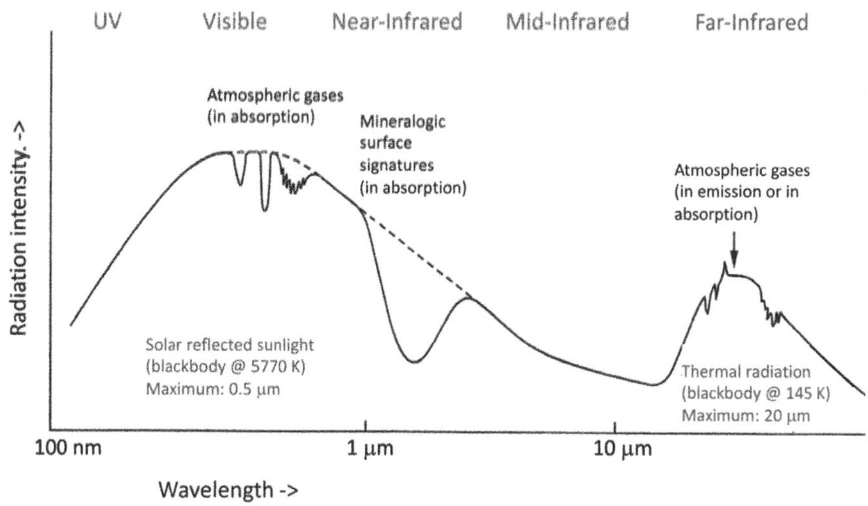

Figure 6.2. *Diagram representing the spectrum of a planet from the visible to the mid-infrared*

COMMENTS ON FIGURE 6.2.– *The reflected solar component dominates in the visible range; lines of atmospheric gases present on the planet appear in absorption in front of the solar continuum. A mineralogical signature from the planetary surface also appears in absorption between 1 and 3 μm. The thermal component dominates in the mid-infrared, with a continuum corresponding to the spectrum of a black body at the planet's effective temperature. The signatures of atmospheric gases appear in emission or absorption, depending on whether the lines are formed in the stratosphere or the troposphere (source: based on (Lissauer and de Pater 2015)).*

For objects in the Solar System, the pumping source is essentially the solar flux. In the case of fluorescence, the pumping rate in s^{-1} is expressed as a function of the solid angle at which the Sun is viewed, its brightness temperature T_B, the respective statistical weights of the upper and lower transition levels, and the Einstein coefficient of spontaneous emission for the transition. The intensity of the line is the product of the pumping rate and the number of atoms (or ions, or molecules) on the target line, multiplied by the energy $h\nu$, ν being the frequency of the transition under consideration (see section 2.1.6).

Fluorescence spectra are most often observed in the UV and the visible ranges: this is due to the shape of the solar spectrum, which is the primary source of excitation. This is the case for ionospheres and the upper atmospheres of giant planets; it is also the case for comets. A particularly remarkable emission is the Ly α line of atomic hydrogen, at 121.6 nm. In the case of comets, lines of molecular fluorescence are observable in cometary infrared spectra. Such emissions have been observed from Halley's comet as well as from other bright comets, particularly Hale-Bopp (see section 2.1.6.4).

6.2.1.2.2. Spectral imaging

Since the beginning of the 21st century, the combined use of high-resolution spectrometers with bi-dimensional infrared detectors has allowed for the emergence of a new, particularly powerful technique, imaging spectroscopy. A high-resolution spectrometer combined with a slit forms an image on a two-dimensional detector for which one dimension is spatial and the other is spectral. The second spatial dimension is obtained by the motion of the slit across the sky (in the case of a ground-based telescope) or by the motion of the satellite along its orbit (in the case of a space mission). Imaging spectroscopy is particularly well-suited to the mapping of species in the planetary atmosphere and the study of their spatio-temporal variations. Among the instruments of this type, we should mention Crires at the Very Large Telescope (operating between 1 and 5 μm at the ESO, Chile) or Texes at the *Infrared Telescope Facility* (operating between 5 and 25 μm at the Mauna Kea Observatory in Hawaii).

The auroral activity of the magnetospheres of magnetized planets can be studied in a very complementary manner via imaging spectroscopy in different spectral ranges. First, the UV/visible range indicates the response of species in the upper atmosphere that are neutral to auroral precipitation. The UV range, which is only accessible from space with instruments like the Hubble Space Telescope or spectral imaging equipment on-board exploration probes, makes it possible to observe the Ly α line of hydrogen at 121.6 nm and the Werner/Lyman bands of H_2 between 80 and 160 nm for giant planets, as well as emissions of oxygen lines from their satellites (see Figure 3.35, Volume 1). Imaging at these wavelengths provides good angular resolution, while spectroscopy makes it possible to measure the energy radiated and to probe the energy of the precipitating electrons.

The near-infrared range probes the response in the ionized upper atmosphere, particularly the rotational-vibrational emission of the H_3^+ molecule between 3 and 5 microns, which is associated with heated regions and has been detected on Jupiter, Saturn, and Uranus. The far-infrared range, between 7 and 14 microns, probes the emission of atmospheric hydrocarbons located at the lowest altitudes, which have been observed on Jupiter. In this range, high-resolution observations from ground-based telescopes are complementary to the more sensitive measurements from on-board spectral imaging equipment.

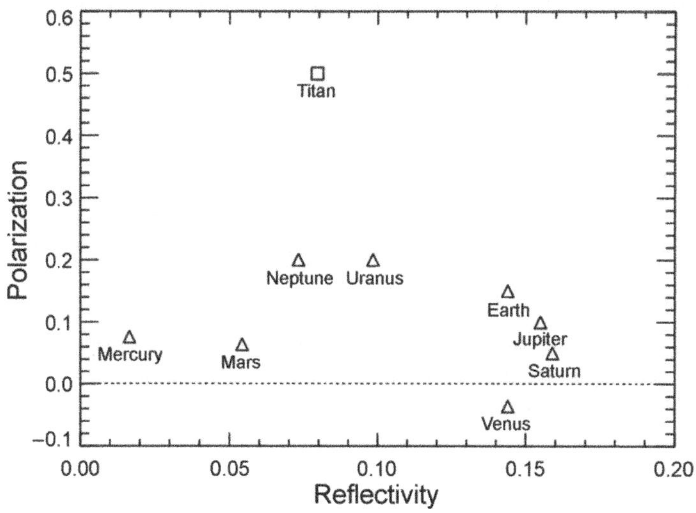

Figure 6.3. *Degree of polarization as a function of reflectivity for planets in the Solar System, at a phase angle of 90°, in the visible red range (R) (source: based on Schmid* et al. *2005))*

Finally, the X-ray range, also only accessible from space, has revealed on Earth and Jupiter the "bremsstrahlung" emission produced by radiation from the braking of high-energy electrons (hard X-rays) and emissions following exchanges of charges in the upper atmosphere with high-energy ions (soft X-rays). We should note the important contribution of the Chandra and XMM-Newton space telescopes in this area.

6.2.1.2.3. Polarimetry

Unlike direct sunlight, the solar radiation reflected by planets exhibits a degree of polarization that can be measured to deduce some of the physical properties of their atmosphere (Rayleigh diffusion, aerosol diffusion) or of their surface. The degree of polarization depends strongly on the phase angle (Sun-planet-observer angle) and is greatest for angles on the order of 90°. Figure 6.3 shows the degrees of polarization of different planets in the Solar System for a phase angle of 90°, as a function of their reflectivity. In the case of Mercury and Mars (see sections 4.2.4.1 and 4.2.4.2, Volume 1), the degree of polarization tells us about properties of the rocky or dusty surface. In the case of Venus, polarization is due to scattering by the cloud layer. In the case of the Earth, the high degree of polarization is due to atmospheric Rayleigh scattering. This is also at work in the polar regions of Jupiter, while the high degree of polarization on Titan (see section 1.3.1) is due to aerosol scattering.

6.2.1.3. *The stellar occultation method*

When an object in the Solar System (planet, satellite, or asteroid) passes in front of a sufficiently bright star (visual magnitude on the order of 10 or less), it is possible to study its properties by analyzing the light curves from the star at immersion and emersion, at the beginning and end of the object's transit in front of the star.

If the object lacks an atmosphere, we observe a relatively abrupt fall and rise in the signal according to the laws of diffraction; the exact measurement of the moments of immersion and emersion provide information about the object's size. By combining observations carried out from several points on Earth, we can reconstitute its shape. Observations are performed in the visible or the near infrared range.

When the observed object is a planet with an atmosphere, the stellar light is deviated by atmospheric refraction; the effect is more notable the higher the atmospheric density (see Figure 6.4). By inverting the light curve, it is possible to work backward to the atmospheric refraction index; if the mean molecular mass is also known, the temperature profile as a function of altitude can be obtained. This method was used successfully in the case of the occultation of the star β Sco by Jupiter in 1971. Employed again in 1977 for a stellar occultation by Uranus, it allowed for the first identification of Uranus's rings (see section 1.5.5); in 1984, it revealed the present of arcs around Neptune. These results were confirmed during the flybys of Uranus (1986) and Neptune (1989) by the Voyager 2 probe; the arcs of Neptune were revealed to be rings with variable density (see section 1.5.6).

With the progressive improvement in the precision of the ephemeris of solar system objects, the observational technique of stellar occultation has gained in importance in the past 20 years. This is likely to continue in the coming years with the release of the catalogs of stellar positions provided by the European Space Agency's Gaia astrometric mission, which has been operating since 2013. A particularly spectacular result was the discovery of rings around distant small bodies, the Centaur Chariklo in 2014 and the trans-Neptunian object Haumea in 2017.

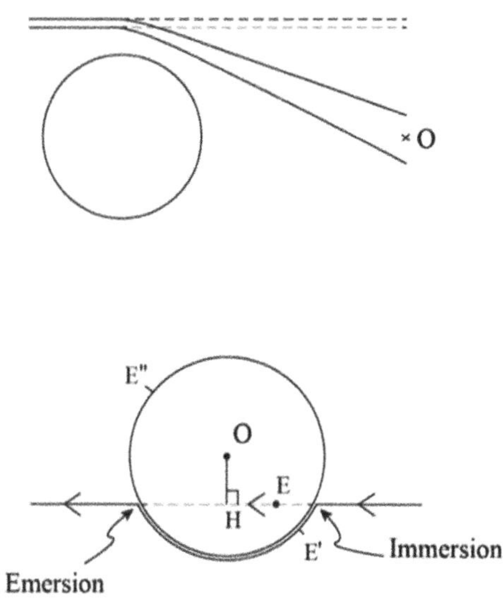

Figure 6.4. *Analysis of an atmosphere by the stellar occultation method: refraction by a planetary atmosphere; O is the center of the planet, E is the star, whose flux is observed at E and E'; OH is the impact parameter (source: (Encrenaz* et al. *2003))*

6.2.1.4. *Observation of the Solar System at radio wavelengths*

Objects in the Solar System are observable in the radio spectral range using both spectroscopy and imaging. This section focuses on observations carried out by low-frequency ground-based radio telescopes, above the terrestrial ionospheric cutoff frequency of 10 MHz. Radio measurements carried out under 10 MHz by space probes are dealt with along with *in situ* plasma measurements in section 6.2.2.3.

6.2.1.4.1. The Sun

While the solar cycle can be monitored from optical observations (flux, number of spots) and the solar magnetic field probed by polarimetric measurements from the Hanle or Zeeman effect, ground-based monitoring of solar eruptions and solar wind activity is mainly carried out in the radio range. At centimeter and decimeter wavelengths, the emission comes principally from synchrotron radiation from relativistic electrons, orbiting at high velocity around magnetic field lines in eruptive loops, at frequencies that are multiples of the local cyclotron frequency f_{ce}.[1] At longer, meter and decimeter wavelengths, beams of electrons accelerated at several thousand km s^{-1} excite the corona and the interplanetary medium along their path, at the fundamental or the first harmonic of the local electron plasma frequency f_{pe}[2], generating so-called *type III* radio bursts that drift rapidly from high to low frequencies. Corona mass ejections, accompanied by shock waves that propagate at 1,000–2,000 km s^{-1} through the corona and the solar wind, produce analogous emissions, referred to as *type II bursts*, but whose frequencies change much more slowly than the previous waves (see Chapter 3 and Figure 3.9, Volume 1). These emissions can be studied through imaging thanks to instruments like the European Lofar interferometer or the Nançay radio heliograph, or at high spectral and temporal resolution by compact phased networks like UTR-2 in Ukraine or the Nançay Decameter Array.

6.2.1.4.2. The planets

Radio astronomy also plays an important part in the study of planets and other objects in the Solar System. The electromagnetic radiation that has been observed has a variety of origins and provides different kinds of diagnostics. They are distributed primarily in three ranges, illustrated in the case of Jupiter by Figure 6.5.

---

1 The electron cyclotron frequency is the natural gyration frequency of electrons around a magnetic field (also called the Larmor frequency). It is written as: $f_{ce} = eB/2\pi$ or $f_{ce}[\text{kHz}] \approx 2\,800\ B[\text{gauss}]$, where B is the magnetic field, e the electron charge, and m the electron's mass.

2 The (electron) plasma frequency is the frequency of natural oscillations of electrons in plasma after any kind of perturbation. It is written as: $f_{pe} = (1/2\pi)\ (Ne^2/\varepsilon_o m)$ or $f_{pe}[\text{kHz}] \approx 9\ N[\text{cm}^{-3}]^{1/2}$, where N is the electron density, e is the electron charge, m is the electron mass, and ε_o is the dielectric permittivity of the vacuum. Radio waves with a frequency of $f \lesssim f_{pe}$ are reflected by plasma and cannot propagate.

At decameter to kilometer wavelengths, magnetized planets are sources of powerful intermittent auroral radio waves. Such a non-thermal radiation is produced by the cyclotron motion of weakly relativistic electrons in the magnetosphere, in the auroral regions surrounding the planet's magnetic poles, at frequencies close to f_{ce} (see Chapter 3, Volume 1). Only Jupiter has a magnetic field strong enough to emit frequencies observable from the ground. The existence of Jupiter's magnetic field was revealed through the detection of intense bursts of decametric radio waves in 1955. Jovian decametric emissions were measured routinely by different phased arrays such as those hosted by the Clark Lake and University of Florida radio observatories, the decameter arrays at Nançay or at Kharkov in Ukraine, and complemented by periodic high-sensitivity observations from other instruments such as LWA1. Auroral radio emissions from other planets at kilometer wavelengths were subsequently detected by space exploration probes. Planetary lightning, such as those associated with giant atmospheric storms on Saturn, has a wideband spectrum that is also observable in the decameter range by sufficiently sensitive radio telescopes.

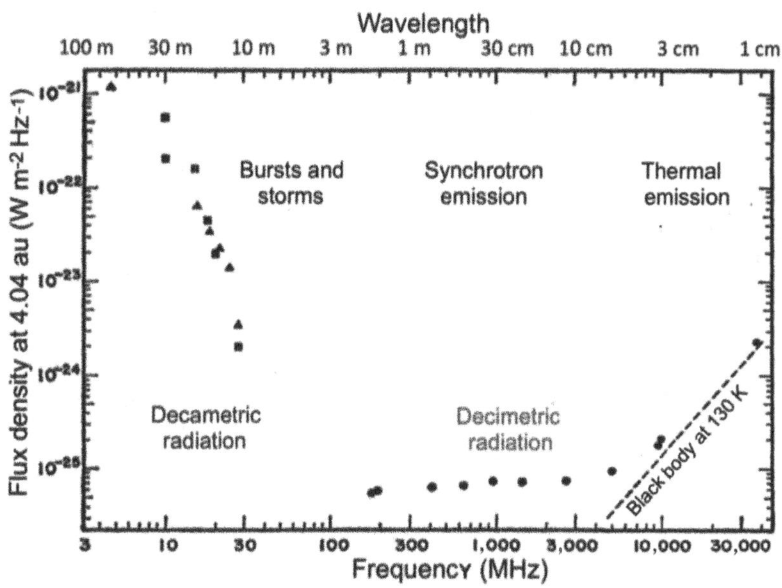

Figure 6.5. *Radio emissions from Jupiter*

COMMENTS ON FIGURE 6.5.– *On the right in the figure, centimeter and millimeter-wave radiation correspond to the atmospheric thermal emission. Decimeter-wave emission at the center is dominated by synchrotron radiation from electrons in the radiation belts. Finally, on the left, decameter, hectometer, and kilometer-wave radiation corresponds primarily to the auroral emissions driven by the cyclotron motion of electrons in the magnetosphere above the magnetic poles (source: based on (Carr and Gulkis 1969)).*

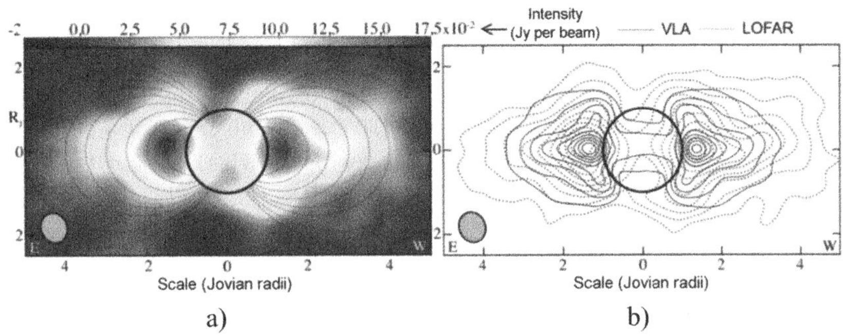

Figure 6.6. *Images of synchrotron radiation from Jupiter's radiation belts. For a color version of this figure, see www.iste.co.uk/encrenaz/solar2.zip*

COMMENTS ON FIGURE 6.6.– *Images obtained with Lofar observations in (a) between 127 and 172 MHz and in (b) Lofar (red contours) and with the VLA at 6 cm (black contours), associated with trapped high-energy electrons, ranging from a few MeV to several dozen MeV, respectively. Jupiter's optical disk is indicated by a circle, and the angular resolution is indicated by a yellow ellipse. These images show the topology of the magnetic field (source: based on Santa-Costa 2009; Girard* et al. *2016)).*

The second component, which is dominant in the centimeter and meter-wave range, corresponds to synchrotron radiation produced by high-energy electrons orbiting in a spiral along the planetary magnetic field lines . This is the case for belts of charged particles around the Earth (Van Allen belts), which also exist around Jupiter. Observation of the latter in 1958 using imaging thus enabled remote mapping of the dynamical topology of the Jovian magnetic field. Figure 6.6 provides an image of Jupiter's synchrotron radiation with recent Lofar and VLA observations.

The highest frequency component, which is dominant in the millimeter and centimeter-wave ranges, is the thermal radiation from the object (planet, comet, asteroid, trans-Neptunian object). As a first approximation, it corresponds to the radiation of a black body at the effective temperature T_{eff} of the object. In the radio frequency spectrum ($\lambda > 300$ µm), according to the Rayleigh-Jeans law, radiance is proportional to the temperature and to the square of the observation frequency ν.

Observation of the radio spectrum thus provides a direct measurement of the object's temperature at frequency ν, provided that its angular size is known. In 1958, the observation of the thermal radiation of Venus in the centimeter-wave range revealed the very high temperature of its surface (see sections 4.1.1 in Volume 1 and 6.1 in this volume). In the case of trans-Neptunian objects, simultaneous measurement in the visible and radio domains makes it possible to determine both the diameter and the temperature of the object (a hypothesis having been made about its albedo).

6.2.1.4.3. Heterodyne spectroscopy

This technique makes it possible to obtain a very high spectral resolving power within a restricted spectral band for the millimeter and centimeter wave ranges. It uses the interference from the incident wave with a wave produced at a neighboring frequency by a local oscillator. The interference is produced in a mixer from which a lower intermediate frequency (IF) is emitted, where the spectral analysis is performed. This can be done by using filters, by acousto-optical spectrometry, or most often by a digital correlator. The bandwidth for the IF stage can reach several GHz. The spectral resolving power obtained can exceed 10^6.

Heterodyne spectroscopy has contributed important discoveries in the field of planetology. We should note the first observation in 1977 of the (1-0) line of CO at 115 GHz in the atmospheres of Mars and Venus, the discovery of HCN in Halley's comet in 1986, the observation of HDO at 225 GHz on Mars and Venus in 1991, and the discovery of SO_2, then, later, SO and NaCl in the atmosphere of Jupiter's satellite Io. In the case of Mars and Venus, high spectral resolution has been used successfully to measure, via the Doppler effect on lines of CO, the velocity fields in the upper atmosphere of these two planets. With arrays of antennas allowing for high spatial resolution, we can obtain resolved maps for minor atmospheric species in planetary disks.

6.2.2. *Methods of space exploration*

The close exploration of planets and satellites by flybys or probes in orbit, by landers and rovers, and by atmospheric probes has opened a new chapter in the research methods for physical and chemical properties of bodies in the Solar System.

The strategy for exploring the Solar System includes five successive steps (see Figure 6.7):

– phase 0 consists of the observation of the object from the ground or from telescopes in Earth orbit;

– phase 1 involves flybys near the body;

– phase 2 includes the remote exploration by a spacecraft in orbit around the body;

– phase 3 involves descent to the object's surface by a lander/rover and/or, for bodies with an atmosphere, an atmospheric probe;

– phase 4 includes the return of samples to Earth collected by astronauts or robotic vehicles.

Most interplanetary missions install scientific instruments at the destination that allow them to carry out their observations and experiments. *In situ* exploration and analyses of bodies in the Solar System are carried out with atmospheric probes, landers, rovers, and orbiters, each of which has a specialized payload. Instruments on-board space probes are miniaturized, highly reliable versions of the instruments used in observatories and laboratories on Earth.

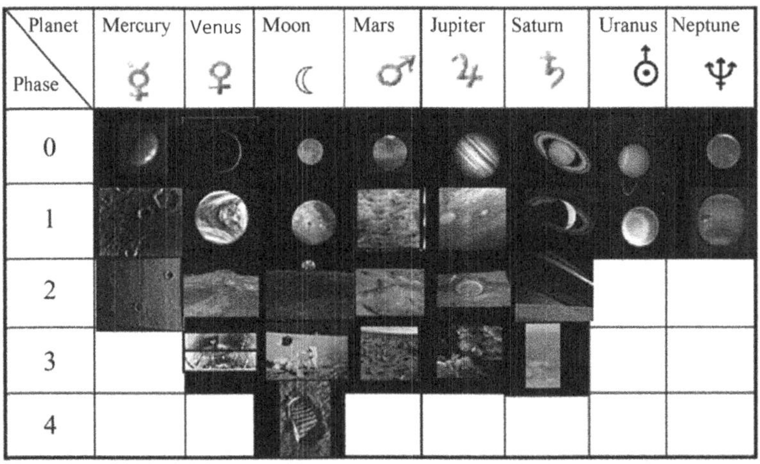

Figure 6.7. *State of exploration for bodies in the Solar System. For a color version of this figure, see www.iste.co.uk/encrenaz/solar2.zip*

COMMENTS ON FIGURE 6.7.– *Phase 0 has been carried out for all planets and their systems of satellites; phase 1 has been carried out for all planets and their major satellites, as has phase 2, with the exception of Uranus and Neptune; Mercury, Uranus, and Neptune are not in phase 3; phase 4 concerns the Moon (explored by humans who collected and brought back samples to Earth); a comet and an asteroid have been sampled by robotic probes (that brought the collected matter back to Earth).*

There are different categories of scientific space instruments: instruments for direct and remote detection, which can be active or passive.

Instruments for direct detection, also called contact scientific instruments, record the features of phenomena in their immediate vicinity.

Remote detection instruments create an image or a spectral characterization of phenomena observed remotely. Instruments that only receive or process light, particles, or other existing phenomena are called passive.

Active instruments have their own means to detect remote objects. A typical example is radar, which generates impulses of radio waves that are broadcast toward a surface, then receives their reflections from this surface, to create images or to study the surface's features.

Active instruments of direct detection enter into contact with the object to be analyzed and provide a source of energy to probe them. X-ray and/or alpha particle spectrometers on the Martian rovers, for example, contain a source of alpha particles that the rover projects at a variety of rock samples to determine their composition by measuring the energy spectra of retro-diffused alpha particles and/or X-rays sent back by the sample.

Interplanetary dust detectors are examples of passive instruments for direct detection: they measure the properties (number, velocity, mass, charge, and direction of flight) for dust particles that penetrate the instrument.

Imaging and spectroscopy systems are classic examples of passive instruments for remote detection.

Box 6.1. *Categories of space instruments*

6.2.2.1. *Atmospheric probes*

All of the space missions that have carried out a flyby or remained for some time in orbit around a body with an atmosphere have probed its gaseous envelope with the remote detection instruments they have on-board. The methods of observation are the same as those that are used from the ground, with the major advantage of being able to obtain a high spatial resolution thanks to the proximity of the object being observed.

Measurements carried out directly in the atmosphere of a body in the Solar System have been conducted on Venus, Mars, Jupiter, and Titan.

The properties of Venus's atmosphere were measured *in situ* during the eight missions of the Venera mission conducted by the Soviet space agency from 1969–1986, and by four atmospheric probes that NASA deployed as part of the Pioneer Venus mission in 1978.

In 1976, NASA's first two landers, Viking 1 and 2, carried out their missions on the Martian soil, and their data remained unique for 20 years. Beginning in 1997, NASA sent a dozen missions that brought rovers to Mars: the study of the Martian atmosphere and meteorology were always one of the main objectives of all of these missions.[3]

The Galileo mission (NASA) on July 13, 1995 included a probe that plunged into the atmosphere of Jupiter on December 7 and sent back data for 58 minutes while it descended more than 110 km through the progressively deeper layers of the Jovian atmosphere, up to a pressure of 22 bars ($22\ 10^5$ Pa).

The Cassini-Huygens mission (NASA, ESA, ASI) released the Huygens probe on December 25, 2004. On January 14, 2005, the probe descended through the atmosphere of Titan, a descent that lasted for 2 hours and 30 minutes.

All of the atmospheric probes, landers, and rovers are transported on a space vehicle, called a "bus", which generally enters into orbit around the body that is the mission objective, then carries out separation and targeting maneuvers for the probe on a direct impact trajectory. After releasing the probe, the bus acts as a radio relay for receiving data collected and transmitted by the probe's telemetry and transmitting it back to Earth. In addition to technical assistance for the probe, the orbiter participates in the exploration of the target and its vicinity with its payload made up of several remote detection instruments (imaging and spectral imaging devices, spectrometers, altimeters, magnetometers, etc.).

The descent of probes through the atmosphere takes place in three successive phases: entry, parachute descent, and landing. During the entry phase, the probe and its payload are protected by an aerodynamic thermal shield that slows the probe by several orders of magnitude while absorbing the heat generated by friction from increasingly dense atmospheric layers. In this phase, the instruments are blocked by the shield and thus are not exposed to the environment to be studied; they do not take measurements. If there is a system of accelerometers on board to measure the deceleration a of the probe due to aerodynamic breaking caused by the shield, it is possible to reconstruct the density profile of the upper atmosphere:

3 A complete list of missions exploring the Solar System is available at the following address: https://en.wikipedia.org/wiki/List_of_Solar_System_probes.

$$\rho(z) = -2(\frac{m}{AC_D})(\frac{a(z)}{V_r^2(z)}) \qquad [6.3]$$

where m is the mass of the probe, A the area of the thermal shield, C_D (measurement before launch) the probe's drag coefficient, and V_r the descent velocity of the probe relative to the atmosphere.

By using the atmospheric gas state equation and assuming a perfect gas in hydrostatic equilibrium, we can obtain pressure and temperature profiles:

$$p(z) = 2\int_{z_e}^{z} g(z)(\frac{m}{AC_D})(\frac{a(z)}{V_r^2(z)})dz \qquad [6.4]$$

$$T(z) = \frac{\mu p(z)}{\rho(z)\mathrm{R}} \qquad [6.5]$$

where z_e is the probe's altitude at entry, $g(z) = g(0)(r_c/z)$ with $g(0)$ the gravitational acceleration to the surface of the body of radius r_c, μ the molecular mass, and R the perfect gas constant.

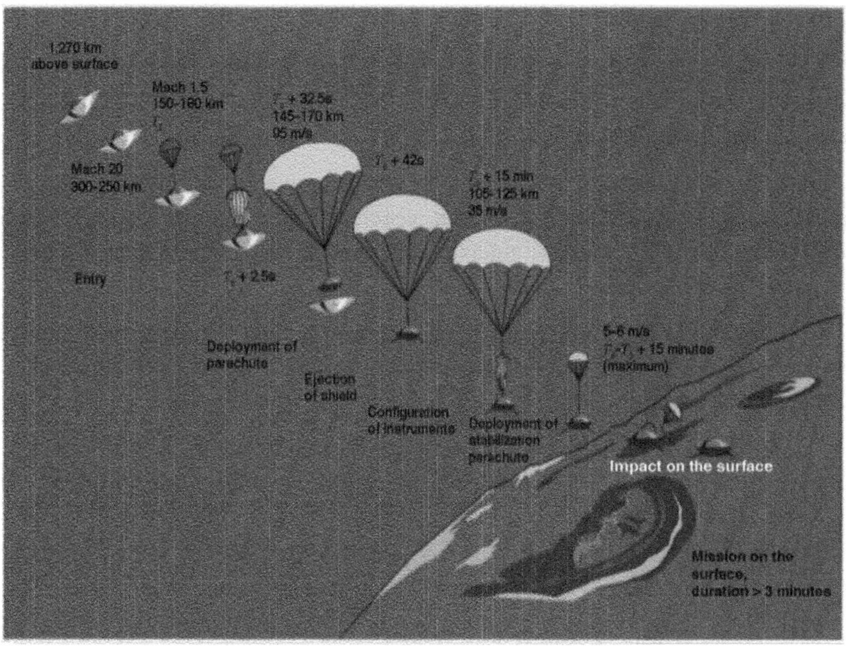

Figure 6.8. *Phases of the Huygens probe mission in the atmosphere of Titan (source: ESA). For a color version of this figure, see www.iste.co.uk/encrenaz/solar2.zip*

The parachute phase begins with the deployment of a pilot chute whose job is to detach the probe's rear cover, which protected it from heat during entry into the atmosphere, and to release the main parachute. This allows the probe to slow to the velocity at which the descent is to be carried out. The thermal shield is then released, and the various sensors of the on-board instruments are exposed to the atmosphere, in order to carry out the operations planned by the observation program (see Figure 6.8).

The payload of an atmospheric probe consists of several instruments for direct and remote detection:

– a set of sensors that, in direct contact with the atmosphere, measure the fundamental physical values (pressure, temperature, permittivity, sound propagation velocity, etc.);

– mass spectrometers that enable the determination of molecular mass for the gases analyzed as well as their identification and quantification; they are often connected to chemical chromatographs of atmospheric gases that separate the various substances present in a homogeneous mixture, which then can be analyzed by the mass spectrometer;

– imaging systems that, in addition to the images taken during descent, provide a panoramic view of the region located under the probe, allowing for characterization of the landing or impact site;

– spectrometers that provide signatures in absorption in the different regions of the electromagnetic spectrum (ultraviolet, visible, infrared), which makes it possible to detect some of the mineralogical components of the surface and/or the atmospheric constituents;

– radar or laser for altimetry, through measurement of the time interval between sending a signal toward the surface and the detection of the reflected signal by instruments on the probe;

– nephelometry to study the localization and composition of the clouds by measuring the concentration of particles in the atmosphere from light diffused at 90° from the incident light.

6.2.2.2. *Chemical analysis*

Knowledge of the chemical and mineralogical composition of the surface of a planetary body is fundamental for understanding its origin and, especially, its evolution. The analytical methods in use are based on the detection of responses from the components that make up the surface to exposure from natural sources of energy (electromagnetic solar radiation or particles of cosmic or solar origin) or that are provoked by *in situ* irradiation of the rocks or the regolith with radioactive sources, lasers, or ion cannons.

Measurements carried out from an orbiter use passive remote detection instruments (spectrometers in the ultraviolet and infrared) to obtain the spectra of solar radiation reflected by the part of the surface covered by the instrument's footprint.

Measurements conducted from lander vehicles or rovers are carried out by active instruments for direct detection (X-ray and gamma-ray spectrometers).

The return of samples from the surface of a celestial body is undoubtedly the method that enables the collection of the most valuable data about their chemical and mineralogical composition: the information that can be obtained by analyzing a few micrograms of extraterrestrial matter in ground-based laboratories is much more complete and precise than what can be obtained by *in situ* analysis. The analysis of extraterrestrial matter in the laboratory uses high-precision instrumentation whose dimensions and complexity are much too great to be installed on a space probe or to be operated remotely.

Between 1969 and 1974, during the six Apollo missions, the astronauts brought back more than 350 kg of rock and soil samples from the Moon's surface, and the Stardust and Hayabusa missions in the first decade of this century collected and brought back to Earth thousands of micrometer-sized particles from comets and asteroids.

6.2.2.2.1. UV, V, IR spectroscopic analysis

Spectra by reflectance in the near infrared, from 0.3 to 5 μm, reveal the presence of species whose spectral features are in absorption and allow for the identification of minerals, rocks, and other geological materials. Ferrous iron (Fe^{2+}) produces deep bands centered on 1 and 2 μm in the spectra of pyroxene, a large band between ~ 0.9 and 1.3 μm, characteristic of olivine, carbonates, and a number of sulfates, as well as a weak band near 1.2 μm in feldspar. Ferrous iron (Fe^{3+}) is responsible for bands at 0.6 and 0.9 μm in oxides and hydroxides, which indicate the effects of aqueous alterations on the minerals present. Depending on its state and mineral structure, water produces bands in the vicinity of 0.96, 1.15, 1.4, 1.9, and 2.9 μm, while structural OH bands are found near 1.4, 2.1–2.5, and 2.7 μm. Other bands are characteristic of carbonates, nitrates, sulfates, phosphates, chlorides, and perchlorates.

The mid-infrared range, between 5 and 20 μm, contains a number of wide bands of molecular transitions associated with modes of rotation-vibration for the most abundant bonds. Thus, a silicate presents a main band, centered around 10 μm, and a secondary band at 18 μm. These vary for SiO and SiO_2, which makes it possible to discriminate between the large families of silicates: the main band for pure SiO is centered at 9.7 μm, and the band for pure SiO_2 is at 10.3 μm.

The spectra for rocks and regolith are affected by the parameters of the samples (for example, the size of grains, porosity, roughness, etc.).

6.2.2.2.2. X-ray spectroscopic analysis

X-ray spectroscopy on-board orbiters is used to analyze the elementary composition of the surfaces of celestial bodies. Matter bombarded by particles or sufficient energy radiation emits by fluorescence X-rays whose energy indicates the electron transitions induced. The energy spectra of this radiation thus provides the composition of the elements in the analyzed soil. X-rays penetrate the soil by several micrometers, and thus allow for analysis of the outermost layer.

X-ray spectroscopy is passive, and in fact uses the solar X-ray radiation that falls on the surface. The distance from the Sun obviously affects the incident solar flux, which is modulated by solar activity. The increase in this activity produces higher energy solar X-rays. This hardening of the incident radiation favors, on the planetary surface, the emission of elements with a large Z (for example, Ca, Ti, Fe) with respect to elements with lower Z (for example, Al, Mg, Si). Observations of the spectra of incident solar X-rays and the planetary emission spectrum should thus be done simultaneously in order to obtain quantitative information about the abundance of different elements on the surface.

6.2.2.2.3. γ spectroscopic analysis

Active instruments for γ spectroscopy are usually mounted on the arms of landers or rovers and use integrated radioactive sources to excite the nuclei of matter on the surface to be analyzed from a very short distance. γ spectrometers on-board orbiters are passive and use excitation by natural sources (radiation associated with major solar eruptions or galactic cosmic radiation).

Gamma rays with characteristic energy are emitted by excited nuclei: natural radioactive elements (K, U, Th) produce lines during decay (for example, the decay of ^{40}K into ^{40}Ar produces the line at 1.46 MeV, and that of ^{238}T into ^{208}Th is emitted at 2.62 MeV).

For stable elements, the excitation of lines is supplied by the bombardment of galactic cosmic rays, which are primarily helium protons and nuclei, with an energy on the order of a GeV and an almost isotropic flux of around 2–3 particles/cm^2. These particles generate nuclear cascades in the bombarded matter. The most efficient particles from these cascades for generating gamma rays are low energy neutrons. The high-energy neutrons produced in this way can be subject to new nuclear collisions and excite stable nuclei to higher levels by inelastic diffusion, in the (n, n'γ) reaction. They can lose energy until they are thermalized and captured, giving rise to a generally complex mixture of captured gamma rays (n, γ).

They also produce gamma rays by forming radioactive nuclei such as ^{26}Al or ^{7}Be by a number of secondary processes. The reaction (n, n'γ) has a large cross-section for all common nuclei at energies above the reaction threshold. Thus, the most abundant elements produce a quantity of gamma particles through this process. The capture cross-section of thermal neutrons depends heavily on the element: it is negligible for ^{16}O, small for ^{12}C, moderately large for hydrogen and ^{56}Fe, and very large for some rare elements such as ^{157}Gd. The neutron cascade penetrates the planetary surface to a depth of several meters. Gamma rays are diffused (with energy loss) or absorbed on a distance scale that is lower by an order of magnitude (a few dozen cm), such that the distribution of equilibrium of the neutrons is found in an area that is quite close to the surface. The thermalization of neutrons depends heavily on the composition of the material, particularly its concentration of hydrogen.

6.2.2.3. *Measuring space plasma*

6.2.2.3.1. A variety of environments

The advent of space exploration made it possible to measure *in situ* the characteristics of the various space plasmas that occupy the heliosphere, beginning with the solar wind after the first missions Luna 1 and 2 and Explorer 10 had demonstrated its existence in the early 1960s. The solar wind was analyzed by a number of probes at heliocentric distances ranging from less than 30 solar radii during the 4th passage at perihelion of the Parker solar probe in 2020, up to several dozen au in the ecliptic plane thanks to the Voyager and New Horizons probes. During the 1990s and 2000s, the Ulysses space probe enabled the observation of the solar wind out of the plane of the ecliptic, particularly above the solar poles. The long-term *in situ* monitoring of the solar wind in the vicinity of Earth has been carried out by space probes such as Wind and ACE since the 1990s.

The ionized and magnetized environments of all the planets in the Solar System, Pluto, and the comets have also been explored by space probes. Of course, it is

Earth's magnetosphere that has been the target of most of *in situ* studies, thanks to dozens of equatorial and polar satellites, a flotilla of probes that has made it possible to take multi-point measurements. We can mention the four Cluster probes in operation since 2000 to study small-scale plasma processes, the five Themis probes launched in 2007 to study substorms, and the four MMS probes launched in 2015 to study the process of magnetic reconnection.

The Pioneer 10 and 11 probes, followed by the Voyager 1 and 2 probes, have made it possible to discover and/or characterize the magnetospheres of the giant planets (see section 5.4, Volume 1). The Voyager mission thus detected in the Jovian magnetosphere the existence of a torus of plasma along the orbit of the volcanic satellite Io. It showed, especially in the case of Saturn, the influence of rings and satellites on the dynamics of the flux of energetic particles trapped in the magnetosphere. The Voyager 2 probe discovered the specific, very inclined configuration of the magnetospheres of Uranus and Neptune and based the study of their dynamics. Jupiter's magnetosphere was re-explored in detail using increasingly powerful instruments thanks to the Galileo equatorial probe at the end of the 1990s, to the remote flybys of Ulysses, Cassini, and New Horizons, and finally to the Juno orbital mission since 2016, whose goal is to characterize the polar magnetosphere. Saturn's magnetosphere was comprehensively explored by the Cassini mission between 2004 and 2017, which discovered that the satellite Enceladus (see section 1.3.2) was active and formed the main source of plasma in the Saturnian magnetosphere.

Cassini also conducted a flyby of Venus and crossed through its ionosphere, which had previously been explored by the Venera and Pioneer Venus probes, then by the Venus Express orbital probe between 2006 and 2015. The Mars Global Surveyor mission in 1997 made possible the first extensive mapping of the Martian magnetic field. Exploration of the Martian plasma environment has been specifically undertaken by the Mars Express orbital probes since 2003 and Maven since 2015. Mercury's mini-magnetosphere was studied for the first time during flybys by Mariner 10 in 1974–1975. The characterization of its dynamics was a goal of the Messenger orbital mission between 2011 and 2015, while its exhaustive study is the object of the Bepi-Colombo mission, currently en route. Finally, we should mention the exploration of Pluto's environment by the New Horizons probe (see section 2.4.2) and the orbital and surface exploration of the comet 67P/Churyumov-Gerasimenko by the Rosetta/Philae mission (see section 2.1.4.2).

6.2.2.3.2. Measurement of plasma

In situ plasma analysis consists essentially of the measurement, for electrons and ions, of the density and temperature at low energy. It also aims at obtaining the energy spectrum and, if possible, the angular distribution of velocities, as well as, for ions, the ionic composition and the energy distribution as a function of charge or of mass. These measurements are carried out by a wide variety of mass spectrometers, in which a constant or variable longitudinal electrostatic field (parallel to the instrument's collimation axis) is generally used to select the type of particle and the energy range to be sampled, while a transverse electrostatic field (or a magnetic field) ensures the deflection of incident particles and the selection of ion species. With improvements in their sensitivity, the early analyzers of currents have evolved toward the counting of individual particles. Modern analyzers now use techniques of incident plasma acceleration (to decrease the energy detection threshold), and of "flight time" measurement (to determine the ion velocity). These detectors are made of plates of micro-channels or solid-state detectors, semi-conductors that produce by cascade a difference of potential proportional to the energy of the incident particle. The angular distribution of the particles is measured by division of the collimation into multiple sectors or by rotation of the satellite or of the analyzer. The challenge is to carry out a complete measurement (over 4π steradians) of the angular distribution of the velocities of the particles over a short time interval, during which the satellite will remain in the investigated region. Satellites that rotate around an axis make it possible to ensure good angular coverage of the measurements.

At low energies, the satellite, which is charged on the surface by the emission of photoelectrons and the collection of plasma, is surrounded by a *sheath* that perturbs the ambient plasma. The measurement techniques thus become more delicate. A variety of probes (Langmuir, electrostatic, mutual impedance) have tried to measure the characteristic tension-current of the satellite immersed in the plasma or its resistivity, and to deduce from this the electron density and temperature, or even the density fluctuations and the drift velocity of the particles in magnetospheric electron and magnetic fields. A powerful, widely used technique is the so-called *quasi-thermal noise* spectroscopy, relying on fluctuations in potential induced on a passive electrical antenna by quasi-thermal agitation of the ambient plasma. While less sensitive to the effects of the sheath because it is probing a larger volume of plasma, this technique provides access to the density and the temperature of both cold and hot electrons and to the velocity of ambient ions.

6.2.2.3.3. Measurement of the magnetic field

The measurement of magnetic fields relies on scalar magnetometers that measure the amplitude $|\boldsymbol{B}|$ of the constant ambient magnetic field and to vectorial magnetometers that enable measurement of $\boldsymbol{B}\cos\theta$, where θ is the angle between the ambient field and the instrument's axis.

The measurements of scalar magnetometers are frequency measurements. The evolution of the techniques in use corresponds to an increase in the sensitivity in Hz nT^{-1}: from the free precession of the proton, which is analogous to the nuclear magnetic resonance (for which the measurement of $\mu_B B/L$, where μ_B is the magnetic moment and L the angular momentum, provides B with a sensitivity of 0.04 Hz nT^{-1}), to the precession of the magnetic moment of the outer electron of the alkaline metals ("Zeeman effect" type measurement, at 3–7 Hz nT^{-1}), and to the variation of the optical properties of ^{4}He gas as a function of the polarization of its electrons (which is equivalent to a measurement of electron gyrofrequency, at 28 Hz nT^{-1}).

Vectorial measurements generally rely on a coil subject to an oscillating current. In *fluxgate* type magnetometers, a magnetically permeable material is excited by this current, and the 2nd harmonic of the excitation, amplified by the projection $\boldsymbol{B}\cos\theta$ of the outer field, is measured in the form of a difference in potential proportional to $\boldsymbol{B}\cos\theta$. Other kinds of instruments include a scalar magnetometer in a coil and operate by canceling the outer field.

6.2.2.3.4. Measurement of waves

On-board measurements of continuous electric fields generally correspond to sampling a difference in potential between two conductive sensors (wires, spheres). All continuous measurements (electrical and magnetic) are of course made difficult by the rotation of the satellite, whenever it spins.

Plasma waves are electrostatic or electromagnetic fluctuations measured *in situ* or after propagation over a short distance, generally guided by the magnetic field or the gradients of magnetospheric plasma, over a wide range of frequencies from the lowest ionic cyclotron frequencies to the electron plasma frequency in the environment. *Radio waves*, which are electromagnetic waves that propagate over long distances, were already addressed in section 6.2.1.4 for ground-based radio telescope measurements. The receptors carried on-board are very sensitive voltmeters, and the sensors are designed to convert the electrical field E (or possibly the magnetic field B) of the wave into a difference of potential ΔV. Electric sensors are dipole antennas made of two spheres, two wires, or a wire and its image by reflection on the conductive body of the satellite, which enables the measurement of $\Delta V = LE$, where L is the characteristic size of the antenna. A single dipole, such as on the Juno probe, enables measurement of the total intensity of incident waves. Two dipoles, such as on the Voyager probes, provide additional information about their polarization. Three dipoles, such as on the Cassini probe, enable, via measurement of the cross correlations of their responses, a complete determination of the wave's polarization and of its direction of propagation (Poynting vector k), enabling *in fine* to conduct imaging as do ground-based radio interferometers (see Figure 3.39,

Volume 1). Magnetic antennas are loops or coils that take advantage of the Faraday effect to measure a difference of potential: $\Delta V = N\, d\varphi/dt$, where N is the number of loops in the coil, $\varphi = \int B dA$ the intercepted magnetic flux and dA is the element of the oriented area of the loop or coil. Reception methods take advantage of heterodyning (conversion to a frequency close to zero before detection, via use of a local oscillator), and a selection of spectra using filters, multi-channels, spectral sweeping, digitization of the shape of the wave, and digital analysis, or a combination of these techniques. In all on-board devices, digitization of the signal occurs as close as possible to the sensors to take advantage of the capabilities of the on-board computers, to reduce instrument noise, and to increase the sensitivity of the measurements.

6.2.2.3.5. Spectral imaging

Spectral imaging applied to space plasma covers a wide range of techniques. In addition to auroral spectral imaging in the UV/visible, IR, X-ray, and radio wavelength ranges already mentioned above (and in a section in Chapter 3, Volume 1), the detection of energetic neutral atoms (ENA, up to several hundred KeV) is a powerful technique for imaging the sources of energetic ions (the torus of Io, the neutral torus of Saturn, Earth's ring current, heliopause, etc.). The latter occasionally cede their energy to a slow ambient neutral atom and transform into an energetic neutral that propagates in a straight line and can be detected remotely. Such instruments go along with ion detectors, with which they share targets, time-of-flight measurements, and measurements of impulsion profile by micro-channels or solid state detectors. They are equipped with devices that block incident ions (electrostatic fields) and UV rays (membranes), which can be confused with neutrals of equal energy. The Cassini probe carried the first energetic neutral atom telescope, called Inca.

Finally, all spatial measurements are limited by the data flow rate between the satellite and the ground, and digital compression techniques used on-board are increasingly sophisticated and useful, particularly since the breakdown of the main antenna on Galileo, which motivated their development in order to save the mission.

6.2.3. *Virtual Observatory and databases*

6.2.3.1. *Virtual Observatory*

The principle of the global Virtual Observatory (VO) is the same as the World Wide Web: to link together astronomical data with compatible standards to allow the data to be compiled and analyzed. Another commonality with the WWW is that VO is not a fixed system but a *way of doing things*. It has been built by researchers and teams that have all observed objects in the sky with original methods and archived the data in different formats. The VO makes it possible for the user to access all of this data without being concerned with its origins, as if it were all on their own machine. The virtual astrophysical observatory makes it possible to "put the Universe online" by combining data from observations collected by space and ground-based telescopes and covering the complete range of wavelengths, from gamma radiation to radio waves, over all time periods. The information thus made available allows astronomers to better understand physical phenomena by bringing together observations at different wavelengths and/or from different time periods.

Before VO, it was very difficult for an astronomer who had observed a phenomenon to know if there had been other observations that could enrich the interpretation of their own observations. To accomplish this, before gaining access to the data, it was necessary to know that it existed, to know by what standards it had been encoded and the observational conditions, and to adapt them in order to incorporate them, along with one's own data, into a single model. The digital archives of the world's observatories are enormous, but are almost impossible to inter-compare. Astronomy is one of the first domains to have accomplished this revolution for a completely scientific reason: astronomy, as well as astrophysics, is built on long-term observation of a number of "astronomical" objects with complex interactions on different scales, from the nanometer-sized grain to galactic clusters. Today, the VO makes it possible for astronomers to have access to an enormous amount of information, hundreds of terabytes of data, obtained by all the telescopes and space probes, as well as the discipline's entire bibliography and all of its modeling data, including synthetic and laboratory spectra. Astronomy is a "remote" science of multiple wavelengths and multiple messengers. The combination of observations from different instruments makes it possible to understand the phenomena at work in the context of potential temporal variability.

Comparison between computer models and these observations is a renewed source of discoveries. Thus, there are more articles based on the Hubble Space Telescope's data archives than on original observations.

These data are primarily from the archives of observations of very large sky surveys (homogeneous information about a large number of objects) and value-added databases, such as the one from CDS.

The CDS, which was created in 1972, initially as the *Stellar Data Center*, now the *Astronomical Data Center of Strasbourg*, is the precursor of the VO, since stars were the first population of astronomical objects whose size necessitated systematic processing. The CDS houses:

– the *Simbad* astronomical global reference database for identifying objects in the sky;

– *Vizier*, the catalog of scientific publication catalogs;

– the *Aladin* interactive software program for access, visualization, and analysis of astronomical images.

The data from scientific articles and large surveys have been homogenized, then connected to the bibliographic database of academic journals, as well as other countries' databases and data models. The CDS thus adds value to the data, then distributes it to the international scientific community. It is made up of a team of about thirty people who are researchers/documentarians/data scientists.

The relevant databases, initially limited to stars, have now been extended to other celestial objects, galaxies, interstellar clouds and nebulae, planets, and small Solar System bodies. The CDS receives more than 800,000 requests per day.

The VO strives for data interoperability and provides software to help researchers find images and spectra, visualize data, explore catalogs, compare data, etc. Data models are expected to be constructed collectively by the community that produces the data. The formats that are used are based on the FITS (*Flexible Image Transport System*) format, the most widely used data format in astronomy. A FITS file contains data and metadata that specifies information, dates, time of exposure, spectral range, size, etc. Data and articles are referenced using an ID. The DOI (*Digital Object Identifier*) reference standard is an identification mechanism that was initially limited to scientific articles and then generalized to all types of objects.

Astronomy was and still is a pioneering discipline. The online data is used by researchers in their daily research work. International cooperation develops standards, links between services, and interoperability, opening a new subfield of astrophysics.

6.2.3.2. *Spectroscopic databases*

For remote sensing of planetary atmospheres, in the calculation of flux emerging from an atmosphere, the equation of radiative transfer (see section 5.1.4, Volume 1) involves optical depth above a given atmospheric level, which itself depends on the absorption coefficient at a given wavelength and a given altitude (see equations [6.1] and 6.2]). This is the sum of the contributions from each atmospheric constituent.

In most cases, the spectral signatures are rotational and vibrational-rotational bands, and the absorption is the sum of the contributions from each of the lines that constitute the band (Hanel *et al.* 2003). The spectroscopic parameters necessary for the calculation (frequency, intensity, broadening coefficient, energy) are listed, with the quantum number of each transition, in spectroscopic databases based on *ab initio* calculation and laboratory measurements. The first databases were established for calculating the transfer in the Earth's atmosphere, within a temperature range that corresponds to Earth's thermal profile. This is the case for the HITRAN database in the United States and the GEISA database in France. Some special cases require different temperature ranges. This is in particular the case for Venus's troposphere, in which the temperature reaches 700 K, and for the giant planets and Titan, where the temperature drops to 50 K. This is also the case for "hot Jupiter" type exoplanets on which the temperature can reach 2000 K.

The GEISA database (Gestion et étude des informations spectroscopiques atmosphériques[4] [Management and study of atmospheric spectroscopic data]) was developed at the Dynamic Meteorology Laboratory (IPSL, Paris, France). It can be accessed online and enables the calculation of radiative transfer line by line and level by level (Jacquinet-Husson *et al.* 2016). The HITRAN (High Resolution Transmission[5]) database was compiled by the Air Force Cambridge research laboratory and updated by the Harvard Smithsonian Center for Astrophysics at Cambridge (Massachusetts, USA) (Gordon *et al.* 2017). In the case of Venus's atmosphere, specific calculations of the absorption of CO_2 at high pressure and temperature have been published (Pollack *et al.* 1993; Scutaru *et al.* 1993). Finally, the ExoMol database has been developed for the calculation of radiative transfer for "hot Jupiters" (Tennyson *et al.* 2016).

6.2.4. *Perspectives of ground-based and space observations*

6.2.4.1. *Ground-based observations*

The largest optical telescopes at current observatories on Earth have mirrors on the order of 10 m in diameter. Some (the four VLT telescopes of the ESO, the two Gemini telescopes) have thin monolithic reconfigurable mirrors; others (the two Keck telescopes, Grantecan, Salt, etc.) have segmented mirrors, formed of the juxtaposition of hexagonal mirrors about a meter in size.

The future of optical ground-based observation lies in giant telescopes, with mirrors that are 20 or 40 m in diameter. Three of this kind of telescope are under construction today: one European, the *Extremely Large Telescope* (ELT, operation

4 See also: https://geisa.aeris-data.fr.

5 See also: https://hitran.org.

planned for 2026/27), and two American, the *Thirty Meter Telescope* (TMT, operation planned for 2030 at the earliest) and the *Giant Magellanic Telescope* (GMT, operation planned for 2029). These instruments will allow for a considerable advance in the variety of domains of interest in the study of the Solar System: the physics of the formation of stars and planets; the discovery and characterization of exoplanets; the chemistry of the surface of objects in the Kuiper belt; the chemistry and meteorology of planetary atmospheres in the Solar System, and the study of exoplanets and the search for potential life there.

For their part, radio astronomers currently have many large radio telescopes with a single antenna and large interferometers: the most recent are, in the millimeter and submillimeter ranges, the global Atacama Large Millimeter-submillimeter Array (ALMA) interferometers on the Chilean Altiplano, made up of 54 antennas of 12 m in diameter and 12 of 7 m, and the French-German-Spanish Northern Extended Millimeter Array (NOEMA), which has 10 antennas of 15 m in diameter, soon to be increased to 12; in the meter and decimeter ranges, the Low Frequency Array (LOFAR) is made up of 50 sets of antennas grouped at 48 European stations. All of these instruments require complex electronics and extremely powerful computers. The Square Kilometer Array (SKA), which is under construction, will be described below. These giant instruments are complementary to optical telescopes, particularly for studying interstellar matter and the formation of planets, as well as their magnetic field.

6.2.4.1.1. The Giant Magellan Telescope (GMT)

The GMT will be installed at the Las Campanas Observatory in Chile. It is expected to begin operations in 2029. It will be made up of seven large rigid adjacent monolithic mirrors: six off-axis of 8.4 meters surrounding a central segment, forming an optical surface of 24.5 meters in diameter. Light will be reflected by the seven primary mirrors onto seven thin secondary mirrors that will send it to a single focus in the primary mirror's central opening. Thanks to adaptive optics, the surface of each secondary mirror will be constantly adjusted to compensate for deformations in the surface of the wave by atmospheric turbulence. The images are thus expected to achieve an angular resolution limited by diffraction that is ten times better than the Hubble Space Telescope. The GMT's first focal instrument will be a cross-dispersed Echelle spectrograph, the GMT-Consortium Large Earth Finder (G-CLEF). This instrument will be placed in a vacuum and connected by fiber optics; it will enable precise measurements of the radial velocity of solar-type stars for the detection and characterization of low-mass exoplanets in orbit around these stars. G-CLEF's bandwidth will extend from 3500 Å to 9000 Å, and will enable studies of the abundance of elements in stars and the observation of faint objects with a large spectral shift.

6.2.4.1.2. The Thirty Meter Telescope (TMT)

This 30 m telescope is being constructed by an international partnership between the state universities of California, Canada, Japan, India, and China. It is expected to be installed on the island of Hawaii, at the top of Mauna Kea, but the decision has been blocked for several years by court battles because the volcano is a sacred site for the native people, and is still unsettled. The telescope would have an altazimuthal mounting, as do all the modern large telescopes, and would have a mobile mass of 2,000 tons, including the instruments. Its main mirror, 30 meters in diameter, will be built of 492 hexagonal reconfigurable facets of 1.4 m. Adaptive optics will correct the effect of atmospheric turbulence by reconfiguring the two mirrors in the path of the optical trajectory. At the beginning of the scientific operation of the TMT, three observational instruments will be available: a Wide-Field Optical Spectrometer (WFOS; 0.3–1.0 μm); an Infrared Imaging Spectrometer (IRIS; 0.8–2.5 μm); and an Infrared Multi-object Spectrometer (IRMS). The construction of new scientific instruments is planned for the first 10 years of operation to complete and expand the TMT's scientific capabilities.

6.2.4.1.3. The Extremely Large Telescope (ELT)

The ELT will be the largest optical and infrared telescope in the world. It is being built by the European Southern Observatory (ESO) on the Cerro Armazones in the north of Chile, not far from the VLT. The main mirror will have a diameter of 39.3 m: it will be made up of 798 hexagonal elements, each 1.45 m in diameter and 40 mm thick. It will use active optics to maintain its ideal form, as does the TMT. The telescope's optics include three aspherical mirrors, M1, M2, and M3, and two plane mirrors, M4 and M5, giving a perfectly stigmatic field of 10' in diameter (see Figure 6.9). Adaptive optics will be carried out by reconfiguring the M4 mirror, which is only 2 mm thick, and the variable inclination of the M5 mirror.

Two platforms with Nasmyth foci, measuring 12 × 29 m, will each be able to accommodate three instruments. There will also be a Coudé focus. For adaptive optics, the ELT will have, as do the other giant telescopes, several artificial stars formed in the upper atmosphere by six lasers positioned next to the telescope. The instruments that will be installed at the beginning of the ELT's operations will be: MICADO (Multi-AO Imaging Camera for Deep Observations), a near-infrared camera (0.8–2.4 μm) with a spatial resolution of 6 to 2 milliarcseconds; HARMONI (High Angular Resolution Monolithic Optical and Near-infrared Integral field spectrograph), an integral field spectrograph in the visible and infrared ranges (0.47–2.45 μm): METIS (Mid-infrared E-ELT Imager and Spectrograph), an imaging spectrograph in the mid-infrared.

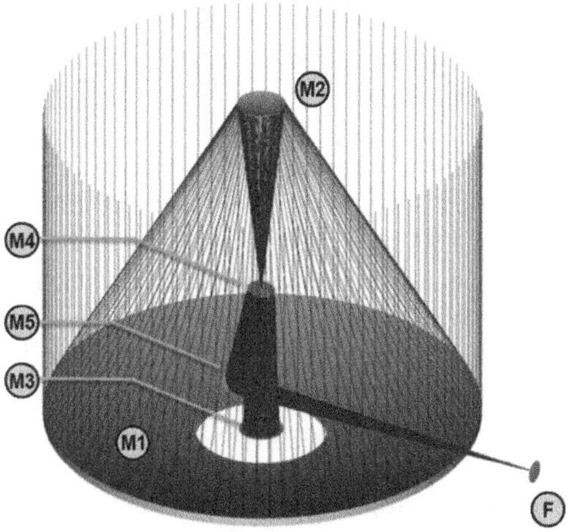

Figure 6.9. *Optical diagram of the E-ELT, with its original anastigmatic configuration, which uses four mirrors upstream from the primary mirror; in blue, the light path from the main M1 mirror to one of the two Nasmyth F foci (source: ESO). For a color version of this figure, see www.iste.co.uk/encrenaz/solar2.zip*

6.2.4.1.4. The Square Kilometer Array (SKA)

The *Square Kilometer Array* is a gigantic radio interferometer whose total surface will reach 1 km^2, hence its name. Thirteen countries are currently involved in this project. SKA is under construction at two sites: one in South Africa, with a first phase of 200 parabolic antennas of 15 m in diameter, and another in Western Australia, which will initially have 130,000 antennas similar to television antennas. It will cover all frequencies between 50 and 350 MHz, covering metric and decametric waves. The progressive implementation of the first phase will extend to 2024, and a second, even more ambitious phase will take place in the 2030s.

6.2.4.2. *Space-based observations*

Many probes traveling to planets and the Moon are operating in 2021: 2001 Mars Odyssey (launched in 2002 by NASA); Mars Reconnaissance Orbiter (2005, NASA); New Horizons (2006, NASA); Akatsuki (2010, JAXA); Mars Science Laboratory with its Curiosity rover (2011, NASA); Mars Atmosphere and Volatile EvolutioN (Maven, 2013, NASA); ExoMars Trace Gas Orbiter (TGO, 2016, ESA); Mars-2020/ Perseverance (NASA), Hope (Emirates) and Tianwen-1 (China), all launched in 2020; Origins, Spectral Interpretation, Resource Identification, Security, Regolith Explorer (OSIRIS-REx, 2016, NASA); BepiColombo (2018, ESA-JAXA); Interior Exploration using Seismic Investigations, Geodesy and Heat Transport (INSIGHT,

2018, NASA). In December 2020, Hayabusa 2 (2014, JAXA) brought back to Earth samples collected from the asteroid Ryugu, and Chang'e 5 (2020, China) brought back lunar samples to Earth.

Among the satellites that study the exoplanets, the most recent are the Transiting Exoplanet Survey Satellite (Tess) from NASA and the Characterising ExoPlanet Satellite (CHEOPS) from ESA, launched respectively in 2018 and 2019 and both currently operational. In the near future we will see the launch in 2021 of the James Webb Space Telescope (JWST) from NASA and ESA, the successor to the Hubble Space Telescope, and then the Wide Field InfraRed Survey Telescope (WFIRST, recently renamed as *Roman*), from NASA, in 2025. These two generalist satellites will of course be very useful for the study of Solar System bodies and exoplanets.

Solar System exploration programs from the main space agencies are already planned for the next twenty years: several missions are in various stages of production and will be launched before the end of the 2020s to conduct observations until the end of the following decade. Two of these missions are satellites dedicated to the discovery of exoplanets (Plato) and the characterization of their atmospheres (Ariel). Others are interplanetary probes: in addition to several Lunar probes, there will be exploration probes to small bodies (Comet Interceptor, in phase 0, Lucy, in phase 1, Psyche, in phase 2), to Mars (ExoMars 2022, in phase 3) and its satellites (MMX, in phases 2, 3, and 4), to the Jupiter system (Juice, in phase 2) and to Titan (Dragonfly, in phase 3). These missions are described below.

6.2.4.2.1. Plato

The scientific goal of ESA's PLAnetary Transits and Oscillations of stars (PLATO) scientific mission (launch planned for 2026) is the detection and characterization of terrestrial exoplanets around bright solar-type stars; the focus is on planets in orbit in the habitable zone. Plato will carry out uninterrupted photometric monitoring in the visible range over a long period of time (months to years) of a large sample of bright stars ($m_v \leq 11$–13). The resulting light curves will be used to detect planetary transits, from which planetary radii will be determined, and also for asteroseismology of host stars, which will make it possible to identify their properties and their age. Since the targets of Plato are bright, we can measure the mass of the detected planets from ground-based observations of variations in their host-star's radial velocity.

6.2.4.2.2. Ariel

ARIEL (Atmospheric Remote-sensing Infrared Exoplanet Large-survey), another ESA mission, is planned for 2028. It will study the atmospheres of exoplanets, how they were formed, and how they have evolved, by examining a diverse sample of around 1,000 extrasolar planets in the visible and infrared ranges. It will be the first mission dedicated to the measurement of the chemical composition and thermal structure of the atmosphere of exoplanets in transit.

6.2.4.2.3. Comet Interceptor

This ESA mission, which includes three space vehicles, will be launched in 2028 with the same rocket as Ariel toward the L2 Lagrange point. Thanks to its own propulsion system, it will be the first mission to visit an Oort cloud comet or an object of interstellar origin that has entered the Solar System. The three vehicles will orbit around the chosen target, thus obtaining a 3D image. Comet Interceptor's destination will not be known during the preparation for the mission; the probe will wait in space for an encounter with a new comet. Its mission has a planned duration of five years.

6.2.4.2.4. Psyche

NASA's Psyche probe will study the asteroid (16) Psyche, which orbits in the main belt at around 3 au. It has an effective diameter of about 235 km; we think it has a metallic core, composed almost completely of iron and nickel, with a few percent of pyroxene with a high concentration of magnesium, and probably hydrated silicates from the impact of chondritic meteorites. The mission is expected to launch in 2022 and will enter orbit around (16) Psyche in 2026. The operations will be very similar to those of the Dawn mission to (4) Vesta, with four increasingly close orbits around the asteroid, to precisely determine its shape and gravitational field.

6.2.4.2.5. Lucy

NASA's Lucy mission will carry out the first exploration of Jupiter's Trojan asteroids, which are essential objects for understanding the history of the Solar System. Lucy will be able to visit six of these primitive asteroids, observing objects that are located near Jupiter's two L4 and L5 Lagrange points. Lucy will be launched in 2021 and will conduct flybys of its targets between 2025 and 2033. The three instruments on-board cover wavelengths from 0.4 to 100 μm; they will provide information about the asteroids' geology, the color and composition of the surface, thermal properties, and other physical properties.

6.2.4.2.6. Mars 2020 and ExoMars 2022

The first objective of these missions is to determine if there was life on Mars and to better understand the history of the water on the planet. In the lineage of Curiosity, one of the goals of NASA's Mars 2020 mission with its rover Perseverance is to select samples for subsequent return to Earth.[6] The ExoMars 2022 mission, developed by ESA and Roskosmos with a planned launch in 2022, will have a rover named Rosalind Franklin, which will include a drill to access the subsurface of Mars and a miniature clean laboratory for researching life.

6.2.4.2.7. MMX

MMX (*Mars Moons eXploration*) is a mission from the Japanese JAXA space agency with extensive international participation. With plans to closely observe Mars's satellites Phobos and Deimos and to return samples for analysis in Earth-based labs, the mission's main scientific objective is to determine the origin of these satellites and to thus try to understand the role of small bodies in the formation of planetary systems in the habitable zone. MMX will be launched in 2024 for a nominal duration of five years. It will carry a mass spectrometer, an X- and gamma- ray spectrometer, a visible and infrared spectrometer, a lidar, two cameras with wide angle and telephoto lenses, a cosmic dust detector, a rover, and a system for collecting samples that will be returned to Earth at the end of the mission (2029).

6.2.4.2.8. Juice and Europa Clipper

Juice (*JUpiter Icy moons Explorer*) is the first ESA mission to the Jupiter system (see section 1.2.1.5). Planned for launch in 2022 and arrival at Jupiter in 2029, the probe will spend at least three years carrying out detailed observations of Jupiter and its three Galilean satellites, Ganymede, Callisto, and Europa. Jupiter's circulation, meteorology, chemistry, and structure will be studied from the top of the clouds to the thermosphere. The study of the Jovian magnetosphere will focus on three-dimensional properties of the magneto-disk and the processes of interaction in the magnetosphere, ionosphere, and thermosphere. The optical and radio auroras and their response to the solar wind will also be observed.

Juice will study the interactions of the Galilean satellites with the magnetosphere, their gravitational interaction, and long-term changes in the tides. Ganymede will be the object of detailed exploration because it represents a natural laboratory for analyzing the nature, evolution, and potential habitability of icy worlds; it also plays a specific role among the Galilean satellites because it has its own magnetic field. The objectives of the observation of Ganymede and, to a lesser extent, Callisto, are thus the characterization of the superficial layers and the detection of potential subsurface oceans; surface mapping (topography, geology, and composition); the study of the physical properties of the icy crusts; the

---

6 See also: https://www.esa.int/Science_Exploration/Human_and_Robotic_Exploration/ Exploration/ Mars_sample_return.

characterization of the distribution of internal mass, dynamics, and the evolution of the interior and the exosphere, the study of Ganymede's magnetosphere, and its interactions with the Jovian magnetosphere.

For Europa, the focus will be on the chemistry essential to life, particularly the detection of prebiotic molecules, the formation of surface features, and the composition of non-icy material. For its part, NASA is preparing the Europa Clipper mission, whose launch is planned in the 2020s. The probe should be able to carry out a series of flybys close to the satellite Europa, after a voyage of several years.

6.2.4.2.9. Dragonfly

NASA's Dragonfly mission will be launched in 2026 and arrive on Titan in 2034 (see section 1.3.4). It will release a drone that will fly over the surface of the satellite. With eight rotors and a nuclear energy source, this 3 m long drone will cover more than 175 km over 2.7 years, first landing on the dunes of Shangri-La, in the area where ESA's Huygens probe landed in 2005. The probe will thus explore the atmosphere of Titan one Saturnian year (29 Earth years) after Huygens, and thus will be able to compare the situations between the two time periods.

The drone's lifetime is in principle only limited by the decay of its plutonium power supply; it could operate for eight Earth years. Dragonfly will visit new sites by using increasingly long flights (up to 8 km) during which it will study the surface both remotely and *in situ* with its scientific instruments and will analyze samples. The on-board instrumentation includes a mass spectrometer to identify the chemical components in surface and atmospheric samples, a gamma and neutron ray spectrometer to analyze the composition of the surface elements under the drone's landing sites, meteorological sensors, a seismometer, and a set of microscopic and panoramic cameras.

A number of other exploration projects for objects in the Solar System have been proposed in response to requests for proposals from space agencies. However, since financial resources are not infinite, these projects must wait until the time when the conditions are favorable for their potential adoption. On the list are several missions that target detailed exploration of bodies in the outer Solar System, such as the systems of Uranus and Neptune, which have only been observed by the Voyager 2 probe, or Enceladus, where particles ejected from its cryogenic geysers would be collected. We also envisage the return of samples from a comet, type E, P, and D asteroids, various sites on Mars, etc. These ideas will be the basis for the space exploration program in our Solar System during the second half of the century.

6.3. Computer simulations

6.3.1. *Dynamics*

As soon as it became possible, the calculations needed for the prediction of the motion of the Moon and the planets were done by computer. Soon after, computer simulations made it possible to study astronomical phenomena that are impossible to produce in the laboratory. Then, the diversity of observations led astronomers to add complexity to the computer simulations. Today, the interweaving of all these reasons, to which have been added the volume of data produced by the simulations themselves, has made astronomy one of the main users of new computing architectures (the set of machines and software).

With the ever increasing flow of data and the increasing spectra and spatial precision, the organization and arrangements of these increasing data flows have given rise to a new field of astrophysics supported by the international Virtual Observatory (see section 6.2.4). The questions raised by these observations are studied by modeling physical systems that involve multiple mechanisms on progressively longer scales of time and space. The astrophysics community is one of the primary users of new generations of massively parallel computers with hybrid architectures, in which some parts of the codes run in a central processing unit and others run in connected processors. The major questions of planetary astrophysics processed by these simulations relate to planetary formation in circumstellar disks, calculation of ephemeris, study of planetary/satellite systems, physics of planetary and interplanetary plasma, and solar physics.

The primary force governing dynamic structures is gravity. Based on the number of objects concerned, their respective sizes, and the mean free path between the objects, the environment will be treated either as a fluid, with hydrodynamic modeling software, like a large number of bodies that have behavior that can be statistically processed with N-corps modeling software, or like a reduced number of bodies in which the behavior of each body is calculated as precisely as possible using mathematical theory. Some software blends these different approaches.

To study questions such as “How do planets form around a star, and how many form?” and “What kinds of radiation and magnetic environments are supplied by a star to the planets in its system?” a wide variety of objects and physical mechanisms need to be taken into account. These questions of the nature of astrophysics require the resolution of other questions that must be addressed first, such as “Which processes control the evolution of a dense collisional environment in the presence of electromagnetic radiation?”

6.3.1.1. *Multi-scale and multi-physics problems*

In the questions asked in astrophysics, the dynamics of the scales of space and time cover several orders of magnitude, and these scales are all linked. Moreover, a number of fundamental physical processes are at work and must be taken into account. These include gravity, magneto-hydrodynamic turbulence, the transfer of radiation, atomic and molecular physics, and non-equilibrium chemistry. Since the energy densities of these processes are often comparable, they must all be taken into account.

To take such a wide range of dynamic scales into account, models use adaptive space-time meshes that make it possible to locally increase the resolution during computation. The model approximates the environment being studied using a mesh. Each cell in the mesh can be "refined", that is to say divided into smaller cells, so that a more detailed description can be achieved locally. This refinement can be applied recursively to each cell, which makes it possible to attain very high resolutions in locations where more precision is needed because of turbulence, for example. This technique is called *Adaptative Mesh Refinement* (AMR).

Moreover, the preparation of observations on instruments in high demand requires the ability to ensure the observability of the relevant phenomena. Computer simulations of a physical process are thus associated with the simulation of the phenomenon to be observed.

6.3.1.2. *Some models*

In planetology, the major questions giving rise to these models are:

– the calculation of planetary ephemeris. The first tables that made it possible to predict the motion of the Sun, the Moon, and the planets only contained empirical extrapolations from observations. Next came predictions based on gravitational theories whose parameters were deduced from observations. Beginning with Newton, when the laws of dynamics became known, it became necessary to take into account all of the gravitational effects that could act on these bodies. Theoretical research has been conducted to model long-term orbital evolution using a linear differential system that connects eccentricities and inclinations. Currently, astronomical ephemeris, produced for example in France by the *Institut de mécanique céleste et de calcul des éphémérides* (IMCCE [Institute for Celestial Mechanics and the Calculation of Ephemeris]), are constructed by digital integration and adjusted based on observations. For example, the distance of Jupiter is today known to within 3 m;

– the study of dynamic systems using formal computation and the application of these tools to the problems of astronomy: in particular, Jacques Laskar has thus studied a number of problems of the stability of bodies in the Solar System[7];

– the process of planetary formation in a disk (see an example of a model in section 6.3.1.4);

– the atmosphere of planets is studied with models called GCMs (see section 6.3.2);

– the physics of interplanetary (solar wind) and planetary (ionosphere and magnetosphere) plasma. The LatHys database brings together a number of results of plasma simulations[8];

– the calculation of planetary spectra based on the simulation of the structure of the planet irradiated by its star. These models are based on the laws of physics and also use thermochemical and spectroscopic databases. The models have been created by comparing synthetic spectra to observations. An example is provided by the PSG (Planetary Spectrum Generator) (Villanueva *et al.* 2018).

6.3.1.3. *Computing machines*

High-end computing technology has evolved toward *massively parallel* architectures on one hand, with multicore processors processing several instructions simultaneously, and *hybrid* architectures on the other hand. In a hybrid architecture, some of the calculations are carried out by an integrated circuit processor, instead of being done by the central processing unit (CPU). These "accelerations" concern mathematical functions, and the co-processors are thus called floating point processing units. In image processing computation, boards are dedicated to image synthesis for the functions of projection, illumination, or layout of the surface. In physics, collisions and gravity may be processed by microprocessors. In optics, ray tracers are dedicated to the computation of light trajectories. The graphics processing unit (GPU) can itself be multi-use and massively parallel.

The EuroHPC (High-Performance Computing, HPC) project provides European researchers with access to *exascale* supercomputers. The term HPC also refers to the science developed around this equipment (hardware, software, etc.).

These *exaflops* supercomputers have a computing power on the order of 10^{18} flops (1 exaflop = 1 billion billion floating point calculations per second). In France, the hybrid Jean-Zay CPU+GPU processor is dedicated to artificial intelligence (AI)

7 See also: https://en.m.wikipedia.org/wiki/Jacques_Laskar.

8 See also: https://dumas.ccsd.cnrs.fr/CNES/insu-01480584v1.

processing and to high-performance scientific computing. Astronomy and astrophysics have always been pioneers in the use of new digital technologies and have supported its development. Hardware also needs to be adapted for massive data processing that requires both intensive computation and fast access to large databases. Some models can also be used in different domains, including cosmology, interstellar medium, protoplanetary disks, and plasma physics. The FargoD[9] model simulates fluid dynamics in rapidly rotating systems (planets and stars), while Heracles simulates fluid dynamics subject to a radiation field, etc.

6.3.1.4. *Models for understanding planetary formation*

Debris disks (see section 4.10) are circumstellar disks where only dust is visible. This dust, with a short lifetime, comes from the largest bodies, planetesimals or planets, hidden in the dust. These disks are also the birthplace of planets, and so, for understanding the process of the formation of these planets, it is primordial to understand the connections between the dynamics of the parent body and the distribution of the generated dust. These models will make it possible to interpret images from the instruments that are observing these dust disks. Simulations should be able to follow the evolution of particles from micrometer to planet size, and to simulate the gravity and constructive and destructive collisions between these bodies. Chemistry and interactions with the central star also need to be taken into account, in order to simulate how this disk will look to instruments.

LIDT-DD, which is described in (Kral *et al.* 2013), is a general, self-coherent model, used for both simulating observations and to extend the development of our theories about the physics of these disks. The models previously developed to model the physics of debris disks were either centered on the physics of collisions or on the dynamics of the grains. This new model makes it possible to monitor both effects simultaneously, which obviously overlap substantially because their characteristic time scales are very similar. This model has made it possible to take into account the effect of collisions when studying disk/planet interactions, but also to study the size distribution of grains in the vicinity of a planet (see Figure 6.10).

This self-coherent model is a hybrid based on two approaches, an N-body model managing dynamics and a statistical "particle in a box" model managing collisions. Other routines have been implemented to make the model capable of operating over long time scales, such as with super-particles (SP) and new ideas for self-regulating the number of SPs in the computer system. This new hybrid model is capable of correctly creating dust in a self-coherent manner in complex systems and comparing the results to observations. Cooperation made it possible to connect the hybrid

9 See also: http://fargo.in2p3.fr/.

model to a model of radiation transfer for optically fine disks (GRaTeR) and thus to be able to produce SEDs (spectral energy distributions) and synthetic images (see Figure 6.11). LIDT-DD makes it possible to model the evolution of planetesimals and the dust in a variety of systems where the dynamics and collisions are significant. LIDT-DD is capable of following the evolution of the collision of fragments as they propagate in the disk.

In particular, the model takes into account the production of new generation fragments from impacts between primordial bodies. Grains from the second, third, or -nth generation can significantly contribute to the disk's luminosity in some regions (see Figure 6.10 where the excess of red points is due to grains created during impact).

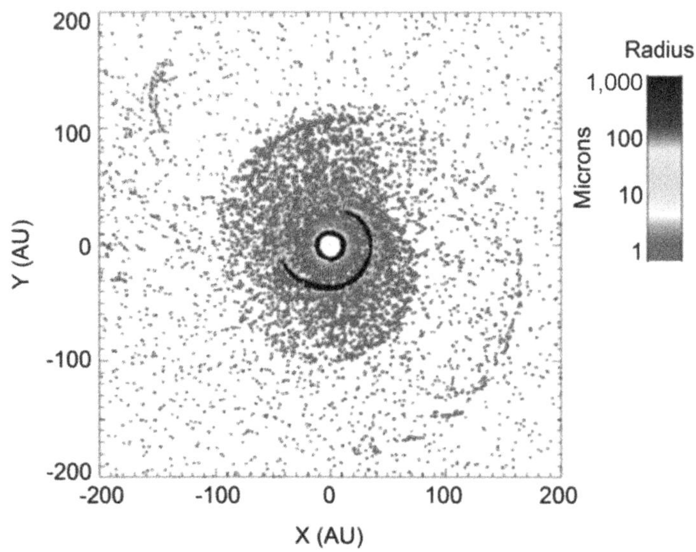

Figure 6.10. *LIDT-DD simulation of a giant collision in a disk of debris, 300 years after the impact (source: (Kral* et al. *2013)). For a color version of this figure, see www.iste.co.uk/encrenaz/solar2.zip*

The first astrophysics question studied by LIDT-DD is the evolution of a planetary system following a violent collision between two subplanetary objects. Indeed, a growing number of observations shows that some disks are too bright to be in collisional equilibrium, given their age, and that violent transitory collisional phenomena could be at work. LIDT-DD is capable of monitoring the evolution of such a complex system in a self-coherent manner. A number of interesting results have been obtained with regard to the brightness of such an event, the asymmetries that develop and resorb over time, and their various time scales. This study also makes it possible to evaluate detectability and to provide predictions of observations for Sphere/VLT or Miri/JWST instruments (see Figure 6.11), which should provide a resolved image of such a system and allow us to see planets as they form.

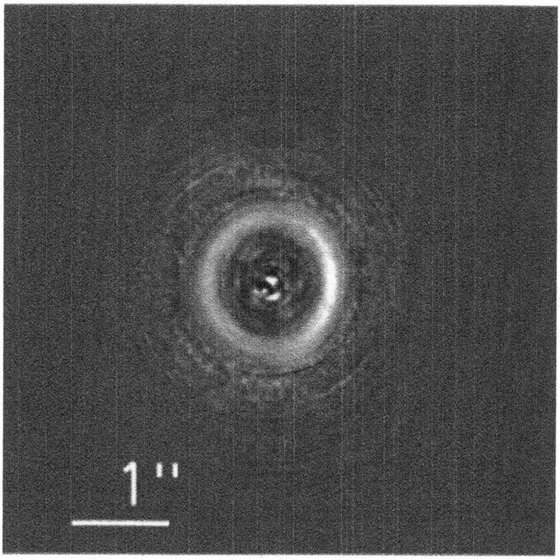

Figure 6.11. *Synthetic Sphere image of an impact in a disk (source: (Kral 2014)). For a color version of this figure, see www.iste.co.uk/encrenaz/solar2.zip*

6.3.2. *Global climate models*

A Global Climate Model (GCM) is a mathematical tool designed to model the atmospheric parameters (vertical profiles of pressure, temperature, various atmospheric components, cloud structure, and wind) and to study their evolution over time. In the case of the Earth, the first atmospheric model dates to 1950. The models have become more complex over the years; there were 23 in 2007 according to the 4th report from the IEGCC (Intergovernmental Expert Group on Climate Change). The most widely used models are ocean-atmosphere connected models that include the contributions of the ocean, the atmosphere, sea ice, and the continents and that analyze their interactions. The next step incorporates Earth system models that add biogeochemical cycles to the preceding factors and, by studying their retroactive effects, make it possible to obtain the most reliable climate projections.

Beginning in the 1990s, planetologists undertook to develop, on the basis of terrestrial models, global climate models that represent other terrestrial planets (Mars and Venus) and other solid surface objects with an atmosphere (Titan, Pluto, Triton). Models for giant planets are also being developed.

6.3.2.1. *Description of a GCM*

6.3.2.1.1. 1D models

As in the case of the Earth, the development of planetary GCMs has become more complex over the years. The first models were radiative-convective transfer models with one dimension (altitude), for a fixed solar incident angle. A pioneering study was carried out by J. Kasting and his team (Kasting *et al.* 1993), which used a 1D model to assess the limits of the habitable zone (HZ) (see section 2.7.7, Volume 1) of a solar-type star for rocky exoplanets with an atmospheric composition of $CO_2/H_2O/N_2$. The study showed that the lower limit of the HZ is fixed by the disappearance of H_2O following photolysis and by the release of hydrogen, while the upper limit is the formation of clouds of CO_2, which increases the albedo and decreases the temperature gradient in the convective zone. The limits found by (Kasting *et al.* 1993) (0.95 and 1.37 au) remain close to the values calculated today. The enormous advantage of 1D models is that they require very short computation times and can model climate change over long time scales.

This approach is particularly useful in the case of modeling of exoplanets, for which very few physical parameters are available. As an example, the LASP (Live Atmospheres-of-Planets Simulator) project developed at the Dynamic Meteorology Laboratory in Paris (Turbet *et al.* 2017) is an online tool for simulating the atmospheric evolution of a rocky (exo)planet. The input parameters are orbital parameters, gravity at the surface, the density of a column of H_2O, the ocean coverage, the amount of CO_2, and the type of star. This tool[10] makes it possible, in particular, to calculate the limits of the habitable zone for different types of stars, and to simulate the evolution of the climate on a terrestrial planet under different environmental conditions (orbit or type of host-star).

6.3.2.1.2. Simplified 2D and 3D models

In the case of rapidly rotating planets, it can be interesting to consider 2D models (altitude and latitude), which make it possible to take into account the variation in sunlight exposure as a function of latitude; in the first place, the variations in latitude for a rapidly rotating plane are much higher than longitudinal variations. This

10 Available at: http://exoplanetes.esep.pro/esep_gcm1d/.

applies particularly to the giant planets, for which there are no topographical variations to take into account. It is also well adapted to the case of the giant exoplanets.

Another simplified approach consists of developing a "simplified forcing" 3D model (Read 2011). This involves using simplified digital models and laboratory experiments to evaluate the role of different entry parameters (radius, rotation, temperature profile) to determine the characteristic scale lengths, the modes of circulation, the types of oscillations, and the processes of instability that are involved in climate change on the terrestrial planets. This approach also makes possible a useful extrapolation for the exploration of rocky exoplanets.

6.3.2.1.3. 3D global models

The construction of a GCM begins with the definition of a three-dimensional mesh representing the planet that is to be modeled (see Figure 6.12). The three dimensions are latitude, longitude, and altitude.

In the case of Mars, the horizontal dimensions of the mesh are typically 1 degree in latitude and longitude, and the altitude dimensions are on the order of a kilometer, clearly below the scale height (8 km at low altitudes). Inside each cell, we calculate the energy assessment by applying the fundamental laws of physics (conservation of mass, energy, and the quantity of motion) in the form of partial differential equations (Navier-Stokes equations) that can be numerically integrated. These equations are resolved inside each cell to determine the atmospheric parameters and their evolution over time: pressure, temperature, ratios of the mixture of minor constituents, cloud structure, wind speed, etc.

There are many applications for GCMs. In the case of Earth, they are continuously compared with observational data acquired from the ground and from space, and thus make it possible to understand the physical mechanisms associated with some aspects of climate change (increases in temperature, drying, etc.). When associated with paleoclimate data (see section 4.3.3.6, Volume 1), they make it possible to study the climate during Earth's history, as well as its recent evolution since the beginning of the industrial era. Above all, they produce climate projections until the end of the 21st century and beyond by studying the response of atmospheric parameters (beginning with temperature) to the increase in the concentration of greenhouse effect gases, primarily carbon dioxide.

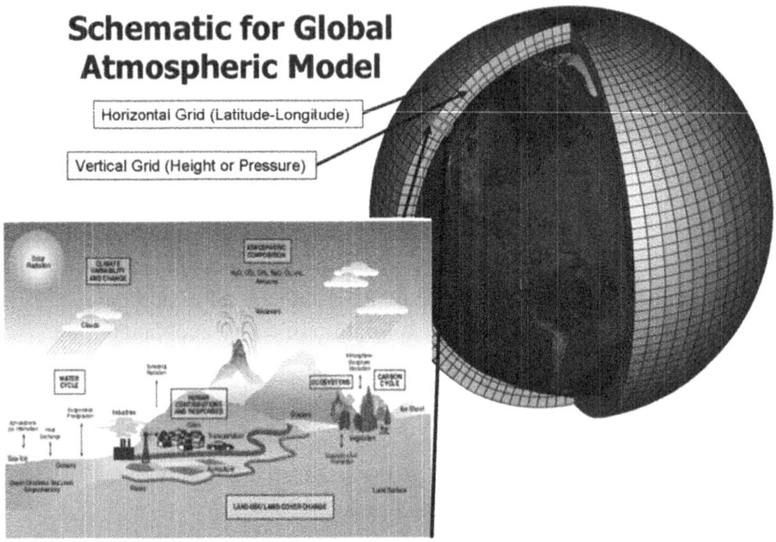

Figure 6.12. *Representative diagram of a GCM. The atmosphere is represented in the form of a three-dimensional mesh as a function of latitude, longitude, and altitude; the fundamental equations of physics are applied to each cell and digitally integrated to calculate the transfers of flux, the wind, and the evolution of atmospheric parameters over time (source: Wikimedia Commons). For a color version of this figure, see www.iste.co.uk/encrenaz/solar2.zip*

6.3.2.2. *Application to Mars*

The application of terrestrial GCMs to the case of Mars dates to the 1990s. In relation to the Earth, the problem is considerably simplified by the absence of oceans and the biosphere. On the other hand, because of the low pressure at ground level, the atmosphere responds to variations in sunlight exposure much more quickly than on Earth, which tends to generate violent winds and dust storms that can surround the entire planet (see section 4.3.3, Volume 1).

The model that is most widely used by the European community is the *Mars Climate Database* developed by a team coordinated by the Dynamic Meteorology Laboratory in Paris.[11] A photochemical module is paired with the GCM to estimate the evolution of the products of photochemistry, particularly O_3 and H_2O_2. A microphysics module processes cloud effects. This database is freely available to the scientific community to calculate a number of atmospheric parameters, at each

11 See also: http://www-mars.lmd.jussieu.fr/.

position on the disk and at each altitude, as a function of the seasonal cycle. These parameters are pressure, temperature, the ratios of the mixtures of minor atmospheric components (H_2O, CO, O_3, H_2O_2) the density of a column of H_2O ice, and the quantity of dust, as well as horizontal and vertical wind. Figure 6.13 shows the agreement obtained between the GCM and data from the TES infrared instrument of the Mars Global Surveyor mission for the distribution of water vapor and the opacity of water ice as a function of the seasonal cycle (Navarro *et al.* 2014).

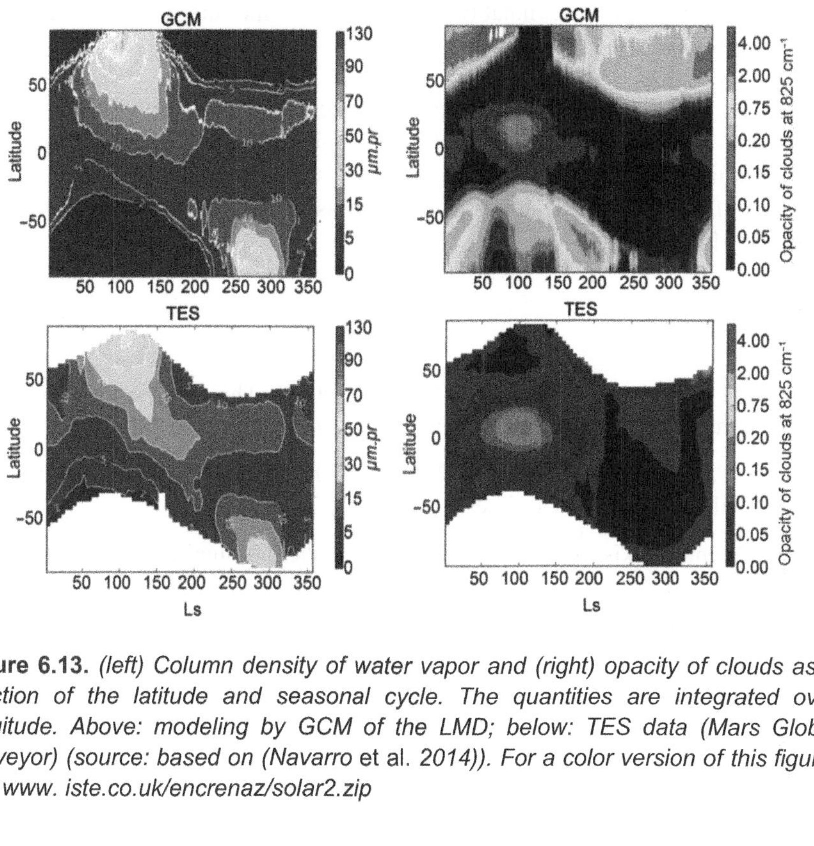

Figure 6.13. *(left) Column density of water vapor and (right) opacity of clouds as a function of the latitude and seasonal cycle. The quantities are integrated over longitude. Above: modeling by GCM of the LMD; below: TES data (Mars Global Surveyor) (source: based on (Navarro* et al. *2014)). For a color version of this figure, see www. iste.co.uk/encrenaz/solar2.zip*

In (Forget and Lebonnois 2013), we find a review of GCMs developed for other Solar System objects as well as an extrapolation for rocky exoplanets.

6.3.2.3. *Other objects in the Solar System and exoplanets*

After Mars, Titan is the first object for which climate models have been developed. With its dense atmosphere, which is rich in molecular nitrogen, Titan's atmosphere presents some analogies to primitive Earth. It exhibits seasonal cycles, with the water cycle replaced by methane and the siliceous ground with hydrocarbon deposits (see Chapter 1 and (Friedson *et al.* 2009; Lebonnois *et al.* 2012)).

The very faint atmosphere of Triton, which is also subject to very significant seasonal effects, has also been modeled using a GCM (Hansen and Paige 1992) to describe the molecular nitrogen cycle. The same studies have been conducted on Pluto to understand the seasonal cycles of CH_4 and N_2, as well as their long-term evolution (Bertrand *et al.* 2018, 2019). The atmosphere of Venus also has been the focus of active research, notably to understand the origin of super-rotation (Lebonnois *et al.* 2016).

Finally, global climate models applied to exoplanets have multiplied in the last few years, whether in reference to the hot Jupiters (Heng and Showman 2016), sub-Neptunians like GJ1214 b (Charnay *et al.* 2015), or rocky exoplanets. One of the goals of these models is to evaluate the limits of the habitable zone (Forget and Leconte 2014).

6.4. References

Bertrand, T., Forget, F., Umurhan, O.M. *et al.* (2018). The nitrogen cycles on Pluto over seasonal and astronomical timescales. *Icarus*, 309, 277–296.

Bertrand, T., Forget, F., Umurhan, O.M. *et al.* (2019). The CH_4 cycles on Pluto over seasonal and astronomical timescales. *Icarus*, 329, 148–165.

Carr, T.D. and Gulkis, S. (1969). The magnetosphere of Jupiter. *Annual Review of Astronomy & Astrophysics*, 7, 577–618.

Charnay, B., Meadows, V., Leconte, J. (2015). 3D Modeling of GJ1214b's atmosphere: Vertical mixing driven by an anti-Hadley circulation. *Astrophysical Journal*, 813(15).

Encrenaz, T., Bibring, J.-P., Blanc, M. *et al.* (2003). *Le système solaire*. CNRS/ EDP Sciences, Paris/Les Ullis.

Encrenaz, T., Bibring, J.-P., Blanc, M. *et al.* (2004). *The Solar System*, 3rd edition. Springer-Verlag Berlin Heidelberg.

Forget, F. and Lebonnois, S. (2013). Global climate models of the terrestrial planets. In *Comparative Climatology of Terrestrial Planets*, Mackwell, S.J., Simon-Miller, A.A., Harder, J.W., Bullock, M.A. *et al.* (eds). University of Arizona Press, Tucson.

Forget, F. and Leconte, J. (2014). Possible climates on exoplanets. *Philosophical Transactions of the Royal Society*, A(372).

Friedson, A.J., West, R.A., Wilson, E.H. *et al.* (2009). A global climate model of Titan's atmosphere and surface. *Planetary and Space Science*, 57, 1931–1949.

Girard, J., Zarka, P., Tasse, C. *et al.* (2016). Imaging Jupiter's radiation belts down to 127 MHz with LOFAR. *Astronomy & Astrophysics*, 587(A3).

Gordon, E.G., Rothman, L.S., Hill, C. *et al.* (2017). The HITRAN 2016 Molecular Spectroscopic Database. *Journal of Quantitative Spectroscopy and Radiative Transfer*, 203, 3–69.

Hanel, R.A., Conrath, B.J., Jennings, D.E. *et al.* (2003). *Exploration of the Solar System by Infrared Remote Sensing*, 2nd edition. Cambridge University Press, Cambridge.

Hansen, C.J. and Paige, D.A. (1992). A thermal model for the seasonal nitrogen cycle on Triton. *Icarus*, 99, 273–288.

Heng, K. and Showman, A. (2015). Atmospheric dynamics of hot exoplanets. *Annual Review of Earth and Planetary Science*, 43, 509–540.

Jacquinet-Husson, N., Armante, R., Scott, N. *et al.* (2016). The 2015 edition of the GEISA spectroscopic database. *Journal of Molecular Spectroscopy*, 327, 31–72.

Kasting, J.F., Whitmire, D.P., Reynolds, R.T. (1993). Habitable zones around main sequence stars. *Icarus*, 101, 108–128.

Kral, Q. (2014). Study of debris disks with a new generation model. PhD thesis, Observatoire de Paris [Online]. Available at: https://ui.adsabs.harvard.edu/abs/2014PhDT.509K/abstract.

Kral, Q., Thébault, P., Charnoz, S. (2013). LIDT-DD: A new self-consistent debris disk model that includes radiation pressure and couples dynamical and collisional evolution. *Astronomy & Astrophysics*, 558(A121).

Lebonnois, S., Burgalat, J., Rannou, P. *et al.* (2012). TITAN global climate model: New 3-dimensional version of the IPSL Titan GCM. *Icarus*, 218, 707–722.

Lebonnois, S., Sugimoto, N., Gilli, G. (2016). Wave analysis in the atmosphere of Venus below 100 km altitude simulated by the LMD Venus GCM. *Icarus*, 278, 38–51.

Lissauer, J.J. and de Pater, I. (2015). *Fundamental Planetary Science*. Cambridge University Press, Cambridge.

Navarro, T., Madeleine, J.-B., Forget, F. *et al.* (2014). Modeling of the Martian water cycle with an improved representation of water ice clouds. *5th International Workshop on the Mars Atmosphere: Modeling and Observations*. Oxford, 13–16 January.

Pollack, J.B., Dalton, J.B., Grinspoon, D. *et al.* (1993). Near infrared light from Venus' nightside: A spectroscopic analysis. *Icarus*, 103, 1–42.

Read, P. (2011). Dynamics and circulation regimes of terrestrial planets. *Planetary and Space Science*, 59, 900–914.

Santos-Costa, D., Bolton, S.J., Sault, R.J. (2009). Evidence for short-term variability of Jupiter's decimetric emission from VLA observations. *Astronomy & Astrophysics*, 508, 1001–1010.

Schmid, H.M., Beuzit, J.-L., Feldt, D. *et al.* (2005). Search and investigation of extra-solar planets with polarimetry. *Direct Imaging of Exoplanets: Science & Techniques, Proc. IAU Colloquium 200*. Nice.

Scutaru, D., Rosenmann, L., Taine, J. *et al.* (1993). Measurements and calculations of CO_2 absorption at high temperature in the 4.3 and 2.7 μm regions. *Journal of Quantitative Spectroscopy and Radiative Transfer*, 50, 179–191.

Tennyson, J., Yurchenko, S., Al-Refaie, A.F. *et al.* (2016). The ExoMol database: Molecular lines lists for exoplanet and other hot atmosphere. *Journal of Molecular Spectroscopy*, 327, 73–94.

Turbet, M., Schott, C., Forget, F. (2017). The LAPS Project: A tutorial, online model to simulate the atmosphere of any terrestrial planet. *EPSC Abstract*, 11.

Villanueva, G., Smith, M.D., Protopapa, S. *et al.* (2018). Planetary Spectrum Generator: An accurate online radiative transfer suite for atmospheres, comets, small bodies and exoplanets. *Journal of Quantitative Spectroscopy and Radiative Transfer*, 217, 86–104.

Appendix
Web links

A list of Internet links that are particularly used in planetology can be found here:

– list of physical and orbital parameters of Solar System objects (with associated tools for trajectory calculation and observability): http://ssd.jpl.nasa.gov/?; https://www.imcce.fr/services/ephemerides/;

– list of orbital parameters of small Solar System bodies, IAU Minor Planet Center: https://minorplanetcenter.net/;

– catalog of exoplanets (with associated tools, histograms, etc.): http://exoplanet.eu/;

– catalog of all stellar objects with positions, motions and physical characteristics (Strasbourg Stellar Data Center): https://cdsweb.u-strasbg.fr/index-fr.gml;

– *Science for Exoplanets and Planetary Systems* digital book (L3 level): http://sesp.esep.pro/fr/index.html;

– list of past and current planetary space missions: https://en.wikipedia.org/wiki/List_of_Solar_System_probes;

– list of the largest ground-based optical telescopes: https://en.wikipedia.org/wiki/List_of_largest_optical_reflecting_telescopes;

– list of the largest radiotelescopes (not completely updated): https://en.wikipedia.org/wiki/List_of_radio_telescopes;

– Martian climate model (with physical atmospheric parameters for any solar longitude, planetary latitude and longitude): http://www-mars.lmd.jussieu.fr/mars/access.html;

– spectroscopic database (provides the molecular spectroscopic data needed for radiative transfer calculations in planetary atmospheres): https://geisa.aeris-data.fr/; https://hitran.org;

– transitions in the submillimeter, millimeter and radio ranges: https://spec.nasa.gov; https://splatalogue.online;

– database for radiative transfer in exoplanets and brown dwarfs: https://exomol.com;

– Planetary Spectrum Generator (calculation of planetary and exoplanetary spectra): https://psg.gsfc.nasa.gov/;

– NASA planetary database: https://pds.nasa.gov;

– ESA planetary database: https://archives.esac.esa.int/psa/;

– Nançay Radio Observatory: http://obs-nancay.fr;

– Planetary auroral spectro-imaging database: https://apis.obspm.fr;

– OV access portal to planetary data: http://vespa.obspm.fr/.

Glossary

Accretion: capture of surrounding matter by the gravitational attraction of an object, for example, a star in formation.

Achondrite: stony meteorite containing less than 35% metal, which comes from the surface of a large asteroid.

Adaptive optics: an observational technique that allows real-time elimination of distortions in the image of a celestial object, due to atmospheric turbulence, by correcting wave surface defects using a thin deformable mirror.

Aerosol: a set of fine particles, solid or liquid, suspended in an atmosphere.

Albedo: fraction of the incident light reflected by a body at a given wavelength.

Ambipolar diffusion: diffusion of charged particles in a quasi-neutral plasma, in other words, in which the charge density is zero at any point, but with microscopic gradients leading to the presence of an electric field.

Antitail: type of cometary tail due to dust, sometimes observed in the direction of the Sun when the Earth crosses the plane of the cometary orbit.

Aphelion: the furthest point from the Sun in a heliocentric orbit.

Asteroid: small solid body in orbit around the Sun. Most are located in the main asteroid belt between the orbits of Mars and Jupiter.

Asthenosphere: ductile part of the Earth's upper mantle, located between the lithosphere and the Earth's lower mantle at a depth of 700 km.

The Solar System 2,
coordinated by Thérèse ENCRENAZ and James LEQUEUX.

Astrometry: measurement of the position, distance and motion of celestial bodies (also called positional astronomy).

Astronomical unit (AU): unit of distance equal to the average Sun-Earth distance, or about 150 million km. Its exact value is 149.6 10^9 m, or 4.848 10^{-6} pc, or 15.812 10^{-6} ly.

Aurora: luminous phenomenon caused by the precipitation of energetic particles from the planetary environment in the upper atmosphere, generally located near the magnetic poles of a planet.

Brown dwarf: "aborted" star, with a mass between 0.01 and 0.08 solar mass (in other words, between 13 and 80 masses of Jupiter), insufficient for the fusion of hydrogen into helium to be triggered in its core.

Centaur: asteroid orbiting the Sun between the orbits of Jupiter and Neptune.

Chondrite: stony meteorite (less than 35% metal) containing chondrules.

Chondrule: elements of undifferentiated meteorites or chondrites. These small submillimeter-sized beads, generally composed of silicates, are inserted into a matrix.

Clathrate: compound formed of gas molecules trapped in a network of host molecules, most often H_2O. The clathrate most often encountered in planetology is methane clathrate (CH_4, $6H_2O$).

Coma: nebulosity surrounding the nucleus of a comet, created by the sublimation of ice and the ejection of dust that accompanies it.

Comet: small body of the Solar System with an often very eccentric orbit, possibly disturbed by the presence of the most massive planets. Far from the Sun, it is made up of a core of ice and rocks. When approaching the Sun, the sublimation of ice leads to the appearance of a coma, a dust tail and a plasma tail.

Conjunction: the moment when the position of a planet, seen from the Earth, is closest to the Sun. Mercury and Venus have lower conjunctions (when the planet is in front of the Sun) and higher conjunctions (when the planet is behind the Sun).

Continental drift: see *plate tectonics*.

Corona: circular volcanic structure bordered by numerous faults. It is found, in particular, on Venus and on Miranda.

Coronagraphy: astronomical technique consisting of masking the central object (generally the Sun or a star) to reveal its immediate environment (the solar corona or possibly a circumstellar disk or exoplanets). This technique is widely used for direct imaging of exoplanets and circumstellar disks.

Coronal hole: dark zone on the surface of the Sun from which the solar plasma escapes in the form of solar wind.

Cosmic abundance: abundance of elements in the Universe. This abundance is commonly related to the quantity of hydrogen, the most abundant element. The other elements are the results of nucleosynthesis. They could also have been produced during the Big Bang, like helium, or later, essentially by stars. For light elements, solar abundances are used; for heavy elements, measured with respect to silicon, the most primitive meteorites are used.

Cryovolcanism: a process similar to volcanism occurring on some icy satellites (Enceladus, Titan, Triton, etc.). The material ejected is water, ammonia or methane under pressure, which immediately recondenses at the surface.

Cyclotron (emission): low-frequency radio electromagnetic radiation produced in the auroral regions of the Earth and giant planets, by energetic electrons (weakly relativistic) circulating along magnetic field lines, at a frequency close to the electron cyclotron frequency.

Dating: method of determining the age of a body in the Solar System. A first method consists of counting the craters present on the surface: the most cratered bodies are the oldest. A second method consists of using samples (lunar or meteoritic) by measuring what remains of long-period radioisotopes of certain elements (uranium, thorium, etc.).

Daughter molecule: dissociation product (molecule, radical, atom or ion) of a cometary parent molecule.

Doppler-Fizeau (effect): variation $\Delta\lambda$ of the wavelength received from a moving source compared to the wavelength emitted λ_0. In the non-relativistic case, $\Delta\lambda/\lambda_0 = v/c$, v being the speed of the source and c the speed of light.

Dwarf planet: according to the definition adopted by the International Astronomical Union in 2006, it is a body of the Solar System that meets the following four criteria: (1) it is in orbit around the Sun, (2) it has sufficient mass to ensure hydrostatic equilibrium (and thus have a quasi-spherical shape), (3) it does not produce a vacuum in its orbital vicinity, and (4) it is not a satellite. There are currently five dwarf planets recognized by the IAU: one asteroid (Ceres) and four TNOs (Pluto, Haumea, Makemake and Eris).

Eclipse: apparent and temporary disappearance, or decrease in brightness, of a celestial object seen from the Earth when another body passes in front of it. Eclipses of the Sun by the Moon are the most common phenomena. This phenomenon is at the origin of the transit method for the detection of exoplanets.

Ecliptic: plane of the Earth's orbit. In the sky, it is the apparent trajectory of the Sun among the stars during the year.

Ecliptic coordinates: coordinates used to locate a point in the sky relative to the plane of the ecliptic, in which the Sun moves according to the seasons. The ecliptic coordinates are the ecliptic longitude λ and ecliptic latitude β.

Edgeworth-Kuiper (belt): see *Kuiper Belt*.

Equatorial coordinates: coordinates used to locate a point in the sky relative to the plane of the Earth's equator. The equatorial coordinates are the right ascension α and declination δ.

Equinox: moment of the year when the Sun crosses the equatorial plane of the planet.

Escape velocity: the minimum velocity that a body must reach to escape the gravitational attraction of the object (planet, satellite, etc.) to which it belongs and move away from it indefinitely.

Exobiology: field of research aimed at studying the physical, chemical and biological processes that may be at the origin of the appearance of life in the Universe (in the Solar System or on exoplanets).

Exoplanet: planet in orbit around a star other than the Sun.

Gegenschein: faint glow along the ecliptic in the opposite direction to the Sun, hence its name (German: opposite glow). It is caused by the scattering of sunlight by interplanetary dust, mostly located near the ecliptic.

Gravitational lens: phenomenon occurring when a massive object is located between a distant object and the observer. The deviation of the light rays of the distant object by the massive object concentrates them on the observer, which produces an amplification of the signal. If the distant object is a star and if the intermediate object is a star or a planet, it is called a gravitational microlens.

Great Red Spot: a very large anticyclone (about 15,000 km longitude extension) located in the southern hemisphere of Jupiter, at a latitude of 22°S. It has been stable since its discovery by Cassini in 1665.

Greenhouse effect: process leading to the heating of a planetary atmosphere when its constituents are transparent to visible solar radiation but absorb infrared radiation. The most efficient greenhouse gases are CO_2, H_2O and CH_4. The solar radiation (maximum in the visible range) heats the surface, which, as it heats up, reflects infrared radiation that the atmosphere absorbs, accelerating the heating process, and the mechanism can run away; the best example is that of the planet Venus.

Habitability: defines the capacity of a star to accommodate life. The criteria of habitability include the presence of liquid water, a source of energy and elements involved in the synthesis of prebiotic molecules (C, H, N, O, P, S).

Habitability zone: theoretical sphere surrounding a star where the temperature on the surface of a planet orbiting this star would be compatible with the presence of liquid water. The more massive the star, the brighter it is, and the further away the sphere is from it; the smaller the star is, the closer it is to it.

Hadley (circulation): planetary convective circulation induced by the variation of insolation as a function of latitude, to which is added another effect (Coriolis force) if the planet is rotating.

Haser (model): model describing the distribution of cometary molecules (parents and daughters) in the coma of a comet as a function of their distance from the nucleus, under the effect of photodissociation due to solar radiation.

Heliosphere: elongated envelope generated by the solar wind. It extends to the heliopause, which defines the boundary between the zone of influence of the solar wind and the interstellar medium. Its elongated shape is due to the motion of the Sun with respect to the surrounding interstellar matter.

Hertzprung-Russell (diagram): diagram representing the luminosity of stars as a function of their effective temperature. It makes it possible to distinguish the different types of stars and to understand their evolution.

Heterosphere: region of the upper atmosphere in which the constituents are no longer mixed, each constituent decreasing as the altitude increases with its own scale height.

Homosphere: region of the atmosphere in which the constituents are uniformly mixed. In the case of the Earth, this region extends from the surface to an altitude of 90 to 100 km.

Horizontal coordinates (also called local or alt-azimuth): coordinates used to locate a point in the sky relative to the horizontal plane where the observer is located. The coordinates are the height h (or its complement, the zenithal distance z) and the azimuth A.

Ice line: in a protoplanetary disk, it is the distance at which simple volatile molecules are in the form of ice. It depends on temperature and is specific to each molecule (H_2O, CH_4, CO). In general, the expression refers to the condensation of water.

Interferometer: instrument made up of several telescopes or radio telescopes that are interconnected, allowing high spatial resolution and mapping astronomical objects. In spectroscopy, the interferometer is a spectrometer that makes it possible to obtain a high spectral resolution (Michelson interferometer or Fourier transform interferometer).

Ionosphere: region of the upper atmosphere in which constituents are ionized under the effect of ultraviolet and solar X-rays.

Kepler's laws: empirical laws stated by Johannes Kepler to describe the motion of the planets around the Sun. They were explained by Newton within the framework of the law of gravity.

Kirkwood gaps: regions located in the main asteroid belt, at certain distances from the Sun, in which there are very few asteroids. They are due to orbital resonances related to the proximity of Jupiter.

Kuiper Belt: ring-shaped zone located beyond Neptune's orbit, considered as the reservoir of periodic comets of low inclination. It is also the reservoir of the Centaurs and certain types of trans-Neptunian objects.

Lagrange points: points located on the orbit of a planet around the Sun or perpendicular to it, or of a satellite around a planet, where the forces of gravity induced by the two bodies balance each other. Three points (L1, L2, L3), located on the axis of the two bodies, are positions of unstable equilibrium; the other two (L4 and L5), located on the orbit of the small body, at 60° upstream or downstream of it, are positions of stable equilibrium.

Lambert's law: law according to which the luminous intensity of an isotropic source is proportional to the cosine of the angle between the emission angle and the surface normal: $I(\alpha) = I_0 \cos \alpha$, I_0 being the emission along the normal axis to the surface.

Late Heavy Bombardment (LHB): episode that occurred at the beginning of the Solar System's history (between 4.1 and 3.9 billion years ago), characterized by a high rate of meteorite impacts on the surfaces of bodies in the Solar System.

Libration: phenomenon of oscillation of a satellite as observed from the planet around which it is rotating. It is due to the eccentricity of the satellite's orbit and the angle that this orbit makes with the equatorial plane of the planet.

Light-year (ly): distance travelled by light in one year. It is equal to $9.461\ 10^{12}$ km.

Lithosphere: rigid envelope surrounding the Earth, comprising the Earth's crust and part of the upper mantle.

Magma: fully or partially molten rock, formed by partial melting of the Earth's crust or mantle. By extension, the term can be applied to similar material in other planets or exoplanets.

Magnetic storm: episode of sudden and intense increase in the Earth's magnetic field.

Magnetopause: the boundary between a planet's magnetosphere, governed by its internal magnetic field, and the interplanetary medium.

Magnetosheath: region of the magnetosphere between the magnetopause and the shockwave.

Magnetosphere: plasma envelope surrounding a star, planet or satellite, in which physical phenomena are governed by the object's intrinsic magnetic field.

Magnitude: logarithmic scale allowing the measurement of the brightness of a star. The magnitude is higher the lower the brightness of the star. If m is the magnitude and B the brightness of the star, we have $m = -2.5 \log B + \text{Cst}$.

Main asteroid belt: main reservoir of asteroids, located between the orbits of Mars and Jupiter.

Mantle: in a solid body, the intermediate layer between the crust and the core.

Mesosphere: the atmospheric layer below the thermosphere. In the case of the Earth, it surmounts the stratosphere; in the case of Mars and Venus, in the absence of a stratosphere, it directly surmounts the troposphere.

Meteor: a luminous trail left in the sky by the fall of an extraterrestrial body that vaporizes by friction in the atmosphere. It is also called a "shooting star".

Meteor shower: a meteor shower (also called shooting stars) is caused by the passage of the Earth through the orbit of a comet, resulting in the fall of many of its cometary dusts into the Earth's atmosphere.

Meteorite: a solid object of extraterrestrial origin that has passed through the Earth's atmosphere without being completely volatilized during impact.

NEO (Near Earth Object): small body in the Solar System whose orbit is close to that of the Earth, with a perihelion at less than 1.3 AU. This class of objects includes about 20,000 asteroids (NEA) and about a hundred short-period comets (NEC).

Nucleosynthesis: formation of the nuclei of different chemical elements by nuclear reactions. The lightest elements (H, He, Li, Be, B) were formed just after the Big Bang (primordial nucleosynthesis). The heavier elements were synthetized in stars.

Nutation: periodic oscillation of the axis of rotation of an object around its mean position, in addition to precession.

Occultation: phenomenon by which a celestial body (star, planet, asteroid, etc.) is entirely or partially masked by an object. Stellar occultations are used to determine the properties of objects in the Solar System that occult them.

Olivine: silicate frequently observed on the surface of the bodies of the Solar System. The most common form is forsterite (Mg_2SiO_4).

Oort cloud: large spherical shell located between 30,000 and 100,000 au (one to two ly from the Sun), supposed to be the reservoir of new comets, formed closer to the Sun like other small bodies, but ejected in this region following gravitational interactions with giant planets.

Opposition: the moment when a planet, seen from the Earth, is exactly in the opposite direction to the Sun. The distance from the Earth to the planet is then minimal. Only Mars and the giant planets, whose orbit is outside that of the Earth, can be seen in opposition.

Orbital resonance: in the three-body problem, orbital resonance occurs when the ratio of the periods of two bodies orbiting a third is a rational fraction. This is the case, in particular, of the Io, Europa and Ganymede satellites around Jupiter.

Parent molecule: molecule sublimated from the ices of the cometary nucleus and released directly into the coma.

Parsec (pc): unit of length used to measure the distance of stars in our Galaxy. It is the distance of a star whose parallax is equal to one arc second. 1 pc = 3.26 ly = 3.08 10^{16} meters.

Perihelion: the point closest to the Sun on a heliocentric orbit.

Permafrost: constituent of the subsoil of a planet or a small body that has the characteristic of being permanently frozen.

Photodissociation: chemical reaction undergone by a molecule under the effect of visible, ultraviolet or higher energy radiation. This radiation can ionize the molecule or dissociate it into several other molecules or atoms, called dissociation products.

Planetary migration: motion of a planet within a protoplanetary disk, as a result of its gravitational interaction with the matter contained in the disk.

Planetesimal: small body a few kilometers in size, formed by the aggregation of dust in the protosolar or protoplanetary disk.

Planetoid: body formed by the agglomeration of several planetesimals following their mutual collisions.

Plasmasphere: toroidal region of the inner magnetosphere consisting of low energy plasma.

Plate tectonics: model explaining the global dynamics of the Earth's lithosphere. It is composed of about fifteen plates with relative movements with respect to each other. It is at the boundary between these plates that most geological phenomena occur (volcanic activity, earthquakes, formation of mountain ranges or oceanic trenches).

Poynting-Robertson (effect): effect by which solar radiation causes a particle of dust to spiral down towards the Sun. It is effective on micrometer-sized grains.

Precession: movement of the axis of a rotating body that describes a cone under the effect of external forces. In the case of the Earth, we speak of the precession of the equinoxes. This term can also refer to the rotation of the orbit of an object.

Protoplanetary disk: circumstellar disk formed of gas and dust, created by the collapse of a nebula rotating on itself. The protostar is formed in the central part, while the planets will form within the protoplanetary disk.

Protosolar (or primitive) nebula: mass of interstellar gas and dust from which the Solar System was formed, following its collapse into a disk.

Pulsar: a neutron star that rotates very rapidly, emitting a beam of radio waves like a lighthouse. It is the ultimate stage in the evolution of some massive stars.

Pyroxene: silicate of formula $XY(SiO_3)_2$, in which X and Y are two cations (examples: Ca, Mg). Pyroxene is a common compound in igneous and metamorphic rocks.

Radiation belts: the inner region of a planetary magnetosphere, in which high-energy particles are trapped by the magnetic field and can emit synchrotron radiation. In the case of the Earth, they are called Van Allen belts.

Radiation pressure: pressure exerted on a surface by the exchange of momentum between the object and the photons emitted by the Sun. It is effective for particles smaller than 0.1 micrometer in size, and tends to push them out of the Solar System.

Red dwarf: star of low mass (between 0.08 and 0.4 solar mass) and relatively faint. Red dwarfs are the most common stars in the Universe.

Reference system: system of celestial coordinates materialized by the position of very distant celestial objects (stars, quasars or galaxy nuclei).

Regolith: layer of fine dust covering the surface of objects without atmosphere or with a tenuous atmosphere, resulting from the pulverization of rocks following meteorite impacts.

Roche limit: distance from a planet below which a satellite begins to break up under the effect of tidal forces induced by the planet. It was established by the French astronomer Édouard Roche (1820–1883).

Scale height: the altitude at which one must rise, in an atmosphere, for the pressure to be decreased by a factor e. The scale height is about ten kilometers on the ground in the atmospheres of telluric planets.

Shooting star: light trail associated with the fall of an interplanetary particle (meteoroid) into the atmosphere. It can be associated with the passage of the Earth in the orbit of a comet; it is then called a swarm.

Small planet: see *asteroid.*

SNC (meteorites): Shergottites, Nakhlites and Chassignites meteorites (named according to the site where they fell) are considered to be of Martian origin, based on the analogy between their chemical, isotopic and petrological characteristics, as well as the observations of the Martian atmosphere made by the Viking probes.

Solar corona: region of the Sun located above the chromosphere that extends over several million kilometers. It is observed with coronagraphs or during total eclipses of the Sun.

Solar wind: plasma flow consisting essentially of electrons, protons and helium nuclei, ejected from the upper solar atmosphere and propagating radially through the Solar System to the confines of the heliosphere.

Stellar parallax: the angle under which one would see, from a star, the half-major axis of the Earth's orbit. Its measurement is used to obtain the distance to the stars.

Stratosphere: atmospheric region above the troposphere and below the mesosphere, in which solar ultraviolet radiation is absorbed, causing the temperature to increase with altitude. The Earth and the giant planets have a stratosphere, but Mars and Venus do not.

Supernova: late evolutionary stage of a massive star, during which all the heavy elements, up to iron, can be synthesized.

Synchrotron (emission): electromagnetic radiation emitted by a highly relativistic charged particle moving in a magnetic field, at the electronic cyclotron frequency and its harmonics.

T-Tauri (phase): phase of intense activity at the beginning of the evolution of a star (around 10 million years). The name refers to the first star on which the phenomenon was observed.

Thermosphere: the upper part of the atmosphere, located above the mesosphere and below the exosphere, with which it forms the heterosphere. The atoms are ionized there under the action of solar ultraviolet radiation.

Tidal effect: see *tide*.

Tide (effect of the): deformation of a star subjected to the gravitational forces of a nearby massive star.

Trans-Neptunian object (TNO): Solar System object in orbit around the Sun beyond the orbit of Neptune.

Trojans (asteroids): families of asteroids located on the orbit of Jupiter near the Lagrange points L4 and L5.

Tropopause: in an atmosphere, the upper limit of the troposphere. It is surmounted by the stratosphere or directly by the mesosphere.

Troposphere: the lower part of an atmosphere, characterized by a transfer of energy by convection and a decrease in temperature as altitude increases.

Van Allen (belts): see *radiation belts*.

Vernal (point): point of intersection (also called gamma point) between the plane of the equator and that of the ecliptic corresponding to the ascending node (passage of the Sun from the southern hemisphere to the northern hemisphere).

Virtual observatory: set of observational data, accompanied by exploitation programs, accessible to any researcher through the Internet.

White dwarf: very dense and not very luminous star, representing the ultimate stage in the evolution of stars with masses less than about eight times that of the Sun.

Zodiacal (cloud): disk of small interplanetary dust, close to the ecliptic, fed by dust from collisions between small bodies (comets or asteroids).

List of Authors

Athena COUSTENIS
LESIA
Paris Observatory
PSL University
Paris
France

Jacques CROVISIER
LESIA
Paris Observatory
PSL University
Paris
France

Thérèse ENCRENAZ
LESIA
Paris Observatory
PSL University
Paris
France

Marcello FULCHIGNONI
LESIA
Paris Observatory
PSL University
Paris
France

Laurent LAMY
LESIA
Paris Observatory
PSL University
Paris
France

James LEQUEUX
LERMA
Paris Observatory
PSL University
Paris
France

Françoise ROQUES
LESIA
Paris Observatory
PSL University
Paris
France

Brigitte ZANDA
Muséum national d'Histoire naturelle
and
Paris Observatory
Paris
France

Index

N

O

P

Q, R

S

T

U, V, W